Carla Gheler-Costa, Maria Carolina Lyra-Jorge, Luciano Martins Verdade (Eds.)
Biodiversity in Agricultural Landscapes of Southeastern Brazil

Carla Gheler-Costa, Maria Carolina Lyra-Jorge,
Luciano Martins Verdade (Eds.)

Biodiversity in Agricultural Landscapes of Southeastern Brazil

Managing Editor: Katarzyna Michalczyk

Language Editor: Andrea Huston

DE GRUYTER
OPEN

ISBN: 978-3-11-048083-2
e-ISBN: 978-3-11-048084-9

Published by De Gruyter Open Ltd, Warsaw/Berlin
Part of Walter de Gruyter GmbH, Berlin/Boston
The book is published with open access at www.degruyter.com.

Library of Congress Cataloging-in-Publication Data
A CIP catalog record for this book has been applied for at the Library of Congress.

Managing Editor: Katarzyna Michalczyk
Language Editor: Andrea Huston

www.degruyteropen.com

Cover illustration: Fábio Henrique Comin

Contents

Part II: **Biodiversity Dimensions**

Part III: **Case Studies**

Preface

The state of São Paulo, Brazil, is one of the most densely populated and developed areas in South America. Such development is evident both in terms of industrialization and urbanization, as well as in agriculture, which is heavily based on sugar cane, *Eucalyptus* plantations and livestock. This intense land use has resulted in great alteration of the original land cover and fragmentation of natural ecosystems. For these reasons, it is almost a paradox that jaguar, a species that requires large areas of pristine forest to exist, is still found in some parts of the state of São Paulo. Is it possible that wild animals can live in coexistence with intense land use? Or perhaps it is the case that such rare encounters with large wild animals in São Paulo will disappear in the near future?

All ecologists are aware of the problem of habitat change caused by humans, but it was not until recently that researchers started to consider that the land used for production could also serve as an important habitat for many different kinds of wild species. This book is about this new approach to conservation. It also highlights the important role that science could and should have in this discussion in order to better understand the problems and propose possible solutions.

In Chapter 1, the authors present a surprising historical review of the processes that have changed the landscape in the state of São Paulo until today. They discuss the main factors that characterize a well-preserved area, listing the pressures that habitats are receiving, and the ecological processes they host. In Chapter 2, the authors found that the majority of the studies related to conservation in São Paulo state do not include local people. According to the authors, the hundreds or thousands of land owners are the real actors in the conservation process and should take an important role in conservation initiatives. The Chapter 3 also evaluates the impact of improving certain aspects relating to conservation. In Chapter 4, the authors discuss how the influence of human activity on wild species and their habitat use are usually neglected by researchers. For example, they discuss the impact of roads on ecological communities and the lack of studies on this topic.

Most of the general public (and even some researchers) tend to think about biodiversity in terms of vertebrates and plants. However, one of the richest organism groups is usually much more difficult to see: soil microorganisms. Chapter 5 deals with the lack of research focus on soil microorganisms. In order to help facilitate new researchers, the authors list all the facilities that new technologies are producing to study this group, as well as examples of how to set up a sampling design and a list of methods and analysis most frequently used by specialists.

A common theme throughout the book is the idea that the productive matrix of the agroecosystem has an important conservation role. The authors also highlight the lack of regulations that should guarantee the conservation of the biodiversity living in agricultural areas. However, it is clear throughout the book that all original cover could or should be manipulated by human activities. In Chapter 6, the authors deal with this issue and conclude that the conservation of patches of original habitats throughout agricultural landscapes is needed in order to increase beta biodiversity.

Human activities not only affect biodiversity and landscapes; they also influence the diversity of agriculture products in the country. In Chapter 7, the authors have found that technology allows for an increase in food production for a limited number of items, but decreases the diversity of food items produced.

Today we know that most biomes have been highly modified, but some of them are still able to preserve a considerable biodiversity. What is not clear is for how long this biodiversity would remain if we stop or disrupt ecological processes by increasing fragmentation and disturbance. In Chapter 8, the authors suggest that the solution is not restricted to biology, but also needs to include human dimensions in order to produce the necessary changes and reach the final goal of biodiversity conservation in agricultural landscapes.

Chapters 9 to 16 focus on case studies where different groups of animals are considered. The general conclusion from these studies is that habitat modification frequently reduces richness, however some species take advantage of those changes, experiencing a great increase in their population. This is important because these few species usually have a negative impact on natural communities, causing loss of biodiversity and concentrating the use of energy in the system into a small group. Another common conclusion from these studies is that specialist species tend to be the first to disappear, while more generalist species experience an increase in their abundance. Interestingly, these changes in natural communities have the ability to also cause negative impacts on humans. For example, in Chapter 11, the authors discuss how the increase in abundance of capybaras have caused car accidents and brought health problems to the people who use the same areas as those wild animals.

In the last Chapter of the book, the authors use remote sensing in order to understand and explain mammal biodiversity in different areas. They conclude that the relationship between the images and the biodiversity found in the area is not enough to produce a proper prediction model. According to the authors, this could be based on the temporal variability of the habitats, generally not considered in this kind of study.

The book includes all aspects needed to understand the current status of biodiversity conservation in Southeastern Brazil. It presents good and clear examples, lists techniques commonly used in different groups of biodiversity and helps to communicate how good science is needed to promote biodiversity conservation.

Dr. Carlos I. Piña

List of Contributors

Preface
Carlos I. Piña

CICyTTP-CONICET, Science and Technology College, FCyT-UADER
Dr. Matteri y España, CP 3105, Diamante, Entre Ríos, Argentina
E-mail: cidcarlos@infoaire.com.ar

Chapter 1
Luciano M. Verdade[1], Carla Gheler-Costa[2], Maria Carolina Lyra-Jorge[3]

[1] University of São Paulo, Department of Tropical Ecosystems Function,
Isotopic Ecology Laboratory/CENA
P.O. Box 96, Piracicaba, SP 13416-000, Brazil
E-mail: lmverdade@usp.br

[2] Sacred Heart University (USC), Department of Health Sciences,
Ecology and Conservation Laboratory (LECO)
Rua Irmã Arminda, 10-50, Jardim Brasil, Bauru, São Paulo 17011-160, Brazil
E-mail: cgheler@gmail.com

[3] Anglo Sorocaba
Avenida Antônio Carlos Comitre, 2277 - Campolim, Sorocaba - SP, 18047-620, Brazil
E-mail: carolina@blumegrupe.com

Chapter 2
Fábio Henrique Comin[1], Carla Gheler-Costa[2]

[1] Sacred Heart University (USC), Department of Exact and Applied Social Sciences
Rua Irmã Arminda, 10-50, Jardim Brasil, Bauru, São Paulo 17011-160, Brazil
E-mail: fhc.eco@gmail.com

[2] Sacred Heart University (USC), Department of Health Sciences,
Ecology and Conservation Laboratory (LECO)
Rua Irmã Arminda, 10-50, Jardim Brasil, Bauru, São Paulo 17011-160, Brazil
E-mail: cgheler@gmail.com

Chapter 3

Alexandre Toshiro Igari[1,2], Leandro Reverberi Tambosi[2]

[1] University of São Paulo, School of Arts, Sciences and Humanities
Av. Arlindo Béttio, 1000, São Paulo, SP - 03828-000, Brazil
E-mail: alexandre.igari@usp.br

[2] University of São Paulo, Biosciences Institute, Ecology Department, Landscape and Conservation Ecology Laboratory
Rua do Matão, Travessa 14, 321, São Paulo, SP - 05508-090, Brazil
E-mail: letambosi@yahoo.com.br

Chapter 4

Milton Cezar Ribeiro[1], Renata de Lara Muylaert[1], Pavel Dodonov[2], Giordano Ciocheti[1], Marcelo Magioli[3], Felipe Martello[1], Alessandro Rocha[1], Bruno Defane Borges[1], Carolina Carvalho[4], Claudia Zukeran Kanda[1], Karen G. Rodríguez-Castro[5]

[1] UNESP, Department of Ecology, Spatial Ecology and Conservation Laboratory
24A Av., 1515, 13506–900, Rio Claro, São Paulo, Brazil
E-mail: (M.C. Ribeiro) mcr@rc.unesp.br, (R.L. Muylaert) renatamuy@gmail.com, (G. Ciocheti) gciocheti@gmail.com, (F. Martello) felipemartello@gmail.com, (B.D. Borges) bruno_borges01@hotmail.com, (C.Z. Kanda) mclaudiakand@gmail.com

[2] Santa Cruz State University (UESC), Ecology Applied to Conservation Laboratory
Jorge Amado Road, km 16, Salobrinho, 45662-900, Ilhéus, Bahia, Brazil
E-mail: pdodonov@gmail.com

[3] University of São Paulo, "Luiz de Queiroz" Agriculture College, Wildlife Ecology, Management and Conservation Laboratory
Av. Pádua Dias, 11, Piracicaba, SP 13418-900, Brazil
E-mail: marcelo.magioli@gmail.com

[4] UNESP, Department of Ecology, Conservation Biology Laboratory
24A Av., 1515, 13506–900, Rio Claro, São Paulo, Brazil
E-mail: carolina.carvalho@ymail.com

[5] Federal University of São Carlos (UFSCar), Department of Ecology and Evolution, Molecular Biodiversity and Conservation Laboratory
Rodovia Washington Luis, km 235, 13565-905, São Carlos, São Paulo, Brazil
E-mail: karengiselle2004@gmail.com

Chapter 5

Dennis Goss-Souza[1], Clovis Daniel Borges[1], Lucas William Mendes[1], Acacio Aparecido Navarrete[1], Dilmar Baretta[2], Tsai Siu Mui[1]

[1] University of São Paulo, Department of Agroindustrial and Food Production/CENA
Avenida Centenário, 303, Caixa Postal 96, Piracicaba, SP 13416-000, Brazil
E-mail: tsai@cena.usp.br

[2] Santa Catarina State University, Department of Soil, Soil Laboratory/CEO
Rua Beloni Trombeta Zanin, 680E, Chapecó, SC 89815-630, Brazil
E-mail: dilmarbaretta@gmail.com

Chapter 6

Eduardo A. Athayde, Leonardo F. Cancian, Leonor Patricia C. Morellato

University of São Paulo State, Biosciences Institute, Botanic Department, Phenology Laboratory
Avenida 24 A, 1515, CEP 13506-900 Rio Claro, SP, Brazil
E-mail: (E.A. Athayde) eanversa@yahoo.com.br, (L.F. Cancian) lfcancian@yahoo.com.br, (L.P.C. Morellato) pmorella@rc.unesp.br

Chapter 7

Giancarlo Conde Xavier Oliveira

University of São Paulo, “Luiz de Queiroz” Agriculture College, Department of Genetics
Av. Pádua Dias, 11, Piracicaba, SP 13418-900, Brazil
E-mail: gcxolive@gmail.com

Chapter 8

Cristina Yuri Vidal[1,] Julia Raquel Mangueira[1], Fabiano Turini Farah[2], Débora Cristina Rother[2], Ricardo Ribeiro Rodrigues[2]

[1] University of Campinas, Vegetal Biology Post-graduation Program, Biology Institute
Rua Monteiro Lobato, 255, CEP 13083-862, Campinas, SP, Brazil
E-mail: (Cristina Yuri Vidal) yurialoha@gmail.com

[2] University of São Paulo (USP), Biological Sciences Department, Laboratory of Ecology and Forest Restoration (LERF)
Av. Pádua Dias, 11, Piracicaba, SP 13418-900, Brazil
E-mail: (Ricardo Ribeiro Rodrigues) rrresalq@usp.br

Chapter 9

Flavio de Barros Molina[1,2], Bruno de Oliveira Ferronato[3], Franco Leandro de Souza[4]

[1] University of São Paulo, Museum of Zoology, Department of Herpetology
Avenida Nazaré, 481, São Paulo SP, 04263-000, Brazil
E-mail: fbmolina@uol.com.br

[2] University of Santo Amaro, Department of Biology
Rua Prof. Enéas de Siqueira Neto, 340, São Paulo SP, 04829-300, Brazil

[3] University of Canberra, Institute for Applied Ecology
Canberra, ACT 2601, Australia
E-mail: brunoferronato@hotmail.com

[4] Mato Grosso do Sul Federal University, Biological Science Center
Campo Grande, MS, 79070-900, Brazil
E-mail: franco.souza@ufms.br

Chapter 10

Rodney Kozlowiski de Azevedo[1], Vanessa Doro Abdallah[2]

[1] Sacred Heart University (USC), Department Exact and Applied Social Sciences, Ichtioparasitology Laboratory
Rua Irmã Arminda, 10-50, Jardim Brasil, Bauru, São Paulo 17011-160, Brazil
E-mail: azevedork@gmail.com

[2] Sacred Heart University (USC), Department Health Sciences, Ichtioparasitology Laboratory
Rua Irmã Arminda, 10-50, Jardim Brasil, Bauru, São Paulo 17011-160, Brazil
E-mail: vanessaabdallah@ig.com.br

Chapter 11

Alex Augusto de Abreu Bovo[1], Katia Maria P. M. B. Ferraz[1], Luciano M. Verdade[2], José Roberto Moreira[3]

[1] University of São Paulo, Department of Forest Engineering, Wildlife Ecology, Management and Conservation Laboratory/ESALQ
Caixa Postal 09, Av. Pádua Dias, 11, Piracicaba, SP, 13418-900, Brazil
E-mail: (Katia Maria P. M. B. Ferraz) katia.ferraz@usp.br

[2] Univesity of São Paulo, Department of Tropical Ecosystems Function, Isotopic Ecology Laboratory/CENA

P.O. Box96, Piracicaba, SP 13416-000, Brazil
E-mail: lmverdade@usp.br

[3] Brazilian Agricultural Research Corporation (EMBRAPA)
PqEB - Av. W5 Norte (final), Brasilia, DF 70770-917, Brazil
E-mail: jose-roberto.moreira@embrapa.br

Chapter 12

Karina L. Silva-Brandão[1], Danilo Bandini Ribeiro[2], Mariana Monteiro de Brito[3], Alberto José Arab Olavarrieta[3], André V. L. Freitas[4]

[1] University of São Paulo, Center of Nuclear Energy in Agriculture, Laboratory of Plant Inbreeding, DVPROD
Av. Centenário, 303, Piracicaba, SP 13400-970, Brazil
E-mail: (K.L. Silva-Brandão) klsilva@gmail.com

[2] Federal University of Mato Grosso do Sul, Centro de Ciências Biológicas e da Saúde
Caixa Postal 549, Campo Grande, MS 79070-900, Brazil
E-mail: biodbr@yahoo.com.br

[3] Federal University of ABC, Centro de Ciências Naturais e Humanas
Av. dos Estados, 5001, Santo André, SP, 09210-580, Brazil
E-mail: (M.M. de Brito) mmbrito.bio@gmail.com, (A.J. Arab Olavarrieta) albertoarab@gmail.com

[4] Campinas State University, Biology Institute, Department of Animal Biology
P.O. Box 6109, Campinas, São Paulo CEP 13083-970, Brazil
E-mail: baku@unicamp.br

Chapter 13

Pedro Gerhard[1], Luciano M. Verdade[2]

[1] Brazilian Agricultural Research Corporation - Satellite Monitoring (EMBRAPA)
Av. Soldado Passarinho, 303, Campinas, SP, CEP 13070-115, Brazil
E-mail: pedro.gerhard@embrapa.br
[2] University of São Paulo, Department of Tropical Ecosystems Function, Isotopic Ecology Laboratory/CENA
P.O. Box 96, Piracicaba, SP 13416-000, Brazil
E-mail: lmverdade@usp.br

Chapter 14

Paula C. Lopes[1], Thiago S. Marques[2], Paula S. Martin[3], Carla Gheler-Costa[4], Luciano M. Verdade[2]

[1] University of São Paulo, Department of Zoology, Entomology and Acarology Laboratory/ESALQ
Av. Pádua Dias, 11, Piracicaba, SP 13418-900, Brazil
E-mail: pclopes83@gmail.com

[2] University of São Paulo, Department of Tropical Ecosystems Function, Isotopic Ecology Laboratory/CENA
P.O. Box96, Piracicaba, SP 13416-000, Brazil
E-mail: (T.S.Marques) thiagomq@yahoo.com.br, (L.M. Verdade) lmverdade@usp.br

[3] Itaiti, Environmental Consulting
Av. Lacerda Franco, 1147 - Cambuci, São Paulo - SP, CEP 15360-000, Brazil
E-mail: paula.sanchesmartin@gmail.com

[4] Sacred Heart University (USC), Department of Health Sciences, Ecology and Conservation Laboratory (LECO)
Rua Irmã Arminda, 10-50, Jardim Brasil, Bauru, São Paulo 17011-160, Brazil
E-mail: cgheler@gmail.com

Chapter 15

Marli Penteado[1], Carlos Yamashita[2], Thiago S. Marques[3], Luciano M. Verdade[3]

[1] Environment Ministry, Chico Mendes Institute of Biodiversity Conservation (ICMBio)
Esplanada dos Ministérios - Bloco B, CEP 70068-900 - Brasília/DF, Brazil
E-mail: marli.penteado@icmbio.gov.br

[2] Environment Ministry, Brazilian Institute of Environment and Renewable Natural Resources (IBAMA)
Esplanada dos Ministérios - Bloco B, CEP 70068-900 - Brasília/DF, Brazil
E-mail: carlos.yamashita@ibama.gov.br

[3] University of São Paulo, Department of Tropical Ecosystems Function, Isotopic Ecology Laboratory/CENA
P.O. Box 96, Piracicaba, SP 13416-000, Brazil
E-mail: (T.S.Marques) thiagomq@yahoo.com.br, (L.M. Verdade) lmverdade@usp.br

Chapter 16

Maria Cecília Lunardelli[1], Maria Carolina Lyra-Jorge[2], Carla Gheler-Costa[3]

[1] Lunardelli & Antunes Ltda
Rua Emilio Cornelsen, 500/403, Curitiba, PR, CEP 80540-220, Brazil
E-mail: lunardellister@gmail.com

[2] Anglo Sorocaba
Avenida Antônio Carlos Comitre, 2277 - Campolim, Sorocaba - SP, 18047-620, Brazil
E-mial: carolina@blumegrupe.com

[3] Sacred Heart University (USC), Department of Health Sciences, Ecology and Conservation Laboratory (LECO)
Rua Irmã Arminda, 10-50, Jardim Brasil, Bauru, São Paulo 17011-160, Brazil
E-mail: cgheler@gmail.com

Chapter 17

Thiago P. de C. e Timo[1], Júlio César Costa[2], Carla Gheler-Costa[3], Maria Carolina Lyra-Jorge[4], Mariana R. Urbano[5], Luciano M. Verdade[6]

[1] University of São Carlos– UFSCar, Campus Sorocaba, Environmental Sciences Departament (DCA)
Rodovia João Leme dos Santos (SP-264), Km 110, Bairro do Itinga, Sorocaba, 18.052-780, São Paulo, Brazil
E-mail: thiago.timo@gmail.com

[2] Costa e Faria Agroenvironmental Consulting
P.O. Box: 91, 13.610-970, Leme - São Paulo, Brazil
E-mail: manacus@gmail.com

[3] Sacred Heart University (USC), Department of Health Sciences, Ecology and Conservation Laboratory (LECO)
Rua Irmã Arminda, 10-50, Jardim Brasil, Bauru, São Paulo 17011-160, Brazil
E-mail: cgheler@gmail.com

[4] Anglo Sorocaba
Avenida Antônio Carlos Comitre, 2277 - Campolim, Sorocaba - SP, 18047-620, Brazil
E-mail: carolina@blumegrupe.com

[5] University of Londrina (UEL), Exact Sciences Center
Rodovia Celso Garcia Cid, CEP 86.051-980, Londrina – PR, Brazil
E-mail: mrurbano@uel.br

[6] University of São Paulo, Department of Tropical Ecosystems Function, Isotopic Ecology Laboratory/CENA
P.O. Box96, Piracicaba, SP 13416-000, Brazil
E-mail: lmverdade@usp.br

Part I: **Human Dimensions**

Luciano M. Verdade, Carla Gheler-Costa, Maria Carolina Lyra-Jorge

1 The Multiple Facets of Agricultural Landscapes

1.1 Agriculture and History

When the first of our ancestors collected wheat grains before they matured and fell on the ground, they started a process – based on the manipulation of the evolutionary process by the selection for the non-dehiscence of the seeds – that changed the surface of the earth. The unprecedented richness generated by food storage – only possible by synchronized harvest of non-dehiscent seeds – supported two major drivers of human history: urbanization and ubiquity. The combination of both led to a splendid development of religion, philosophy, arts, science and technology, besides eventual eruptions of density-dependent infectious diseases and a permanent (though variable in intensity, in space and time) state of belligerence (Verdade et al., 2014). This process called – *lato sensu* – agriculture has been supporting what we generally call civilization, including our history, culture, religions and socioeconomics.

The use of fire allowed the replacement of even tropical forests by agricultural fields during millennia (Pyne, 1998). Considering the population size of its native inhabitants and their crops, the Atlantic Forest must have been cut down for slash and burn agriculture at least once in its entire area before the arrival of the first Europeans (Dean, 1995). The slash-and-burn system relied on the rotation of agricultural fields in order to restore soil fertility (Palm et al., 1996). However, the increasing human population demanded a continuous expansion in agricultural areas, which is limited in a Malthusian world. The lack of extra arable lands in certain areas stimulated the use of exogenous fertilizers in order to stop rotation and keep permanent the already existing agricultural fields (Zanden, 1991; Juo & Manu, 1996). Organic materials such as manure and wastes were then used for centuries resulting on a considerable increase of soil organic matter in certain regions of the world (Glaser & Woods, 2004). However, only the use of industrial fertilizers and other agrochemicals (e.g. insecticides and herbicides) from the middle 20th century on allowed an exponential increase in productivity and a rapid expansion of agriculture over regions with infertile soils in what has been romantically called the “Green Revolution” (Conway & Barbier, 1988).

As a result, agroecosystems replaced pristine ecosystems worldwide originating agricultural landscapes with their own history, geography and ecology, shared by many generations of humans and millions of other living species in their Darwinian struggle for life. Of course, such struggle was unevenly balanced among them which resulted in the extinction of many (Dirzo et al., 2014) and the adaptation of some (Verdade et al., 2011) species to this brave (and apparently still unbalanced...) new world.

In Southeastern Brazil, cattle and sugarcane arrived with the first Europeans in the early 16th century. Pastures for livestock production currently cover approximately 200 Mha with 30% considered degraded or under low productivity systems (Ferraz &

de Felício, 2010; Sparovek et al., 2010, 2012). Sugarcane experienced historical cycles of expansion and decline for the production of sugar, *aguardente* (i.e. liquor), and more recently, ethanol, with the ethanol trade expecting to cover up to 14 Mha in Brazil in 2016 (Verdade et al., 2012). Tree species of the genus *Eucalyptus* have been introduced in Southeastern Brazil in the late 19th century as a surrogate for native wood trees that by then were being literally burnt down (Dean, 1995). Currently, *Eucalyptus* plantations are the main silvicultural crop in Brazil covering approximately 3 Mha (IEA, 2014) to produce paper, celluloses and charcoal (Stape et al., 2004).

The food, fiber and biofuels produced by these crops became major commodities in the modern economy and Brazil became a major player in the so called "agribusiness" (Martinelli et al., 2010; Chapter 3, this book). These crops currently cover vast areas of Southeastern Brazil as agroecosystems (i.e. landscape matrices) (Fig. 1.1) and furnish prime matter for three of the most important Brazilian agroindustrial sectors. Whatever happens with these sectors in terms of land use change and agricultural practices will have a deep influence in the history, geography and ecology of their respective landscapes (Chapter 2, this book). These multiple facets of agricultural landscapes should be considered by public policy concerning their conflicting missions of producing domesticated species and conserving wild species on a finite land (Verdade et al., 2014a).

Fig. 1.1: Agroecosystems: A) Pasture; B) Sugarcane plantation; C) *Eucalyptus* plantation and mixed agricultural landscape (D) containing all of them in Southeastern Brazil.

1.2 Agriculture and the Wilderness

As a consequence of agricultural expansion, land use change and agricultural practices have been impacting not only wildlife patterns of distribution and abundance but also the evolutionary processes that mold these patterns (e.g. Thompson, 1998; Palumbi, 2001; Nacci et al., 2002; Galetti et al., 2013; Snell-Rood & Wick, 2013). Therefore, the quantity and quality of the agroecosystem will determine both the diversity of patterns (e.g. β-diversity) and the complexity of processes (e.g. demography, trophic relations, adaptation) concerning the wildlife on an agricultural landscape as a whole. However, their local wildlife (i.e. the wild species that use or dwell on them) gives some conservation value to agricultural landscapes; therefore, they merit conservation efforts (Verdade et al., 2014a).

1.2.1 Agricultural Impacts on the Wildlife

Agriculture expansion impacts wildlife in two major ways: land use change (LUC) and intensification. LUC can occur directly over pristine ecosystems (dLUC) or indirectly by the replacement of one agroecosystem by another and its consequent shift to pristine areas (iLUC). Agriculture intensification is usually associated with an increase in productivity on an agroecosystem or the replacement of the agroecosystem by a more productive one. By the agronomical point of view, low productive agroecosystems are usually called marginal or degraded land. LUC and intensification usually affect not only biodiversity patterns of abundance and distribution, but also the ecological and evolutionary processes that mold those (Verdade et al., 2014a) (Tab. 1.1).

LUC results in direct habitat loss and the consequent increase on local extinction rate and decrease on α- to γ-diversity, depending on the spatial scale on which it occurs, based on island biogeography theory (MacArthur & Wilson, 1967) and the subsequent approach of forest fragmentation (Harris, 1984). dLUC has historically happened in Europe and China and more recently in North America (Diamond, 1997). Currently it still occurs in Southeast Asia (Koh et al., 2011), Africa (Bekunda et al., 2009), and Southern South America (Sala et al., 2009). iLUC in the Amazon has been associated with the sugarcane expansion in Southeastern Brazil (Lapola et al., 2010), but deforestation in the Brazilian Amazon appears to be more related to timber (mostly illegal) extraction followed by pasture implantation for livestock – and more recently soybean – production (Fearnside, 2008).

In terms of environmental security, the agricultural expansion over degraded or marginal lands is preferable in relation to LUC (Murphy et al., 2011; Joly et al., 2014; Mendenhall et al., 2014). However, the consequent increase of agrochemical use associated with agricultural intensification can result in local impacts on wildlife (Verdade et al., 2012; Phelps et al., 2013) (Tab. 1.1). Environmental contamination by agrochemical residues can result in a decrease in α-diversity (e.g. Meche et al., 2009),

Tab. 1.1: Agricultural impacts on wildlife.

Driver	Impact	Patterns affected	References	Processes affected	References
Land use change (LUC)	Habitat loss	γ-diversity	Metzger et al., 2010	Extinction rate, ecophysiology	Koh, 2007; Vynne et al., 2014
Agriculture intensification	Environmental contamination	α-diversity	Meche et al., 2009	?	Piña et al., 2009; Ferronato et al., 2009a
	Water eutrophication	α-diversity	Liu et al., 2013	?	Liu et al., 2013
	Invasion of exotic species	β-diversity?		Trophic relations, biodiversity homogenization	Jaksic et al., 2002; Magnusson, 2006; Ferronato et al., 2009b
	Barriers	β-diversity?		Dispersal	Verdade et al., 2002b; Ferronato et al., 2009a; Prevedelo & Vieira, 2010; Gheler-Costa et al., 2013; Chapter 4 (this book)
	Alterations in spatial-temporal heterogeneity	β-diversity?		Trophic relations?	
		Biomass?		Adaptation?	

but also in unpredictable impacts on the ecophysiology and reproductive biology of wild species (e.g. Ferronato et al., 2009; Piña et al., 2009; Vynne et al., in press). Water eutrophication by the use of chemical fertilizers can also affect α-diversity and many ecological processes, including its potability (e.g. Liu et al., 2013). Agricultural intensification has also been associated with the introduction of invasive species of plants and animals (Mooney & Hobbs, 2000; Jaksic et al., 2002) with their innumerous environmental impacts such as biodiversity homogenization (Magnusson, 2006), spread of diseases (Tompkins et al., 2011; Lafferty, 2014) and competition with native species (Ferronato et al., 2009b). Even environmental-friendly management practices associated with agroecosystems can have deleterious effects on wildlife. As an example, *Eucalyptus* plantations "in line" to avoid soil erosion can affect dispersal capacity of small marsupials in Southeastern Brazil (Prevedelo & Vieira, 2010).

Last but not least, agroecosystems are usually less heterogeneous in space, but more heterogeneous in time, as compared to pristine ecosystems (Verdade et al., 2014a). Such variation in spatial-temporal heterogeneity creates "pulses" of biomass production on the landscape, possibly affecting the use of space and trophic relations of wild species. In addition, their effects on evolutionary processes are unpredictable and should be prioritized by future studies.

1.2.2 Wildlife Management in Agricultural Landscapes

In some regions of the world (e.g. the "Montado" areas of the Iberian Peninsula) biodiversity friendly agroecosystems became the most diverse remaining environments. For these reasons, agroecosystems (i.e. the matrices of agricultural landscapes) should no longer be considered as "non-habitats", as they are habitats for some species. Instead, we should improve agricultural practices – besides regulate land use change based on zoning – in order to increase agroecosystem carrying capacity as well as their α- and β-diversity and, consequently, decrease local species extinction (Driscoll et al., 2013). We should also consider not only ecosystem services provided by biodiversity friendly agricultural landscapes (e.g. carbon sequestration, water "production" and pollination), but also the maintenance of the evolutionary process upon which both wild and domestic species depend to adapt to an increasingly human-induced changing world (Faith & Pollock, 2014; Verdade et al., 2014a).

For the reasons above, it is not only necessary to include agriculture in the context of wildlife management, but also wildlife management in the context of agriculture. There are only four alternatives of action for wildlife management at the population level, as proposed by Caughley (1994) and adapted by Verdade et al. (2014b): biological conservation, control, sustainable use and monitoring. The first aims at increasing depleted populations; the second aims at decreasing excessive populations; the third aims at establishing sustainable yields of valuable species; and, the last aims at checking possible changes in population status (Tab. 1.2). Complementarily to the management at the population level, the landscape (limited, preferably, by drainage, see Magnusson, 2001) should be considered as the ideal unit of administration, including for restoration programs (see Chapter 8, this book), as they are recognized as such at the public policy level (Verdade et al., 2014b).

Agricultural landscapes of Southeastern Brazil present a reasonable diversity of plants (Chapter 6, this book), invertebrates (Chapter 12, this book), fish (Chapter 13, this book), amphibians (Chapter 14, this book), reptiles (Marques et al., 2013, Chapter 9, this book), birds (Chapter 16, this book) and mammals (Dotta & Verdade, 2007, 2009, 2011; Gheler-Costa et al., 2012; Martin et al., 2012; Chapter 16 and 17, this book). Those species demand management actions in order to increase the agricultural landscapes' carrying capacity and consequently their β-diversity (Verdade et al., 2014a). However, wild species can cause damage to agricultural crops by herbivory (e.g. Ferraz et al.,

2003) and to livestock by predation (e.g. Palmeira et al., 2008; Azevedo et al., 2010). In such cases it is usually necessary to prevent the contact between the wild species and the crop or the livestock. Otherwise, land owners tend to blame – and, therefore, overkill – the former.

Tab. 1.2: Wildlife management in agricultural landscapes (adapted from Verdade et al., 2014a).

Category	Examples	References
Conservation biology	Biodiversity in agricultural landscapes	Plants (Chapter 6, this book), invertebrates (Chapter 12, this book), fish (Chapter 13, this book), amphibians (Chapter 14, this book), reptiles (Marques et al., 2013; Chapter 9, this book), birds (Chapter 15, this book), mammals (Dotta & Verdade, 2007, 2009, 2011; Gheler-Costa et al., 2012; Martin et al., 2012; Chapter 16 & 17, this book)
Control	Livestock predation	Palmeira et al., 2008; Azevedo et al., 2010
Sustainable use	Capybaras	Verdade & Ferraz, 2006, Ferraz et al., 2009; Chapter 11, this book
Monitoring	Limiting factors	Verdade et al., 2014b
	Conceptual basis (diversity of patterns vs complexity of processes)	Trophic process (Verdade et al., 2011), use of space (Magnusson, 1999), adaptation (Rosalino et al., 2013, 2014)
	Innovation	Methodological: Mammals (Ferraz et al., 2009, 2010; Verdade et al., 2012a; Lyra-Jorge et al., 2014), birds (Penteado et al., 2014), reptiles (Sarkis-Gonçalves et al., 2004) Technological: Molecular markers (Palomares & Adrados, 2014), stable isotopes (Marques et al., 2014)
	Governance	Lindenmayer & Likens, 2010; Pezzini et al., 2012; Magnusson et al., 2013, 2014

As some species can become abundant in anthropogenic environments, they could possibly be included in hunting programs. However, hunting is legally restricted in Brazil, even preventing researchers' access to hunting data (Verdade & Seixas, 2013). For this reason, capybaras became "plagues" in agricultural landscapes and even urban areas in Southeastern Brazil with expensive but rather ineffective control (Verdade & Ferraz, 2013). Alternatively, the species could possibly be the best option of medium to large mammals for a "didactic" sustainable use program (Verdade & Ferraz, 2006; Moreira et al., 2012; Verdade & Ferraz, 2013).

Notwithstanding, as most species are not endangered, nor harmful or valuable, the most important management action is monitoring them in order to detect as soon as possible whether they change their status. However, the implementation of a large-scale long-term biodiversity monitoring programs is currently limited at three levels: conceptual, innovation and governance (Verdade et al., 2014b). Possibly, the major concept yet to be improved is – not so surprisingly – what to monitor. The vast majority of monitoring programs still rely on "one size fits all" indicators – usually specialist-specific – based on patterns of diversity (see Lindenmeyer & Likens, 2010 for an extensive review). However, we should possibly look for ecological and evolutionary processes whose complexity would mold such diversity of patterns (Verdade et al., 2014b). Among them, the use of space (Magnusson, 1999), trophic relations (Verdade et al., 2011) and adaptation (Rosalino et al., 2014) are possibly the best options to begin with, as they are measurable and can possibly be reduced to indexes or relatively simple algorithms. In addition, traditional field methods – usually developed for pristine temperate environments –need to be adapted and checked from time to time in order to be effective in tropical anthropogenic environments. Such adaptation and checking usually requires creativity from researchers. Simple but effective initiatives have been carried out for mammals (Pinto et al., 2006; Ferraz et al., 2009, 2010; Verdade et al., 2012b; Lyra-Jorge et al,. 2014), birds (Penteado et al., 2014) and reptiles (Sarkis-Gonçalves et al., 2004). Lastly, governance should be improved in order to create facilities to hold an interoperable data bank of such processes as indicators and make them available and legible for policy makers.

1.3 Agriculture and the Future

The future of human kind seems inevitably urban and agriculture continues to support urbanization. Considering that the human population has not yet stabilized, we can suppose that agricultural landscapes will become increasingly widespread (Laurance et al., 2014). Agricultural sectors are quite adaptive. They learn from their past. Therefore, we can expect adaptive changes during such growth.

Sugarcane and cattle production extensively depended on slavery and destruction of pristine ecosystems during almost four of their five centuries in Brazil. On the other hand, silviculture is more recent and technological. In order to keep producing food, fiber and energy in the future these sectors will have to adapt to the multi-functionality of agricultural landscapes, which will keep their presumptive mission of producing domesticated species, but will have to add a second, although crucial mission of conserving wild species and their evolutionary processes (Martinelli et al., 2010; Verdade et al., 2014a). This will require continuous advances in its theoretical basis, technology and governance.

Acknowledgements: This study was funded by the Biota Program/FAPESP (Proc. 2006/60954-4). Luciano M. Verdade holds a CNPq Productivity Scholarship (309468/2011-6).

Bibliography

Azevedo, F.C.C., Concone, H.V.B., Pires-da-Silva, A., & Verdade, L.M. (2010). Puma (*Puma concolor*) predation on a water buffalo (*Bubalus bubalis*). Mammalia, 74, 431-432.

Bekunda, M., Palm, C.A., de Fraiture, C., et al. (2009). Biofuels in developing countries. In R.W. Howarth & S. Bringezu (Eds.), Biofuels: Environmental Consequences and Interactions with Changing Land Use. Proceedings of the Scientific Committee on Problems of the Environment (SCOPE) International Biofuels Project Rapid Assessment, 22-25 September 2008, Gummersbach Germany (p.249-269). Ithaca NY, USA: Cornell University. (http://cip.cornell.edu/biofuels/).

Conway, G.R., & Barbier, E.B. (1988). After the green Revolution: Sustainable and equitable agricultural development. Futures, 20 (6), 651-670.

Dean, W. (1995). With Broadax and Firebrand: The Destruction of the Brazilian Atlantic Forest. Berkeley, CA, USA: University of California Press.

Diamond, J. (1997). Guns, Germs and Steel: The fates of Human Societies. New York: W.W. Norton.

Dirzo, R., Young, H.S., Galetti, M., et al. (2014). Defaunation in the Anthropocene. Science, 345, 401-406.

Dotta, G., & Verdade, L.M. (2007). Trophic categories in a mammal assemblage: diversity in an agricultural landscape. Biota Neotropica, 7 (2), 287-292.

Dotta, G., & Verdade, L.M. (2009). Felids in an agricultural landscape in São Paulo, Brazil. CATnews, 51, 22-25.

Dotta, G., & Verdade, L.M. (2011). Medium to large-sized mammals in agricultural landscapes of South-eastern Brazil. Mammalia, 75, 345-352.

Driscoll, D.A., Banks, S.C., Barton, P.S. et al. (2013). Conceptual domain of the matrix in fragmented landscapes. TREE, 28 (10), 605-613.

Faith, D.P. & Pollock, L.J. (2014). Phylogenetic diversity and the sustainable use of biodiversity. In: L.M. Verdade, M.C. Lyra-Jorge & C.I. Piña (Eds.), Applied Ecology and Human Dimensions in Biological Conservation (p.35-52). Heidelberg, Germany: Springer-Verlag, DOI: 10.1007/978-3-642-54751-5_3.

Fearnside, P.M. (2008). The roles and movements of actors in the deforestation of Brazilian Amazonia. Ecology and Society, 13 (1), 1-18.

Ferraz, J.B.S., & de Felício, P.E. (2010). Production systems: An example from Brazil. Meat Science, 84, 238-243.

Ferraz, K.P.M.B., Lechevalier, M.-A., Couto, H.T.Z., & L.M. Verdade (2003). Damage caused by capybara (*Hydrochaeris hydrochaeris*) on a cornfield in São Paulo, Brasil. Scientia Agrícola, 60 (1),191-194.

Ferraz, K.M.P.M.B., Ferraz, S.F.B., Moreira, J.R., et al. (2007). Capybara (*Hydrochoerus hydrochaeris*) distribution in agroecosystems: a cross-scale habitat analysis. Journal of Biogeography, 34 (2), 223-230.

Ferraz, K.M.P.M.B., Peterson, A.T., Scachetti-Pereira, R. et al. (2009). Distribution of capybaras in an agroecosystem, Southeastern Brazil, based on ecological niche modeling. Journal of Mammalogy, 90 (1), 189–194.

Ferraz, K.M.P.M.B., Manly, B., & Verdade, L.M. (2010). The influence of environmental variables on capybara (*Hydrochoerus hydrochaeris*: Rodentia, Hydrochoeridae) detectability in anthropogenic environments of southeastern Brazil. Population Ecology, 52, 263–270.

Ferronato, B.O., Genoy-Puerto, A., Piña, C.I., et al. (2009a). Notes on the hematology of free-living *Phrynops geoffroanus* (Testudines: Chelidae) in polluted rivers of southeastern Brazil. Zoologia, 26 (4), 795-798.

Ferronato, B.O., Marques, T.S., Guardia, I. (2009b). The turtle *Trachemys scripta elegans* (Testudines, Emydidae) as an invasive species in a polluted stream of southeastern Brazil. Herpetological Bulletin, 109, 29-34.
Galetti, M., Guevara, R., Côrtes, M.C., et al. (2013). Functional extinction of birds drives rapid evolutionary changes in seed size. Science, 340, 1086-1090.
Gheler-Costa, C., Vettorazzi, C.A., Pardini, R., Verdade, L.M. (2012). The distribution and abundance of small mammals in agroecosystems of Southeastern Brazil. Mammalia, 76, 185-191, DOI: 10.1515/mammalia-2011-0109.
Gheler-Costa, C., Sabino-Santos, Jr. G., Amorim, L.S., et al. (2013). The effect of pre-harvest fire on the small mammal assemblage in sugarcane fields. Agriculture, Ecosystems and Environment, 171, 85-89.
Glaser, B. & Woods, W. (2004). Explorations in Amazonian Dark Earths in Time and Space. Heidelberg, Germany: Springer-Verlag.
Harris, L.D. (1984). The Fragmented Forest: Island Biogeogrphy Theory and the Preservation of Biotic Diversity. Chicago: The University of Chicago Press.
Instituto de Economia Agrícola – IEA (2014). Área e produção dos principais produtos da agropecuária do estado de São Paulo. http://www.iea.sp.gov.br/out/banco/menu.php. Accessed August 2, 2014.
Jaksic, F.M., Iriarte, J.A., Jiménez, J.E., & Martínez, D.R. (2002). Invaders without frontiers: cross-border invasions of exotic mammals. Biological Invasions, 4, 157-173.
Joly, C.A., Metzger, J.P. & Tabarelli, M. (2014). Experiences from the Brazilian Atlantic Forest: ecological findings and conservation initiatives. New Phytologist, 203, 459-473.
Koh, L.P. (2007). Potential Habitat and Biodiversity Losses from Intensified Biodiesel Feedstock Production. Conservation Biology, 21 (5), 1373–1375.
Koh, L.P., Miettinen, J., Liew, S.C. & Ghazoul, J. (2011). Remotely sensed evidence of tropical peatland conversion to oil palm. PNAS, 108 (12), 5127–5132.
Lafferty, K.D. (2014). Biodiversity loss and infectious diseases. In L.M. Verdade, M.C. Lyra-Jorge & C.I. Piña (Eds.), Applied Ecology and Human Dimensions in Biological Conservation (p.73-89). Heidelberg, Germany: Springer-Verlag, DOI: 10.1007/978-3-642-54751-5_5.
Lapola, D.M., Schaldach, R., Alcamo, J., et al. (2010). Indirect land-use changes can overcome carbon savings from biofuels in Brazil. PNAS, 107 (8), 3388–3393.
Laurance, W.F., Sayer, J., & Cassman, K.G. (2014). Agricultural expansion and its impacts on tropical nature. TREE, 29 (2), 107-116.
Lindenmayer, D.B., & Likens, G.E. (2010). Effective Ecological Monitoring. Collingwood, Australia: CSIRO Publishing.
Liu, Y., Duan, M., & Yu, Z. (2013). Agricultural landscapes and biodiversity in China. Agriculture, Ecosystems and Environment, 166, 46–54.
Lyra-Jorge, M.C., Gheler-Costa, C., Piña, et al. (2014). Wildlife surveys in agricultural landscapes: Terrestrial medium- to large-sized mammals. In L.M. Verdade, M.C. Lyra-Jorge & C.I. Piña (Eds.), Applied ecology and human dimensions on biological conservation (p.133-147). Heidelberg, Germany: Springer-Verlag, DOI: 10.1007/978-3-642-54751-5_9.
MacArthur, R.H., & Wilson, E.O. (1967). The Theory of Island Biogeography. New Jersey, USA: Princeton Univerty Press, Princeton.
Magnusson, W.E., Braga-Neto, R., Pezzini, F., et al. (2013). Biodiversidade e monitoramento ambiental integrado. Áttema Editorial. Santo André, SP, Brasil.
Martin, P.S, Gheler-Costa, C., Lopes, P.C., et al. (2012). Terrestrial non-volant small mammals in agro-silvicultural landscapes of Southeastern Brazil. Forest Ecology and Management, 282,185-195, http://dx.doi.org/10.1016/j.foreco.2012.07.002.
Magnusson, W.E. 1999. Too early for space ecology? Bulletin of the Ecological Society of America, 80 (1), 87-88.

Magnusson, W.E. (2001). Catchments as basic units of management in conservation biology courses. Conservation Biology, 15 (5), 1464-1465.
Magnusson, W.E. (2006). Homogeneização biótica. In C.F.D. Rocha, H.G. Bergallo, M van Sluys & M.A.S. Alves (Eds.), Biologia da Conservação: Essências (p.211-229). São Carlos, SP, Brasil: RiMa Editora.
Magnusson, W.E., Braga-Neto, R., Pezzini, F., et al. (2013). Biodiversidade e monitoramento ambiental integrado. Áttema Editorial. Santo André, SP, Brasil.
Magnusson, W.E., Lawson, B., Baccaro, F., et al. (2014). Multi-taxa surveys: integrating ecosystem processes and user demands. In L.M. Verdade, M.C. Lyra-Jorge & C.I. Piña (Eds.), Applied Ecology and Human Dimensions on Biological Conservation (p.177-187). Heidelberg, Germany: Springer-Verlag. DOI: 10.1007/978-3-642-54751-5_12.
Marques, T.S., Lara, N.R.F., Bassetti, L.A.B., et al. (2013). Population structure of *Mesoclemmys vanderhaegei* (Testudines, Chelidae) in a silvicultural system in southeastern Brazil. Herpetology Notes, 6, 179-182.
Martinelli, L.A., Naylor, R., Vitousek, P.M., & Moutinho, P. (2010a). Agriculture in Brazil: Impacts, costs, and opportunities for a sustainable future. Current Opinion in Environmental Sustainability, 2, 431-438.
Martinelli, L.A., Joly, C.A., Nobre, C.A., & Sparovek, G. (2010b). A falsa dicotomia entre preservação da vegetação natural e a produção agropecuária. Biota Neotropica, 10, 323-330.
Meche, A., Martins, M.C., Lofrano, B.E.S.N., et al. (2009). Determination of heavy metals by inductively coupled plasma-optical emission spectrometry in fish from the Piracicaba River in Southern Brazil. Microchemical Journal, 94, 171–174.
Mendenhall, C.D., Karp, D.S., Meyer, C.F.J., et al. (2014). Predicting biodiversity change and averting collapse in agricultural landscapes. Nature, 504, 213-217.
Metzger, J.P., Lewinsohn, T.M., Joly, C.A., et al. (2010). Brazilian Law: Full Speed in Reverse? Science, 329, 276-277.
Mooney, H.A., & Hobbs, R.J. (2000). Invasive species in a changing world. Washington, D.C.: Island Press.
Moreira, J.R., Verdade, L.M., Ferraz, K.M.P.M.B., & Macdonald, D.W. (2012). The Sustainable Management of Capybaras. In J.R. Moreira, K.M.P.M.B. Ferraz, E.A. Herrera & D.W. Macdonald (Eds.), Capybara: Biology, Use and Conservation of an Exceptional Neotropical Species (p.283-302). New York: Springer, DOI: 10.1007/978-1-4614-4000-0_17.
Murphy, R., Woods, J., Black, M., & McManus, M. (2011). Global developments in the competition for land from biofuels. Food Policy, 36, S52-S61.
Nacci, D.E., Gleason,T.R., & Munns, Jr. W.R. (2002). Evolutionary and ecological effects of multigenerational exposures to anthropogenic stressors. Human and Ecological Risk Assessment, 8 (1), 91-97.
Palm, C.A., Swift, M.J., & Woomer, P.L. (1996). Soil biological dynamics in slash-and-burn agriculture. Agriculture, Ecosystems and Environment, 58, 61-74.
Palumbi, S.R. (2001). Humans as the world's greatest evolutionary force. Science, 293, 1786-1790.
Palmeira, F.B.L., Crawshaw Jr., P.G., Haddad, C.M., et al. (2008). Cattle depredation by puma (*Puma concolor*) and jaguar (*Panthera onca*) in Northerh Goiás, Central-western Brazil. Biological Conservation, 141 (1), 118-125.
Penteado, M., Silva, W.R., & Verdade, L.M. (2014). Wildlife surveys in agricultural landscapes: Birds. In L.M. Verdade, M.C. Lyra-Jorge & C.I. Piña (Eds.), Applied ecology and human dimensions on biological conservation (p.149-158). Heidelberg, Germany: Springer-Verlag, DOI: 10.1007/978-3-642-54751-5_10.
Pezzini, F.F.P.H., Melo, A., Oliveira, D.M.S., et al. (2012). The Brazilian Program for Biodiversity Research (PPBio) Information System. Biodiversity & Ecology, 4, 265-274, DOI: 10.7809/b-e.00083.

Phelps, J., Carrasco, L.R., Webb, E.L., et al. (2013). Agricultural intensification escalates future conservation costs. PNAS, 110 (19),7601-7606.
Piña, C.I., Lance, V.A., Ferronato, B.O., et al. (2009). Heavy metal contamination in *Phrynops geoffroanus* (Schweigger, 1812) (Testudines: Chelidae) in a River Basin, São Paulo, Brazil. Bulletin of Environmental Contamination and Toxicolology, 83, 771–775.
Pinto, G.R.M., Ferraz, K.M.P.M.B., do Couto, H.T.Z. & Verdade, L.M. (2006). Detectability of capybaras (*Hydrochoerus hydrochaeris*) in forested habitats. Biota Neotropica, 6 (1), 1-5.
Prevedello, J.A., & Vieira, M.V. (2010). Plantation rows as dispersal routes: A test with didelphid marsupials in the Atlantic Forest, Brazil. Biological Conservation, 143, 131–135.
Pyne, S.J. (1998). Forged in fire: History, land, and anthropogenic fire. In W. Balée (Ed.), Advances in Historical Ecology (p.62-103). New York: Columbia University Press.
Rosalino, L.M., Martin, P.S., Gheler-Costa, C., et al. (2013). Allometric relations of Neotropical small rodents (Sigmodontinae) in anthropogenic environments. Zoological Sciences, 30, 585-590, doi:10.2108/zsj.30.585.
Rosalino, L.M., Lyra-Jorge, M.C., & Verdade, L.M. (2014). Adaptation and evolution in hanging environments. In L.M. Verdade, M.C. Lyra-Jorge & C.I. Piña (Eds.), Applied ecology and human dimensions on biological conservation (p.53-71). Heidelberg, Germany: Springer-Verlag, DOI: 10.1007/978-3-642-54751-5_4.
Sala, O.E., Sax, D., Leslie, H. (2009). Biodiversity consequences of biofuel production. In R.W. Howarth & S. Bringezu (Eds.), Biofuels: Environmental Consequences and Interactions with Changing Land Use (p.127-137). Proceedings of the Scientific Committee on Problems of the Environment (SCOPE) International Biofuels Project Rapid Assessment, 22-25 September 2008, Gummersbach Germany. Ithaca NY, USA: Cornell University, http://cip.cornell.edu/biofuels/.
Sarkis-Gonçalves, F., Castro, A.M.V., & Verdade, L.M. (2004). The influence of weather conditions on caiman night-counts. In Crocodiles. Proceedings of the 17th Working Meeting of the Crocodile Specialist Group (pp.387-393). Gland, Switzerland: IUCN - The World Conservation Union.
Snell-Rood, E.C., & Wick, N. (2013). Anthropogenic environments exert variable selection on cranial capacity in mammals. Proceedings of the Royal Society B, 280,1-9.
Sparovek, G., Barretto, A., Klug, I., et al. (2010). A revisão do Código Florestal Brasileiro. Novos Estudos, 88 (1),181-205.
Sparovek, G., Berndes, G., Barreto, A.G.O. P., & Klug, I.L.F. (2012). The revision of the Brazilian Forest Act; Increased deforestation or a historic step toward balancing agricultural development and nature conservation. Environmental Science & Policy, 16, 65-72.
Stape, J.L., Binkley, D., Ryan, M.G. (2004). *Eucalyptus* production and the supply, use and efficiency of use of water, light and nitrogen across a geographic gradient in Brazil. Forest Ecology and Management, 193, 17-31.
Thompson, J.N. (1998). Rapid evolution as an ecological process. TREE, 13 (8), 329-332.
Tompkins, D.M., Dunn, A.M., Smith, M.J., & Telfer, S. (2011). Wildlife diseases: From individuals to ecosystems. Journal of Animal Ecology, 80, 19-38.
Van Zanden, J.L. (1991). The first green revolution: The growth of production and productivity in European agriculture, 1870-1914. Economic History Review, 46 (2), 215-239.
Verdade, L.M., & Ferraz, K.M.P.M.B. (2006). Capybaras (*Hydrochoerus hydrochaeris*) on an anthropogenic habitat in Southeastern Brazil. Brazilian Journal of Biology, 66 (1b), 371-378.
Verdade, L.M., & Ferraz, K.M.P.M.B. (2013). Capivaras de Piracicaba: o bom, o mau e o feio.. In A.M. Meira, J.A. Monti, K.M.P.M.B. Ferraz, M. Cooper, R.B. Caramez & W.B.C. Delitti (Eds.), Febre Maculosa: Dinâmica da Doença, Hospedeiros e Vetores (p.143-161). Piracicaba, SP, Brasil: ESALQ.
Verdade, L.M., & Seixas, C.S. (2013). Confidencialidade e sigilo profissional em estudos sobre caça. Biota Neotropica, 13 (1), 21-23.

Verdade, L.M., Zucoloto, R.B., & Coutinho, L.L. (2002). Microgeographic variation in *Caiman latirostris*. Journal of Experimental Zoology, 294 (4), 387-396.

Verdade, L.M., Rosalino, L.M., Gheler-Costa, C., Pedroso, N.M., & Lyra-Jorge, M.C. (2011). Adaptation of mesocarnivores (Mammalia: Carnivora) to agricultural landscapes of Mediterranean Europe and Southeastern Brazil: A trophic perspective. In L.M. Rosalino, & C. Gheler-Costa (Eds.), Middle-Sized Carnivores in Agricultural Landscapes (p.1-38). New York: Nova Science Publishers.

Verdade, L.M., Gheler-Costa, C., Penteado M., & Dotta, G. (2012). The Impacts of sugarcane expansion on wildlife in the state of São Paulo, Brazil. Journal of Sustainable Bioenergy Systems, 2, 138-144, DOI:10.4236/jsbs.2012.24020.

Verdade, L.M., & Ferraz, K.M.P.M.B. (2013). Capivaras de Piracicaba: o bom, o mau e o feio. In A.M. Meira, J.A. Monti, K.M.P.M.B. Ferraz, M. Cooper, R.B. Caramez & W.B.C. Delitti (Eds.), Febre Maculosa: Dinâmica da Doença, Hospedeiros e Vetores (p.143-161). Piracicaba, SP, Brasil: ESALQ.

Verdade, L.M., Lyra-Jorge, M.C., & Piña, C.I. (2014). Redirections in conservation biology. In: L.M. Verdade, M.C. Lyra-Jorge & C.I. Piña (Eds.), Applied Ecology and Human Dimensions in Biological Conservation (p.3-17). Heidelberg, Germany: Springer-Verlag, DOI: 10.1007/978-3-642-54751-5_1.

Verdade, L.M., Penteado, M., Gheler-Costa, C., et al. (2014). The conservation value of agricultural landscapes. In: L.M. Verdade, M.C. Lyra-Jorge & C.I. Piña (Eds.), Applied ecology and human dimensions on biological conservation (p.91-102). Heidelberg, Germany: Springer-Verlag, DOI: 10.1007/978-3-642-54751-5_6.

Vynne, C., Booth, R.K., & Wasser, S.K. (2014). Physiological implications of landscape use by free-ranging maned-wolves (*Chrysocyon brachyurus*) in Brazil. Journal of Mammalogy, 95 (4), 696-706.

Fábio Henrique Comin, Carla Gheler-Costa

2 The Role of Family Farms in Environmental Conservation in Agricultural Landscapes

2.1 Introduction

This is an integrated ecological and social vision study that seeks to understand how social and economic practices of farmers relate to environmental issues with a view to conservation. The central theme of this study addresses human dimensions and biodiversity conservation in agricultural landscapes, understanding the relationship between the role of environmental research and the participation of family farming in the Corumbataí river basin in the state of São Paulo, Brazil.

We know that changes in land use and land cover are among the major forces causing environmental changes around the world, and agriculture is still one of the human activities that cause the greatest impact on natural environments, either by habitat destruction, the introduction of exotic species or pollution (Turner & Meyer, 1994). Martinelli and Filoso (2009) consider that agriculture only exists where ecosystems are able to maintain their basic operating functions; therefore, the environmental services provided by ecosystems in general should be valued. According to Geist and Lambim (2002), anthropogenic actions also cause indirect damages to natural resources, which are based on economic, political, technological, cultural and demographic factors. Therefore, understanding the dynamics of human dimensions in agricultural landscapes from the standpoint of environmental multifunctionality is possibly the best way to conserve biodiversity in these landscapes.

In recent decades, the Corumbataí river basin suffered changes in land use that affected agricultural production practices of small-scale farmers and increased the practice of agriculture focused on "agribusiness". Concurrently with the landscape changes that resulted on an agrarian system designed for the production of commodities, the relevance of environmental issues also increased, and the region became host to numerous scientific investigations regarding the environment. Within this context, humans have been primarily responsible for the landscape changes, and research that seeks to understand the cause/effect relationship of the degradation and/or the conservation of natural resources in agricultural landscapes is needed.

Recently, many studies have argued that ecosystem services are related to ecological processes and also to evolutionary processes. According to Younés (2001), because humans play a key role in the alteration of ecosystem processes, the integration of human activities on ecosystems is crucial to consolidate efforts for the conservation of biodiversity. To Brondizio and Moran (2008), understanding the processes that mediate the perception of environmental changes in both, individual behavioural responses and the local community, is the key element in human and natural systems research. There are good examples in the scientific debate that integrate the social and natural sciences (Thomas, 1956; Turner II et al., 1990; Moran, 1990; Berlin, 1992; Begossi, 1993; Vitousek et al., 1997; Berkes et al., 2003; Drew, 2006; Liu et al., 2007; Glikman et al., 2011; Sponarski et al., 2014).

In some ways, this study had strong links with multidisciplinary research proposals because it is linked to the research project "Socio-environmental changes in the state of São Paulo and perspectives for conservation" (Biota/FAPESP). Thus, this study broadly joined natural sciences with social sciences. This project showed that about 80% of the agricultural matrix in the state of São Paulo is responsible for maintaining biodiversity and is worthy of conservation efforts. In this sense, some studies in ecology have pointed to a new paradigm in environmental conservation in agricultural landscapes and recognize the need to "look at the whole" and not just isolated issues or issues specific to each area of knowledge (Perfecto et al., 2009; Joly et al., 2010). Specifically, in the Corumbataí river basin, Gheler-Costa et al. (2012), Dotta and Verdade (2007, 2009, 2011), Gerhard and Verdade (Chapter 13, this book) and Penteado and Verdade (Chapter 15, this book) show that not only should the forest fragments and water resources be prioritised in scientific research, but all of the processes of land use and occupation should be included because these processes influence the local biota.

This axiom relies on recognising that the lack of socio-environmental studies in agricultural landscapes limited environmental conservation and that such studies are essential to environmental conservation. In order to investigate this question, it is necessary to understand that in a socially occupied territory, residents are protagonists in the management and conservation of biodiversity process. This accentuates the need for applied scientific research directed at the relationships and interactions between human dimensions in agricultural landscapes, which may contribute to the establishment of socio-environmental nature conservation policies.

In Brazil, the Corumbataí river basin is one of the most studied regions. Since the 1980s, it has been the subject of scientific research on the conservation of biodiversity and water issues. Many studies in this basin have contributed to the identification, classification and characterisation of fragile areas important to conservation purposes, especially with regard to water attributes. However, these studies, conducted over almost three decades in the Corumbataí river basin, apparently have not contributed to political decisions, given that the region is suffering from pollution, deforestation and rural exodus (Armas et al., 2007; Antoniazzi, 2008; Garden et al., 2008; Nobre, 2008; Borges et al., 2008; Comin et al., 2009; Mingoti & Vettorazzi, 2011; Morais et al., 2012). We found that human dimensions have been usually ignored in research on environmental issues during the last three decades by local researchers and that family farmers should be taken into consideration regarding the environmental conservation in agricultural landscapes.

In Brazil, family farming accounts for approximately 30% of the GDP (Gross Domestic Product) of the agricultural sector because approximately 30% of all agricultural land is under the possession of family farmers, who contribute approximately 30-50% of food security at a global level (Altieri, 2004). In São Paulo, family farms cover an area of more than 2 million hectares, and in the Corumbataí river basin, specifically in the cities studied (Analândia, Ipeúna and Corumbataí), family farms comprised of an area of approximately 7,000 hectares, which produces food and environmental services. The family farm includes forest agricultural production, fishing, ranching and aquaculture that are managed and operated by a family, and it is predominantly dependent on family labour, both by women and men. Family farms grow traditional food crops and contribute to the protection of agricultural biodiversity and the sustainable use of natural resources. In the Corumbataí river basin, family farms are geared to the production of beans, rice, cotton, garlic, corn, milk, meat, poultry, eucalypt, sugar cane, vegetables, processed products and distilled liquor. These products cater mainly to the local demand, which is culturally well accepted and contributes to the economy.

For Cullen Jr et al. (2000), the type of neighbourhood and land use can profoundly affect biological diversity, ecological processes and conservation. In the case of the Corumbataí river basin, we know little about the rural population and how it relates to nature. On a local scale, residents of rural areas in the basin are actors that play an important role in conserving (or not) local ecosystems as they still relate to the natural environment through their farming practices and the use of water resources.

The general assumption of this study is that in the Corumbataí river basin, there is a very large gap between the research conducted for conservation and research with social actors (human dimensions) who live and use this same landscape. The present study aimed to integrate the family farmer in matters of environmental conservation of the agricultural landscape, understanding that there is no environmental conservation without social conservation. In this sense, we assume the relevance of human dimensions in biodiversity conservation in agricultural landscapes.

2.2 Study Area

The Corumbataí river basin is situated in the east-central region of the state of São Paulo, between parallels 22°04'46"and 22°41'28"S and meridians 47°26'23" and 47°56'15"N (Fig. 2.1). The Corumbataí river crosses the municipalities of Analândia, Ipeúna, Itirapina, Charqueada, Santa Gertrudes, Piracicaba, Rio Claro, Iracemápolis, São Pedro and Cordeirópolis, and is an example of the need for environmental conservation, given the importance of its waters for public water supply as well as its ecological functions. The Corumbataí river has five sub-basins: High, Medium and Low Corumbataí, Passa-Cinco and Rio Claro IPEF (2001) and the following main tributaries: Cabeça, Passa-Cinco and Claro (Nobre, 2008).

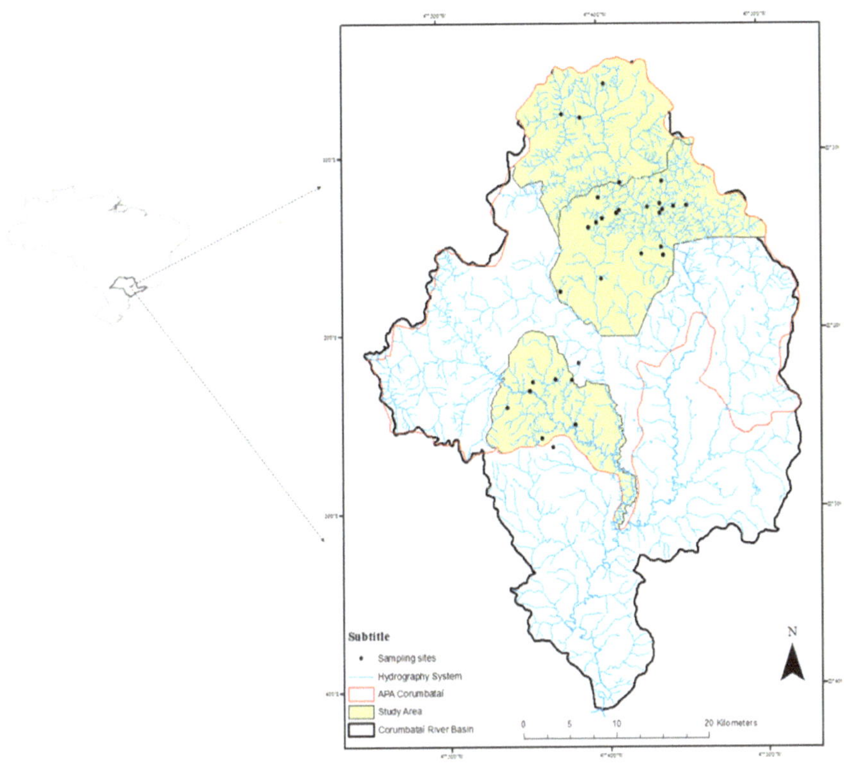

Fig. 2.1: Study sites and sampling sites in the Corumbataí APA.

The climate is characterised as subtropical, with well-defined seasons that are dry in the winter, with average temperatures of approximately 17°C, and rainy in the summer, with average temperatures of 22°C (Ceapla, 2013). The relief of the basin has the presence of elongated hills and springs, medium and large hills and perennial and intermittent valleys. The basin geomorphological characteristics are prevalent in the Cenozoic era, with headwaters in the Serra Geral cuestas, typical of the Paraná sedimentary basin, presenting four geological formations, including the Pirambóia, Rio Claro, Irati and Corumbataí (Koffler, 1993, 1994).

According to Dean (1977), in the early 19th century this region was almost completely covered by forest and Cerrado. Rodrigues (1999) highlights the five types of forest formations: i) seasonal dry forests; ii) semideciduous upland, forests that are not susceptible to drought; iii) swamp forest; iv) riparian forest, and v) Cerrado. According to Valente (2005), current land use includes grassland (42.29%), sugarcane (27.77%), native forest and Cerrado (11.97%), planted forest (5.69%), annual and perennial agriculture (4.96%), urban areas, road networks, mining and water courses (7.32%).

In the upper portion of the Corumbataí river basin is the Environmental Protection Area (APA) of Corumbataí Botucatu and Tejupa, a conservation unit established by State Decree 20960 on June 8, 1983, and recognised in 1985 by the Federal Law of 18 July 2000, which established the National System of Conservation Units. The APA Corumbataí spreads over 335,205 hectares, and the predominant economic activities are the cultivation of sugar cane, eucalypt plantations, agriculture, livestock and the mining of sand, gravel and clay. The Corumbataí river basin is the universe of inference of this study, and the sample universe is the Corumbataí Environmental Protection Area (APA Corumbataí). This region can be considered a representative region for the entire Corumbataí river basin because it is in these areas where research on local populations was conducted in this study, as they require more attention in relation to environmental conservation (Ferraz et al., 2009) (Fig. 2.1).

2.3 Development

To verify the scientific production in the Corumbataí river basin, two sources of data were explored: i) library collections of regional scientific literature and ii) an inventory of publications in indexed scientific journals. The searches were conducted in digital libraries and databases of peer-reviewed journals; 42 keywords related to social and environmental issues consistent with the subject of the proposed study were used, and the search period was from 1950 to 2012: native forest, eucalypt, sugarcane, grassland, pesticides, pollution, fragmentation, biodiversity, fauna, flora, vertebrates, invertebrates, fish, mammals, birds, reptile, amphibian, insect, palaeontology, conservation unit, environmental protection area, environmental management, environment, environmental, environmental education, sustainability, conservation, water, soil, family agriculture, sociology, anthropology, rural, rurality, rural sociology, agriculture, agricultural production, economy, Passa-Cinco, Cabeça, Ipeúna, and Analândia; all terms were truncated with Corumbataí.

For the socioeconomic and environmental survey of family farmers, interviews were performed in the field that were constructed from the previous classification and categorisation of the properties of the APA Corumbataí. During the fieldwork, we adopted a specific methodology called "Diagnostic Analysis of Land Systems", available by agreement INCRA/FAO (2010). It includes helpful research approaches and participatory management (Blumenthal & Jannink, 2000) and action research, which focus only on investigations that empower local communities (Chambrers, 1994). Some aspects related to the origin of production systems on ideological and symbolic levels (Douglas, 1969; Shalins, 1976) were also explored. Thus the questionnaires and interviews were used to obtain oral reports about production units, and each of these interviews was recorded with the permission of the interviewees.

2.4 Analytical Methodology

For the review of the production (scientific and non-scientific) in the river basin, the data were organised according to their origin, being divided as follows: Thesis/ Dissertation (doctor and master), scientific peer-reviewed articles and other (maps, books, monographs, conference abstracts). The division of data into three categories was carried out to distinguish between scientific materials (peer reviewed and published) and academic materials, such as theses and others. The results were conducted so that we could ascertain the number of studies in the river basin that were actually made public, i.e. sent to journals and analysed by peers and then published. We considered doctoral thesis, master's dissertations, monographs and conference abstracts as academic and non-scientific texts.

A frequency rate of records was established to standardise the sampling effort among the different databases analysed because some records may occur in more than one database. To establish this rate, we used the following formula:

$$\text{Frequency Index} = \frac{\text{Number of records from the database}}{\text{Total number of records collected}}$$

For socio-environmental diagnoses we have built a framework of analysis with 4 main themes: i) Which are they? ii) Who are they? iii) How do they live? and iv) Do they have relationships with biodiversity conservation? Thus, we sought to establish rural development in progress, and the trends of family farming in the region. Therefore, from the framework, each interview/questionnaire was systematised, establishing the analytical methodology. Thus, the relevant questions related to "socio-economic-environmental" data were analysed in the following ways: 1) the data were divided into categories; 2) compared within the pre-defined categories; and 3) regarding family agriculture and the environment, the notion of environmental multifunctionality was used as an analytical tool.

Specifically, for analyses of family farming, we used the work of Hugues Lamarche (1993, 1998) that was conducted in Canada, France, Tunisia, Poland and Brazil under the titles "Family Farming: a multiform reality" and "Family Farming: Myth to reality". These studies provide a comprehensive and thorough study of the family farm in the contemporary world and its many facets. Further analysis of this diagnosis made with family farmers were based on the criteria of multifunctionality, which refers to the set of services and products created by farming that have direct or indirect impacts on the economy, society and nature (Bonnal et al., 2003; Bonnal et al., 2009). Adopting analysis of multifunctionality implies recognising the economic, environmental and social functions of family farming in the Conrumbataí river basin in addition to enabling the understanding of the human dimensions related to strategies of social reproduction and environmental conservation.

2.5 Results

2.5.1 Scientific Production in the Corumbataí River Basin

In the sampling universe, regarding the term Corumbataí and its 42 complementary keywords, we found 3,026 records from 1950 to 2012. We observed a consistent increase since 2000 with a tendency to increase even further in the next decade, demonstrating the evolution of scientific literature (Tab. 2.1).

Tab. 2.1: Evolution of bibliographic production for 60 years in the Corumbataí river basin.

Decade	Number of scientific literature
1950	2
1960	0
1970	5
1980	8
1990	36
2000	136
2010	10
2012	98

Given the concern to demonstrate that human dimensions are disregarded in research in the Corumbataí river basin in relation to other areas of knowledge, the set of 42 keywords used was satisfactory because it covered the entire universe of interest. Among these keywords, "Corumbataí" followed by "environment", "water", "soil", "vegetation" and "fish" were keyword pairs related to the natural sciences that appeared in most searches. The results demonstrate that human dimensions are not usually considered in research conducted in the Corumbataí river basin, because the keywords "anthropology", "sociology", "rurality", "rural" and "family farmers", had fewer than five matches in our search (Fig. 2.2).

2.5.2 Family Farms: Socioeconomic and Environmental Characterisation

In the APA Corumbataí, there are 49 family farms and 151 people, 74 of whom are directly involved in agricultural activities. We established three types of family farmers using analogies to the models proposed in the literature (Lamarche, 1998; Carmo, 1999; Guanzirolli & Cardim, 2000; Abramovay, 2003; Gonçalves & Souza 2005; MDA, 2009):

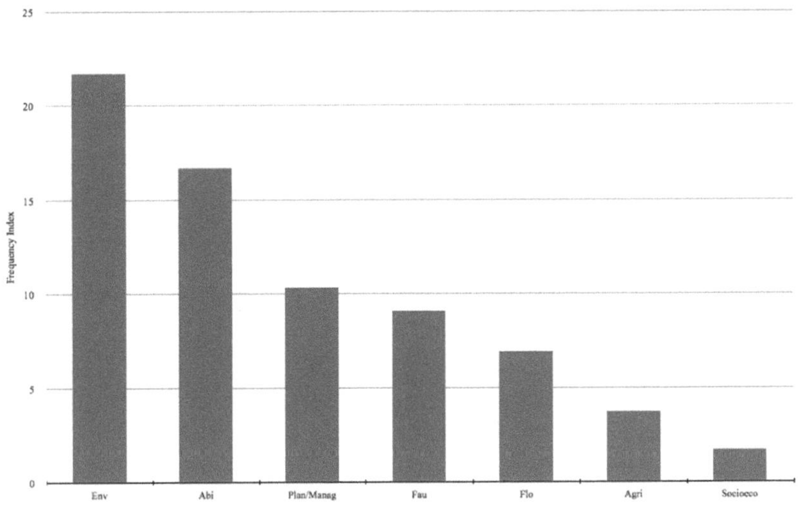

Fig. 2.2: Frequency index of keywords for some areas (Env - Environment; Abi - Abiotic; Plan/Manag - Planning and Management; Fau - Fauna; Flo - Flora; Agri - Agriculture; Socioeco - Socioeconomy).

i) Family farmers who identify themselves as such, who are competitive and use technology, and who have the ability to adapt to the environment have greater autonomy from the market and are in the process of growth. In this group, which includes productive units where the exploitation of the land directly by the owner is not the only possible land use, are those who lease land for sugarcane cultivation and/or have integrated farms with cooperatives or private firms. This type of producer reveals features that Lamarche (1998) calls the business model, or the dependent system, where dependence is exercised both in technological as well as financial terms;
ii) The second type includes content producers and entrepreneurs who are family farmers who identify themselves as such and are less capitalised and less reliant on technology, but who are able to think in terms of agricultural maintenance in the field and approach the model of subsistence agriculture. These characteristics categorise these producers as very familiar and happy, using mainly family labour. The future of these proprieties are thought of in terms of reproduction and continuity of the model and, therefore, a family business, as proposed by Lamarche (1998) and;
iii) The third type includes family farmers who are marginalised and less competitive, dependent on policy support and who will most likely fail, further accentuating the desertification of rural areas. With little land structure and occupancy of unsuitable land, mostly in cattle raising activities, been moderately familiar, while Lamarche (1998) categorises them as resigned or frustrated. They do not fit the models of

subsistence farming, as they are among those who are classified as having low agricultural diversity.

Relating to environmental issues, we have observed in productive unit visits the existence of 76 water sources and found that 91% of these have permanent preservation areas (APP). Among these, 33% are typical and function as native riparian vegetation, 33% have exotic vegetation, such as sugarcane and grass, 24% have native and exotic vegetation, such as grass and isolated trees, and 9% have no vegetation (exposed soil), and the soil is compacted by livestock use.

All of the farmers declare that the production units investigated have at least one area with native forest fragments, often small and basically formed by secondary forest (abandoned pasture). The presence of even small and/or isolated forest fragments adds possibilities for environmental adaptation in the agricultural landscape as well as for some proposed conservation declaration for the region, such as ecological corridors. Generally, data indicate the high sensitivity of these forest habitats and freshwater environments on agricultural properties.

2.6 Discussion

The results indicate that in the Corumbataí river basin there is a strong evolution of ecological research on biodiversity conservation. These results show that in the fragile rural spaces of APA Corumbataí, family farming presents social, economic and ecological characteristics that are related to and influence processes of nature conservation. Despite the gains obtained by studies conducted in the last sixty years in the Corumbataí river basin, we observe that there is a huge gap in the research regarding human dimensions. We evaluated that the dominant scientific-technical approaches in the field of natural resource conservation, in addition to studies in the physical, biotic and abiotic environment, must incorporate the human dimension in the preservation of biodiversity.

According to Martins (2006), the farmers identify themselves with the perception and appropriation of resources, access, and ecosystem management practices. Family farmers respect the APP areas and keep the Legal Reserves (areas of natural vegetation that the farmers are legally required to have on their properties by the Brazilian Forest Code (Metzger et al., 2010). Because of this respect, such people should be observed as beneficial to local environmental conservation. According to Gliessman (2000), the transition of landscapes altered to more sustainable agroecosystems is complex and requires a redesign of the components of agroecosystems so that they function based on a new set of social and ecological processes. Therefore, we will now treat the issue of environmental multifunctionality of family farming in APA Corumbataí.

Recognising the multifunctional approach to production units means breaking with the exclusive hegemony of economic "tools", which may facilitate the farmers' understanding of their complex relationship with nature. The functions of environmental multifunctionality from productive units will enable the understanding of human dimensions related to strategies of social reproduction and environmental conservation (Bonnal et al., 2003). For some authors, the family farm is diversified in its processes of interaction and use of the environment; therefore, it has a great proximity to sustainability (stability, resilience and environmental equality) (Veiga, 1996; Moreira, 2008). Thus, the emphasis on the multifunctionality of agricultural families and their relationships with the environment can also strengthen the conservation of natural resources in the region.

Thus, in this work, considerations of environmental multifunctionality of agriculture families represents an attempt to recognise the productive units not only as producers of agricultural or social goods but also to extend the scope of its environmental functions to include river basin environmental conservation. Therefore, it is possible to develop a theoretical framework on environmental multifunctionality of family farming in the agricultural landscape as a way of understanding the dynamics of the rural population in the face of nature conservation.

To Brondizio & Moran (2008), understanding the processes that mediate the perception of environmental changes and the individual local community behavioural responses is a key element in the research of human and natural systems. There are good examples in the scientific debate that integrate social and natural sciences, bringing together these two fields (Thomas, 1956; Moran, 1990; Turner II et al., 1990; Berlin, 1992; Begossi, 1993; Vitousek et al., 1997; Berkes et al., 2003; Drew, 2006; Liu et al., 2007; Glikman et al., 2011; Sponarski et al., 2014). These authors share the idea that the social and human dimensions can be absorbed by environmental conservation, providing an integrated view of reality. There are, in other parts of the world, experiences joining the characteristics of natural resources, interactions of social actors, governance and political participation, information exchange, and economic issues with room for both conflicts and synergies between conservation and rural development (Romero et al., 1997; Moran et al., 2002).

These studies demonstrate how people can work together to improve the quality of rural life, guarantee the rights of workers, allow rural democracy, conserve biodiversity and maintain cultural traditions. In Brazil, there are studies that focus on participatory and integrated management considering humans as part of the ecosystem (Ribeiro & Galizoni, 2002; Berkes, 2005; Seixas, 2005; Vieira et al., 2005; Moran, 2011). However, specifically in the Corumbataí river basin, the hegemony of positivist research without a systemic vision of socio-environmental problems is still prevalent. According to Moran (2011), although these issues may divide social and natural scientists due to their emphasis on the environment or people, it is important to understand that there are many reasons, not just biological, that can change the directions of research on biodiversity conservation, including aesthetic, cultural and economic factors.

Moran et al. (1996) remind us that the physical sciences highlighted the role of human actions, and this encouraged the community of social scientists (economists, among them) to join in the effort to understand global environmental change. Thus, research on the human dimensions challenges most of the social sciences to develop new forms of analyses of change in land use. This is also true for ecology, where many things have been done to incorporate and improve ways of thinking in the conservation of agricultural landscapes (Moran et al., 2011), such as green economy, crops certification, payment for environmental services and the concept of sustainability.

2.7 Final Considerations

From these results, we demonstrated that the family farmers investigated in the studied agricultural landscapes respect the APP areas and maintain legal reserves and have a direct relationship with conservation. In addition, there is great agricultural diversity, both in the plants and animals produced by family farmers on their properties. In this sense, we can consider the productive units of agricultural landscapes as small islands in an ocean of production of monocultures and pastures, and this is the principle on which we can base the features of the agricultural landscape or environmental multifunctionality, as suggested in this study. By bringing the family farm to the sphere of environmental conservation, we recognise that the agricultural landscape is socially occupied and that the family farmers have clear links to ecosystem services and conservation.

The results presented allow us to observe that agricultural practices and the ways of life of family farmers in the region are favourable for biodiversity conservation, as these farmers have an important role in environmental conservation by maintaining fragments of APP and RL. Moreover, the results also indicate that in the Corumbataí river basin, there is a consistent improvement of ecological and conservation research and that in fragile rural areas of APA Corumbataí, the inhabitants have social, economic and ecological characteristics that relate to and influence the nature conservation. Therefore, to identify, through this case study, the incorporation of the human dimensions of this relationship and the challenges of environmental conservation, this study took an interdisciplinary techno-scientific model as a trend of research studies of conservation in agricultural landscapes. Thus, combining the natural sciences with social sciences, we suggest that this is one of the goals that interdisciplinary research on nature conservation in agricultural landscapes should prioritize.

Acknowledgments: This study was funded by FAPESP (06/60954-4). FHC was funded by a doctoral fellowship by CNPQ. We would like to thank Prof. Dr. Luciano Martins Verdade by his judicious and important revision of this manuscript. We would like to thank the farmers in the Corumbataí river basin.

Bibliography

Abramovay, R. (2003). O futuro das regiões rurais. Porto Alegre: Editora da UFRGS.

Altieri, M.A. (2004). Linking ecologists and traditional farmers in the search for sustainable agriculture. Frontiers in Ecology and Environment, 2, 1, 35–42.

Antoniazzi, L.B. (2008). Oferta de serviços ambientais na agricultura. Dissertação (Mestrado em Economia Aplicada) - Escola Superior de agricultura "Luiz de Queiroz". Piracicaba: Universidade de São Paulo,.

Armas, E.D.; Monteiro, R.T.R.; Antunes, P.M., et al. (2007). Diagnóstico espaço-temporal da ocorrência de herbicidas nas águas superficiais e sedimentos do rio Corumbataí e principais afluentes. Química Nova, 30, 5, 1119-1127.

Begossi, A. (1993). Ecologia Humana: um enfoque das relações homem-ambiente. Interciência, 18, 121-132.

Berkes, F. (2005). Sistemas sociais, sistemas ecológicos e direitos de apropriação de recursos naturais. In P.F. Vieira, F. Berkes, C.S. Seixas (Eds.), Gestão integrada e participativa de recursos naturais: conceitos, métodos e experiências (pp.47-72). Florianópolis: Secco/APED.

Berkes, F., Colding, J., & FOLKE, C. (2003). Navigating Social-Ecological Systems: Building Resilience for Complexity and Change. Cambridge Univ. Press: Cambridge, UK.

Berlin, B. (1992). Ethobiological classification: principlis of classification of plants and animal in tradicional societies. Princeton: Princeton University Press.

Blumental, D., & Jannink, J. L. (2000). A classification of collaborative management methods. Conservation Biology, 4, 2, 11-13.

Bonnal, P., Piraux, M., & Fusillier, J.L. et al. (2003). Approche de la multifonctionnalité de l'agriculture à la Réunion: les modèles agricoles, la relation agriculture-emploi et la perception des CTE par les acteurs locaux. Paris: MAAPAR et CIRAD.

Bonnal, P., & Maluf, R. S. (2009). Políticas de desenvolvimento territorial e multifuncionalidade da agricultura familiar no Brasil. Política e Sociedade, 14, 211-250.

Borges, A.K.P., Tauk-Tornisielo, S.M., & Domingos, R.N., et al. (2008). Performance of the constructed wetland system for the treatment of water from the Corumbataí River. Brazilian Archives of Biology and Technology, 51, 6, 1279-1286.

Brondizio, E.S., & Moran, E.F. (2008). Human Comensions of Climate Change: The Vulnerabilty of Small Farmers in the Amazon. Philosophical Transactions of the Royal Society, 363, 1803-09.

Carmo, R.B.A. (1999). A Questão Agrária e o Perfil da Agricultura Brasileira.
http:// www.cria.org.br/gip/gipaf/itens/pub/sober. Accessed 10 March 2014.

Ceapla (2013). Atlas ambiental da Bacia do Corumbataí. http://igce.rc.unesp.br/ceapla. Accessed 15 March 2013.

Chambers, R. (1994). The origens and practice of participatory rural appraisal. Word Development, 22, 953-969.

Comin, F.H., Gheler-Costa, C., & Verdade, L. et al. (2009). Relações e conflitos na conservação ambiental da bacia do Rio Passa-Cinco, São Paulo, Brasil. OLAM Ciência e Tecnologia, 9, 2, 254-258.

Cullen Junior, L., Bodmer, R.E., & Padua, C.V. (2000). Effects of hunting in habitat fragments of the Atlantic Forests, Brazil. Biological Conservation, 95, 49-56.

Dean, W. (1977). Rio Claro: um sistema brasileiro de grande lavoura – 1820-1920. Rio de Janeiro: Paz e Terra.

Dotta, G., & Verdade, L.M. (2009). Felids in an agricultural landscape in São Paulo Brazil. Cat News, 51, 6, 24-27.

Dotta, G., & Verdade, L.M. (2007). T rophic categories in a mammal assemblage: diversity in an agricultural landscape. Biota Neotropica, 7, 2, 287-292.

Dotta, G., & Verdade, L.M. (2011). Medium to large-sized mammals in agricultural landscapes of South-eastern Brazil. Mammalia, 75, 345-352.
Douglas, M. (1969). Purity and Danger. London: Routledge and Kegan Paul.
Drew, J.A., & Henne, A.P. (2006). Conservation biology and traditional ecological knowledge: Integrating academic disciplines for better conservation practice. Ecology and Society, 1, 2, 34.
Ferraz, S.F.D., Paula, F.R., & Vettorazzi, C.A. (2009). Incorporation sustainability indicators on site selection for Forest restoration in the Corumbatai River basin. Revista Árvore, 33, 5, 937-947.
Gheler-Costa, C., Vettorazzi, C.A., & Pardini, R. et al. (2012). The distribution and abundance of small mammals in agroecosystems of southeastern Brazil. Mammalia, 76, 185 – 191.
Gliessman, S.R. (2000). Agroecologia: processos ecológicos em agricultura sustentável. Porto Alegre: Editora da Universidade.
Glikman, J.A.,, J.J., Vaske, Bath, A.J. et al. (2011). Residents' support for wolf and bear conservation: the moderating influence of knowledge. European Journal Wildlife Research, 58:295–302.
Gonçalves, J.S., & Souza, S.A.M. (2013). Agricultura familiar: limites do conceito e evolução do crédito. Instituto de Economia Agrícola. http://www.iea.sp.gov.br/out/verTexto.php?codTexto=2521>. Accessed 15 March 2013.
Guanziroli, C., & Cardim, S.E. (2000). Novo retrato da agricultura familiar: O Brasil redescoberto. Brasília: Projeto de Cooperação Técnica FAO/INCRA.
Instituto de Pesquisa e Estudos Florestais (IPEF) (2001). Plano Diretor: Conservação dos recursos hídricos por meio da recuperação e da conservação da cobertura florestal da bacia do Rio Corumbataí. Piracicaba: Vitor's Design.
Instituto Nacional de Colonização e Reforma Agrária (INCRA/FAO) (2010). Guia Metodológico – análise – diagnóstico de sistemas agrários. Brasília: Incra/FAO.
Joly, C.A., Rodrigues, R.R., & Metzger, J.P. et al. (2010). Biodiversity Conservation Research, Training, and Policy in São Paulo. Science, 328, 1358-1359.
Lamarche, H. (1993). A agricultura familiar: comparação internacional. Uma realidade multiforme. v.1. Campinas: Unicamp.
Lamarche, H. (1998). A agricultura familiar: comparação internacional. Do mito à realidade. v.2. Campinas: Unicamp.
Geist, H.J. & Lambim, E.F. (2002). Proximate causes and underlying driving forces of tropcial deforestation. BioScience, 52, 2, 143-150.
Liu, J. (2007). Complexity of Coupled Human and Natural Systems. Science, 317, 1513-1516.
Martinelli, L.A., & Filoso, S. (2008). Expansion of sugarcane ethanol production in Brazil: Environmental and Social Challenges. Ecological Applications, 18, 885-898.
Martins, R.C. (2006). Representações sociais, instituições e conflitos na gestão de águas em territórios rurais. Sociologias, 8, 5, 288 – 325.
Metzger, J.P., Lewinsohn, T.M., Joly, C.A., et al. (2010). Brazilian Law: Full Speed in Reverse? Science, 329, 276-277.
Ministério do Desenvolvimento Agrário (MDA) (2013). Programa Nacional de Agricultura Familiar. http://www.mda.gov.br/portal/saf/programas/ pronaf.html. Accessed 16 March 2013.
Mingoti, R., & Vettorazzi, C.A. (2011). Relative reduction in annual soil loss in macro watershelds due to the relief and forest cover. Engenharia Agrícola, 31, 6, 1202-1211.
Morais, E.B., Tauk-Tornisielo, S.M., & Ventorini, S.E. (2012). Impacto de atividades agropecuárias na qualidade das águas do rio Cabeça, na bacia do Rio Corumbataí, SP. Holos Environment, 12, 1, 125-163.
Moran, E.F. (1990). A ecologia humana das populações da Amazônia. Rio de Janeiro: ed. Vozes.
Moran, E.F., Packer, A., Brondizio, E.S., et al. (1996). Restoration of Vegetation Cover in the Eastern Amazon. Ecological Economics, 18, 41-54.

Moran, E.F., Brondizio, E. S., & McCracken, S. (2002). Trajectorias of land use: soils, succession, and crop choice. In C. H. Wood & R. Porro (Eds.), Deforestation and land use in the Amazon (pp. 102-114). Gainesville: University of Florida Press.

Moran, E. F., & Osgtrom, E. (2005). Seeing the Forest and the Trees: Human Environment Interactions in Forest Ecosystems. Cambridge: MIT Press.

Moran, F.E. (2011). Meio Ambiente e Ciências Sociais: interações homem-ambiente e sustentabilidade. São Paulo: Senac.

Nobre, M.F. (2008).O zoneamento ecológico-econômico como instrumento de planejamento e gestão ambiental: uma proposta para a bacia hidrográfica do rio Corumbataí. Tese (Doutorado em Geociências) - Instituto de Geociências e Ciências Exatas, Universidade Estadual Paulista,"Júlio de Mesquita Filho" Rio Claro.

Perfecto, I., Vandermeer, J., & Wright, A. (2009). Nature's Matrix: Linking Agriculture, Conservation and Food Sovereignty. London: Earthscan.

Ribeiro, E. M., & Galizoni, F. M. (2002). Água, população rural e políticas de gestão: o caso do vale do Jequitinhonha, Minas Gerais. Ambiente & Sociedade, 5, 2, 129-146.

Rodrigues, R.R., & Bononi, V.L.R. (2008). Diretrizes para a conservação e restauração da biodiversidade no Estado de São Paulo.São Paulo: Imprensa Oficial do Estado de São Paulo, 2008.

Ribeiro, M.C., Metzger, J.P., & Martensin, A.C. et al. (2009). The Brazilian Atlantic Forest: How much is left, and how is the remaining forest distributed? Implications for conservation. Biological Conservation, 142, 1141–1153.

Romeiro, A. R., Reydon, B., & Leonardi, M. L. (1997). Economia do meio ambiente. Campinas: Instituto de Economia/EMBRAPA.

Sahlins, M. (1976). Culture and practical reason. Chicago: The University of Chicago Press.

Seixas, C.S. (2005). Abordagens e técnicas de pesquisa participativa em gestão de recursos naturais. In: P.F. Vieira, F. Berkes & C.S. Seixas (Eds.), Gestão Integrada e Participativa de Recursos Naturais (p.73-105). Florianópolis: APED Editora.

Spornarski, C.C., Vask, J.J., Bath, A.J., & Musiani, M.M. (2014). Salient values, social trust, and attitudes toward wolf management in south-western Alberta, Canada. Environmental Conservation, available on CJO2014. doi:10.1017S0376892913000593.

Thomaz Jr., W.L. (1956). Man's Role in Changing the Face of the Earth. Chicago: Univ. of Chicago Press.

Turner II, B.L., Clark, W.C. & Kates, R.W., et al. (1990). The Earth as Transformed by Human Action: Global and Regional Changes in the Biosphere over the Past 300 Years. Cambridge: Cambridge Univ. Press.

Turner, B.L. II, & Meyer, W.B. (1994). Global land use and land cover change: an overview. In W.B. Meyer & B.L. Turner (Eds.), Changes in land-use and land cover: a global perspective (pp. 149-169). Cambridge: Cambridge University Press.

Valente, R.O.A., & Vettorazzi, C.A.(2005). Avaliação da estrutura florestal na bacia hidrográfica do Rio Corumbataí, SP. Scientia Florestalis, 68, 45-57.

Veiga, J. E. (1996). Agricultura familiar e sustentabilidade. Cadernos de Ciência & Tecnologia, 13, 3, 383-404.

Vieira, P.F. (2005). Gestão dos Recursos Comuns para o Ecodesenvolvimento. In P.F.

Vieira, F. Berkes & C.S. Seixas (Eds.), Gestão Integrativa e participativa de recursos naturais: conceitos, métodos e experiências. Florianópolis: Secco/APED.

Vitousek, P.M., Mooney, H.A. & Lubchenco, J., et al. (1997). Human Domination of Earth's Ecosystems. Science, 277, 494-499.

Younés, T. (2001). Ciência da Biodiversidade: questões e desafios. In: I. Garay, B.F. de Souza Dias (Eds.), Conservação da Biodiversidqde em Ecossistemas Tropicais. Petrópolis: Vozes.

Alexandre Toshiro Igari, Leandro Reverberi Tambosi

3 Agribusiness and Socioeconomic Drivers of Land Cover Change in Brazil

3.1 Conceptual Framework of Economic and Social Drivers for Land Use Change

3.1.1 Population Growth, Industrialization and Agricultural Technology

Human history evidenced the advance of croplands over natural areas to meet the needs for food, fibers and fuels (Grigg, 1982). The gloomy predictions of human demographic models pointed out population increase as a determinant driving force for poverty, starvation (Malthus, 1798) and also for depletion of natural resources (Meadows et al., 1972). Population growth (Fig. 3.1 – box at rightmost section) affects rural land use through multiple and sometimes opposite pathways. Changes in economy patterns, from agriculture dominant to industry and services based, induce urban populations to increase faster than rural ones. Migration to developing cities reduces the availability of rural labor (Fig. 3.1 – boxes at bottom left corner), what increases the labor costs for maintenance or expansion of cultivated lands (Rudel et al., 2005).

The investment decisions in farmlands rely on the relative costs of the factors of production (Angelsen et al., 2001), namely land (farmland, natural resources), labor (human work) and economic capital, which includes built capital (facilities, machinery) and financial capital (Fig. 3.1 lower dashed box). The growth of urban population fosters the expansion of cities over surrounding rural areas, what contributes to inflate the rural land prices (Livanis et al., 2006). As well as labor costs, the increase of land price also changes the relative agricultural costs. This scenario favors the investment in labor-saving and land-saving technologies, such as mechanization and fertilization of croplands respectively (Angelsen et al., 2001).

In the 1960 decade, the Green Revolution fostered new agricultural technologies that allowed increasing the productivity of croplands. The productivity increase roughly balanced the supply and the demand of agricultural goods. However, technology by itself, cannot endlessly expand agricultural production, and does not represent the solution for starvation or underdevelopment, given the unequal distribution of resources (Angelsen & Kaimowitz, 2001a).

Technological development can accelerate the depletion of natural resources, since the improvement of land productivity relies on intensive use of fossil fuels, freshwater, minerals, agrochemicals, and more recently on genetically modified organisms (Clay, 2004; MEA, 2005). As a consequence, highly intensified agricultural regions present environmental problems due to water eutrophication, soil compaction and salinization (Clay, 2004; MEA, 2005), and changes in species abundance which triggers pests and diseases outbreaks (Chapter 11 and 12, this book). Despite these significant environmental impacts, some Green Revolution hypotheses argue that adoption of agricultural technologies may increase the productivity in croplands and relieve the pressure of farmland expansion over remaining natural vegetation (Angelsen & Kaimowitz, 2001a; Balmford et al., 2005).

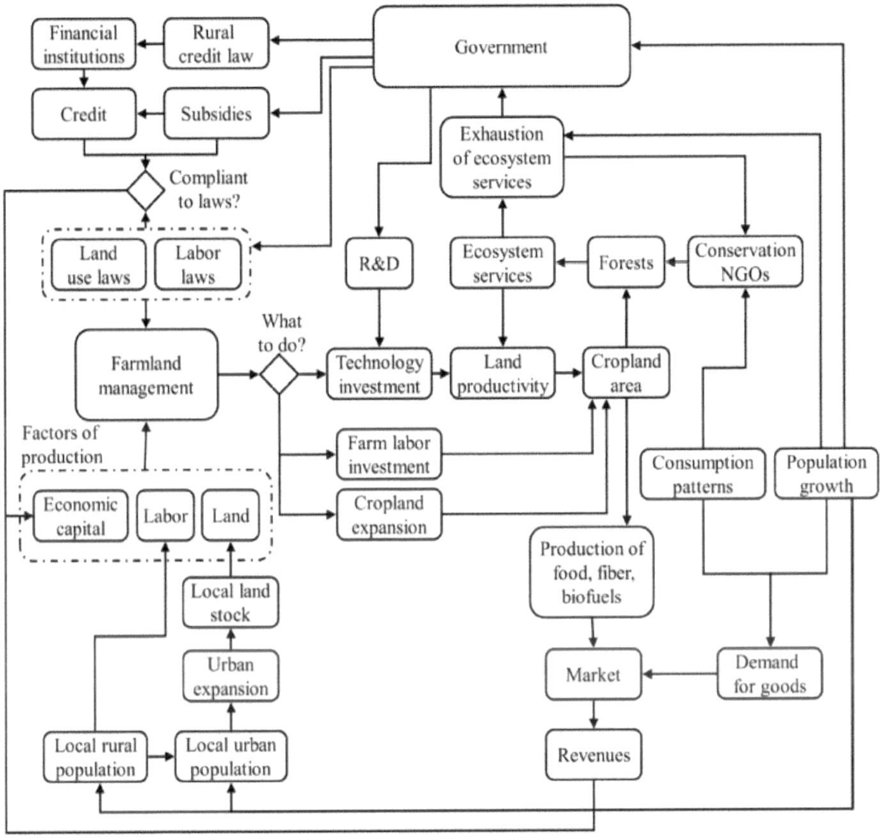

Fig. 3.1: Conceptual Framework of demographic, legal, economic and technological drivers on farm land-use management. Arrows represent the influence of one variable on another.

3.1.2 Development-based Hypotheses for Forest Recovery

3.1.2.1 Forest Transition Hypothesis

Forest Transition (FT) is a controversial hypothesis which states that urban development starts a chain of events that could lead to the retreat of cultivated areas and to forest recovery. Industrialization and urbanization would create new jobs and attract workers to the cities. Rural labor costs would increase and agriculture would concentrate in the more productive lands, allowing forest regrowth in the abandoned and less productive areas. Rudel et al. (2005) identified the scarcity of rural labor as a key driver for increase of forest area in Greece, Ireland and Portugal in the period between 1990 and 2000. Rudel et al. (2005) refer to this sequence of events, from development of the cities to forest recovery in marginal areas, as the "economic development path" in the FT hypothesis.

3.1.2.2 Agricultural Adjustment Hypothesis

Agricultural Adjustment (AA) hypothesis complements the scenario of FT hypothesis, and argues that increased labor costs would foster farmland technology, increasing agricultural production. When technology increases agricultural yields to the level of the demand for agricultural goods, Mather and Needle (1998) predict the concentration of agricultural activities in the more suitable lands, hindering farmland expansion, reducing the cultivated area, and allowing the regrowth of secondary forests in the less productive and abandoned areas.

3.1.3 Caveats on Development-based Hypotheses

Despite demographic (rural exodus) and technological local effects on limiting farmland expansion, the globalized access to markets of agricultural goods (Fig. 3.1 – boxes at bottom right corner) may increase the agricultural revenues and contribute to intensify the pressure toward farmland expansion (Angelsen, 1999). If local production is export-aimed, but does not represent a substantial international market-share, agricultural prices would be weakly reduced by local yield increase. In this scenario, where local sale price is perceived as static at any local yield level, productivity gains would increase farm profitability and stimulate the expansion of cultivated areas, even in less profitable sites (Angelsen et al., 2001). Without enforced regulatory constraints, incentives, taxes or market-based interventions, setting aside agricultural land to natural capital conservation can become unfeasible.

The material-intensive urban consumption patterns (Fig. 3.1 – right side) may leverage the effects of population growth on farmland expansion. The changes in dietary habits of the growing Chinese urban population, replacing locally produced vegetables by pork and poultry products, foster the expansion of Brazilian soybean croplands for animal feeding (Nepstad et al., 2006).

However, globalization of markets and the change of people's preferences may also play a controlling role on farmland expansion over natural areas. The growing concern about the exhaustion of ecosystem services spurred the development of environmentalist Non-Governmental Organizations - NGOs (Fig. 3.1 – boxes at top right). In Brazil, this emerging scenario shows initiatives as the Brazilian soybean roundtable and the Amazonian beef boycott. Both initiatives had as starting point the concerns of distant urban consumers about the environmental impacts of agricultural production (Nepstad & Stickler, 2008). Taking into account the risk of economic losses, transnational companies and NGOs pressed Brazilian farmers to suspend the expansion of soybean cropland on native areas, and also induced the Amazonian ranchers to comply with Brazilian environmental and land use laws.

3.2 The Conceptual Framework in Brazil

3.2.1 Native Vegetation in Private Farms and the Forest Code

Brazil is the world's fifth largest country, with an area of 8.5 million km^2, and around 32% of the land is used for agriculture. The agricultural area increased 14% between 1990 and 2011, while native forested areas declined 10% in the same period (UNSD, 2013). Expressive investments in agriculture development since the middle of the last century led the country to the third place in global exports of agricultural products. In 2012, exports of agricultural goods accounted for US$86 billion, representing 35% of Brazilian total exports (WTO, 2013).

Even so, Brazilian forests still cover 5.2 million km^2 (UNSD, 2013), an area larger than the entire European Union. However, in only five years (2000 to 2005), forest loss in Brazil accounted for 164,000 km^2 (Hansen, 2010), an area equivalent to almost half the Germany territory. Although recent conservation efforts caused the decline of deforestation rate, the conversion of forests to pastures and crops led the country to become the world leader in tropical deforestation (Hansen, 2010).

In the decade of 1960 Brazilian agriculture was losing competitiveness in traditional export commodities as coffee, sugar, cotton and cocoa, and the growing urban population was dealing with frequent shortages of basic agricultural goods as rice, sugar and beef (Igari & Pivello, 2011). To address these issues, the Brazilian National Congress passed two laws, in late 1965, with the objective of enhancing the productivity of agriculture: the New Forest Code (NFC) and the Rural Credit (RC) (Igari & Pivello, 2011). NFC was a land-use law that aimed to restrain the degradation of agricultural soils and freshwater stocks through the definition of mandatory areas of permanent preservation (commonly known as APP) which encompasses native vegetation in riparian areas, hilltops and slopes steeper than 45°. Furthermore a given percentage of the farm could not be converted to agricultural use (later on, these areas were named as Legal Reserves).

3.2.2 Agricultural Intensification and the Rural Credit Law

The RC law was approved two months after the NFC, and stated that a percentage of the deposits in the banking system would have to be invested as rural credit under subsidized interest rate (Fig. 3.1 – boxes at top left corner), in order to modernize

Brazilian agriculture (Igari & Pivello, 2011). Moreover, an intriguing clause in the RC law stated that the subsidized loans could not be conditioned to compliance with the NFC. This decoupling between subsidized loans and compliance to NFC (Fig. 3.1 - decision diamond shape at top left area) may even be interpreted as a governmental financial incentive to degrade the natural capital (Igari et al., 2009).

This early conflict between NFC and RC is reflected in the present scenario. In 2005, public subsidies to rural loans represented three times the total budget of the Ministry of Environment, the institution responsible for enforcing NFC and for the management of most Brazilian federal nature reserves. Moreover, the rural credit loans were equivalent to 44% of the agricultural added value to Brazilian Gross Domestic Product (GDP) in 2008 (Igari & Pivello, 2011).

The financial support provided by RC was preceded by a governmental program to foster domestic production of industrialized goods, which benefited the farm equipment industry. In 1960, imports supplied all farm equipment in Brazilian agriculture, but three years later, there already were six national companies, with total production capacity of 21,600 units per year (Vegro et al., 1997).

The local development of agricultural technology gained *momentum* in 1973, when the federal government created EMBRAPA, the national agency for research on agriculture and cattle raising (Fig. 3.1 – R&D box), which was the main responsible for successful adaptation of temperate agricultural crops to the Brazilian soils and climate (Kaimowitz & Smith, 2001).

3.3 Confronting Development-based Hypotheses to Empirical Data

The broad diversity of land use patterns, of rural/urban population composition, of technological development and economic wealth among the Brazilian states offers a promising opportunity to study the relationship among the hypothesized driving forces for land use change. It is possible to identify a wide range of combinations between the extreme scenarios, from remote barren rural regions, which present large portions of pristine areas under risk of massive deforestation (Hansen et al., 2010), to developed states, where secondary forests seem to recover in rural landscapes (Baptista & Rudel, 2006). Furthermore, Brazilian demographic and agricultural surveys deliver standardized long-term data on land use, demographic and technological characteristics since the 1960 baseline (prior to NFC and RC law) until the present scenario. Moreover, sub-national approaches of land use change allow controlling the variation of government policies present in cross-national studies, what usually entangles the causation effect of the explanatory variables.

In the next section we sought to answer two questions, in order to investigate the validity of the “economic development path” of FT hypothesis (Rudel et al., 2005) and of the AA hypothesis (Mather & Needle, 1998): (a) Does the reduction of rural population lead to the lack of rural labor and then to farmland retraction and forest recovery? (b) Does the technological development in farms lead to reduction of croplands, and then, to forest recovery?

3.3.1 Empirical Data

The state-level data regarding farmland area, forest area inside farms, number of tractors (as a proxy of mechanization), and number of rural workers were extracted from Brazilian agricultural census of 1960 (IBGE, 1960a) and 2006 (IBGE, 2006). Rural and urban population data were taken from demographic census of 1960 (IBGE, 1960b) and 2010 (IBGE, 2011). The 2010 demographic census was the closest survey to 2006 agricultural census.

Rural credit data was attained from Brazilian Central Bank annual reports from 1969 to 2006 (BCB, 1985; BCB, 1999; BCB, 2012), and the average annual RC value was calculated for each state in that period. RC values were converted to US dollars (US$) through the average daily exchange rate of 2006, where US$1.00 = R$2.17 (FED, 2012).

The percentage of farmland in each state was calculated through the division of the total farmland area (in 1960 and 2006) by the official state area (IBGE, 2012). The average percentage of native forests inside farms was attained through the division of total area of native forests inside farms by the total farmland area in each state. The average level of mechanization was estimated by the number of tractors per km^2 of farmland in the state. The average intensity of rural labor in the state was estimated by the number of rural workers per km^2 of farmland. The rural and urban population data were directly extracted from the demographic census in each year.

The half century timeframe was chosen in order to capture the contribution of the RC law on farm mechanization and of NFC on conservation of native vegetation in private farms. Both laws could then foster AA and FT in Brazil. Rudel (1998) predicted a time lag of 15 to 20 years in the "economic development path" of FT hypothesis. This means that land abandonment and forest recovery would take place one generation after the urbanization-led demographic changes. The half century timeframe would then be large enough to capture significant changes in all variables. The broad-scale timeframe and the sub-national assessment were also compatible to the approach of the study conducted by Rudel (2001), in which he identified the positive effects of the Green Revolution on FT at the south of the United States between 1935 and 1975.

3.4 Empirical Results

Total results encompassing the 26 Brazilian states showed an expressive increase of the Brazilian urban population (+127 million) contrasting with a shrinking rural population (-8 million) from 1960 to 2010 (Tab. 3.1). The negative variation of the rural population is reflected on the reduction of the average rural labor intensity

Tab. 3.1: Demographic (urban and rural population), technological (rural labor and mechanization), financial (rural credit) and land-use (farmland in the state and forests inside farmlands) results concerning the 26 Brazilian states in 1960 and their variation (Δ) until the most recent survey (2010 for demographic data and 2006 for all other variables). Highlighted numbers are the three higher (+) and lower (-) values of each column. [A] Corroborate agricultural adjustment hypothesis (+ mechanization → - farmland → + forests in farms). [F] Corroborate "economic development path" of forest transition hypothesis (+ urban population → - rural population → - rural labor → - farmland → + forests in farms).

State		Urban population ($\times 10^3$)		Rural population ($\times 10^3$)		Rural labor (worker × km^{-2})		Mechanization (tractor × km^{-2})		Farmland in the state (%)		Forests in farms (%)		Mean annual rural credit (US$ × $yr^{-1} \times 10^6$)
		1960	Δ 2010	1960	Δ 2010	1960	Δ 2006	1960	Δ 2006	1960	Δ 2006	1960	Δ 2006	1969-2006
Acre	(AC)	**34** $^{(-)}$	**499** $^{(-)}$	125	75	**32** $^{(-)}$	253	**0.0** $^{(-)}$	**1.8** $^{(-)}$	62	**-39** $^{(-)}$	**82** $^{(+)}$	-19	**32** $^{(-)}$
Alagoas	(AL)	421	1877	835	-12	**1903** $^{(+)}$	240	1.7	15.4	69	7	31	**-19** $^{(-)}$	267
Amapá	(AP)	35	**565** $^{(-)}$	**33** $^{(-)}$	35	**39** $^{(-)}$	111	0.2	**0.9** $^{(-)}$	9	-3	61	-14	**5** $^{(-)}$
Amazonas	(AM)	237	2519	123	**602** $^{(+)}$	261	**473** $^{(+)}$	0.0	**2.0** $^{(-)}$	**4** $^{(-)}$	-2	**88** $^{(+)}$	**-38** $^{(-)}$	78
Bahia	(BA)	2049	8056	**3870** $^{(+)}$	46	1030	-233	0.3	9.1	31	**20** $^{(+)}$	25	8	913
Ceará [A]	(CE)	1103	5241	2186	-82	732	**714** $^{(+)}$	0.2	7.0	**74** $^{(+)}$	**-20** $^{(-)}$	29	7	343
Espirito Santo	(ES)	370	2559	799	-216	987	132	1.8	40.0	63	-1	30	-15	257
Goiás	(GO)	517	4904	1077	-494	202	-39	0.7	16.8	57	19	13	8	1189
Maranhão	(MA)	443	3701	2034	392	1158	**-395** $^{(-)}$	0.1	4.6	25	14	25	7	195
Mato Grosso	(MT)	139	2346	182	367	71	4	0.1	8.8	9	**44** $^{(+)}$	27	12	1069
Mato Grosso do Sul	(MS)	208	1890	364	-12	57	14	0.3	12.3	65	19	14	5	859
Minas Gerais [A]	(MG)	**3914** $^{(+)}$	**12800** $^{(+)}$	**6167** $^{(+)}$	**-3285** $^{(-)}$	579	2	1.2	27.0	67	-11	**9** $^{(-)}$	13	2228
Pará	(PA)	624	4573	917	**1474** $^{(+)}$	637	-284	0.7	3.5	4	14	50	-10	246
Paraíba [A,F]	(PB)	695	2144	1296	-368	1359	-63	1.2	6.5	72	-5	**11** $^{(-)}$	**20** $^{(+)}$	189
Paraná	(PR)	1311	7595	2953	**-1420** $^{(-)}$	1128	**-398** $^{(-)}$	4.6	69.8	57	20	25	-7	**3761** $^{(+)}$
Pernambuco [A,F]	(PE)	1828	5221	2252	-506	**2132** $^{(+)}$	**-393** $^{(-)}$	1.7	8.5	60	-5	19	6	448
Piauí	(PI)	292	1759	957	111	393	**482** $^{(+)}$	0.1	3.9	36	2	27	**23** $^{(+)}$	131

State		Urban population (× 10^3)		Rural population (× 10^3)		Rural labor (worker × km^{-2})		Mechanization (tractor × km^{-2})		Farmland in the state (%)		Forests in farms (%)		Mean annual rural credit (US$ × yr^{-1} × 10^6)
		1960	**Δ 2010**	**1960**	**Δ 2010**	**1960**	**Δ 2006**	**1960**	**Δ 2006**	**1960**	**Δ 2006**	**1960**	**Δ 2006**	**1969-2006**
Rio de Janeiro	(RJ)	**5253** (+)	**10214** (+)	1397	-870	875	-105	**5.5** (+)	31.9	69	**-22** (-)	18	-3	347
Rio Grande do Norte [A,F]	(RN)	428	2038	713	-11	812	-36	0.9	12.5	70	-9	11	20	153
Rio Grande do Sul [A,F]	(RS)	2412	6690	2976	-1383	616	-6	**7.0** (+)	**73.9** (+)	**77** (+)	-5	**10** (-)	2	**3920** (+)
Rondônia	(RO)	**31** (-)	1112	**40** (-)	378	140	193	0.3	6.6	**1** (-)	**34** (+)	**80** (+)	**-46** (-)	73
Roraima	(RR)	**12** (-)	**332** (-)	**16** (-)	90	**37** (-)	136	**0.0** (-)	2.6	**4** (-)	4	12	**35** (+)	**14** (-)
Santa Catarina	(SC)	688	4561	1441	-440	967	-21	1.9	**113.8** (+)	62	1	29	-3	1022
São Paulo	(SP)	**8044** (+)	**31508** (+)	**4779** (+)	**-3080** (-)	895	-349	**14.1** (+)	**72.9** (+)	**78** (+)	-10	12	-1	**4648** (+)
Sergipe	(SE)	291	1229	461	87	**1696** (+)	120	0.6	19.6	67	1	16	-4	115
Tocantins	(TO)	63	1094	260	**442** (+)	114	9	**0.0** (-)	6.9	35	17	21	14	178
Mean		**1209**	**4886**	**1471**	**-311**	**725**	**22**	**1.7**	**22.3**	**47**	**3**	**30**	**0**	**872**
SD		**1881**	**6273**	**1576**	**1030**	**600**	**277**	**3.1**	**28.8**	**27**	**18**	**23**	**18**	**1300**
Brazil – total		**31444**	**127026**	**38254**	**-8078**	**626**	**-124**	**2.5**	**22.4**	**29**	**9**	**22**	**6**	**22678**

(rural workers per km²) and on the increase of average mechanization in farmlands (tractors per km²). The mechanization increase was supported by an average annual rural credit of US$22.6 billion. Despite the reduction of rural labor intensity and the increase of mechanization, the national percentage of farmland has increased from 29% to 38% (+9%) between 1960 and 2006. In the same period, there has been an increase (+6%) of the average percentage of native forest area inside farmlands (Tab. 3.1). The standard deviation (SD) was much larger than mean values in 11 of the 13 variables considered in the analysis (Tab. 3.1), what expresses the heterogeneity among states.

The urban population has increased in all 26 states, and in several of them the increase until 2010 was one order of magnitude larger than the entire urban population in 1960 (Tab. 3.1). Meanwhile, the rural population increased in 12 states, but even in these states the increase of urban population was far larger than rural population growth (Tab. 3.1).

Rural labor intensity declined in 12 states from 1960 to 2006 and there has been an overall augment of the average mechanization in farmlands until 2006. Only 3 states (SP, RS and PR - refer to Table 3.1 for states names) accounted for more than 54% of the national rural loans between 1960 and 2006 (marked with "(+)" in Table 3.1). These three states also showed substantial mechanization increase in the same period.

The percentage of farmland declined in 12 states (Tab. 3.1). This farmland reduction does not necessarily means that forests would recover, since farmland could give place to other non-forest land uses, as mining or cities. Furthermore, from these 12 states, 6 states (AC, AP, AM, ES, RJ and SP) also showed decline in the average percentage of forests inside farms.

Despite the increase in the percentage of forests inside farms in 14 states and the overall increase of mechanization (Tab. 3.1), the land-use predictions from AA hypothesis were supported only in six states (marked with "A" in Table 3.1). Furthermore, the predictions from the "economic development path" of FT hypothesis were identified in only four states (marked with "F" in Table 3.1), although the reduction of rural labor intensity has took place in 12 states.

3.5 Alternative Approaches

The technological optimism of the AA hypothesis derives from Borlaug Green Revolution global previsions, in which agricultural technology would deliver expanding yields and save land for forest recovery (Angelsen & Kaimowitz, 2001a). Interpreted at global scale, the Borlaug previsions are plausible whether gains of agricultural productivity match the increase on demand of agricultural goods. Beyond this point, agricultural production would exceed the demand, and further farmland advances would be avoided. Agricultural production would then concentrate in countries where the factors of production (labor, land and capital) are cheaper

(Angelsen, 1999). In this context it is not surprising the occurrence of FT in temperate developed countries, where rural labor and farmland are quite more expensive, while agricultural frontiers expand in tropical regions (Rudel et al., 2005).

The expanding access of local agricultural products to regional and global markets, as well as the growing mobility of workforce toward distant labor opportunities may entangle local land use decisions and broad-scale variables. The results from the southern state of RS, which apparently corroborate both AA and FT hypotheses, can illustrate this point. Kaimowitz and Smith (2001) pointed out the successful technological adaptation of soybean varieties to tropical conditions as the starting point of mechanization in many Brazilian states. Soybean production started in southern states of Brazil in the early 1960s. The initial yield was very constrained by capital needs for mechanized production. Thus, soybean production was largely benefited by subsidized RC, and then, mechanized soybean croplands replaced traditional labor-intensive cultures as coffee in the south of Brazil.

The possible causation effect of RC on variation of mechanization intensity may be interpreted from an alternative perspective to the FT hypothesis. Rezende (2006) argue that Brazilian rural labor costs have really showed an increasing trend in the decade of 1960. However, this trend was not pushed by competition with new urban labor opportunities, as predicted by FT hypothesis. In that time, rural workers were increasingly benefited by recent labor rights which assured better wages and work conditions.

According to Rezende (2006) the government passed the subsidized RC law in 1965 to compensate farmers for the increased labor costs. This new scenario benefited the mechanization of large farms, which were better prepared to deal with RC requirements. Large farms were widely benefited by scale economies derived from substitution of human workforce by mechanization. The foreign demand for soybean spurred a chain of events on local production in which land prices increased, many southern small farmers sold their lands, soybean mechanized crops took place in the more suitable lands and total farming declined.

The social side-effect was a significant unemployment increase in the Brazilian countryside and many workers migrated to cities (Kaimowitz & Smith, 2001). Although this chain of events corroborates FT and AA hypotheses predictions, migration was pushed by rural unemployment instead of availability of better urban jobs. In this scenario, rural unemployment may increase pressure over marginal lands for subsistence agriculture (Barbier et al., 2010) preventing forest regrowth.

Furthermore, in a sub-national scale perspective, many southern small farmers were attracted by cheaper lands and migrated to the expanding agricultural frontier states as MT. New farmers started to produce soybeans, largely expanding the croplands on Cerrado (Brazilian savannah) and on Amazonian forest (Kaimowitz & Smith, 2001), Thus, contrary to AA previsions, the investments on labor-saving technologies and the improvement of workforce mobility lead to the increase of deforestation (Angelsen & Kaimowitz, 2001b; Walker, 2012).

The contrasting land use results from RS and MT illustrate intertwined effects of migration, mechanization, financial constraints and foreign market opportunities. Similarly, despite the evidence of crop productivity increase leading to FT in the south of the United States, Rudel (2001) admitted that governmental incentives for cotton expansion in the west might have contributed to land retirement in southern region, what clearly illustrates the entangled land use change causation chain.

Agricultural frontiers and regions undergoing FT represent contrasting and complementary scenarios of land use change. This process may be interpreted as a global-scale AA (Meyfroidt et al., 2010) in which economic balance between the factors of production determines what, where and how agricultural goods will be produced. Rural labor mobility and trade of agricultural goods are two relevant exchange channels between agricultural frontiers and FT regions (Walker, 2012). The way these variables contribute to local land use dynamics may be dependent on the interactions of driving forces in both regions (Pfaff & Walker, 2010).

Expanding agricultural frontiers would present characteristics that make agriculture more competitive (as cheaper lands and lower labor costs) as well as cropping advantages (suitable soils, freshwater availability and favorable climate). Meanwhile FT would take place where some combinations of these variables make local agriculture perform worse. Then, if transport costs are low, local demand in these FT regions would be supplied by agricultural goods produced in distant agricultural frontiers (Pfaff & Walker, 2010).

Forest Transition would then promote the recovery of natural areas strictly in less profitable agricultural lands, what would constrain the diversity of ecosystem functions and the supply of ecosystem services. At a global perspective, agricultural frontiers expansion would cause the loss of tropical biodiverse biomes (Barbier, 2004; Hansen et al., 2010) while FT would mainly recover temperate secondary forests (Rudel et al., 2005). This market-driven land use change is unlikely to converge to a reasonable land use pattern capable to supply enough ecosystem services.

In central and southern regions of Brazil large extents of flat terrain with fertile soils became intensive monoculture areas, with low native vegetation cover. This human driven homogenization of landscapes reduces the diversity of habitats and leads to changes in species community composition (Chapter 13, 14, 15, this book). Thus, the ongoing global and regional agricultural adjustments (Meyfroidt et al., 2010) would cause as side-effect some human driven “ecological adjustments”, where ecosystems would be displaced or conserved mainly according their agricultural profitability. Walker (2012) concluded that much of the Atlantic Forest recovery in the developed states of São Paulo (Lira et al., 2012) and Santa Catarina (Baptista & Rudel, 2006) may be supported by imports of forest and agricultural products from Amazonian deforestation. Furthermore, some ecosystem services as watershed balance and nutrient cycling can be profoundly affected by long-distance transport of agricultural goods, since local stocks of water and nutrients are incorporated to exported crops and livestock (MEA, 2005). Obviously, other ecosystem services as carbon storage can be compensated worldwide, despite the carbon content differences among world biomes. However, much of local ecosystem services, as pest control, pollination and soil conservation, can be largely damaged in this economic-driven "ecological adjustment" that does not take into account the importance of ecosystem services and natural capital.

Angelsen and Kaimowitz (2001b) summarized the results of studies in Europe, USA, Latin America, Asia and Africa, which presented some alternative perspectives for predictions of AA and FT hypotheses. The authors argue that economic incentives or subsidies remove financial constraints to farm expansion and may foster deforestation. Thus, to avoid environmental losses, agricultural incentives and subsidies would have to be conditioned to broad-scale assessments of environmental impacts. Fearnside (2008) also considers the restriction of incentives and subsides as a key issue to address rural demographic movements and to improve policies for reducing Amazonian deforestation.

Igari et al. (2009) pointed out that subsidized rural credit plays a pivotal role in recent advance of Brazilian capital-intensive agriculture, given historically increased interest rates. Then, RC policy could constitute an important tool to enforce environmental land use laws as the NFC. In July 2008, the Brazilian Central Bank, the institution responsible for ruling rural credit, inserted environmental criteria for rural credit loans in the Amazon biome (BCB, 2008). In 2012, the reform of NFC softened some mandatory standards for conservation of native vegetation, but it also incorporated a clause that conditioned the subsidized rural loans to farmers in compliance with preservation constraints. It represents a late but promising contribution on stopping government subsidized forest loss. Moreover, it also represents an important step to foster restoration of native vegetation in profitable lands. The restoration actions can increase landscape heterogeneity resulting in better conditions for species conservation in agricultural landscapes (Chapter 6, this book), fostering recovery or enhancement of ecosystem services (Chapter 8, this book).

Rudel et al. (2009) have also proposed alternative pathways to AA and FT hypotheses. Performing meta-analysis, the authors identified changes in the main drivers of tropical deforestation. From 1960 to 1985 rural population growth was the main cause of tropical deforestation, mainly through small farms advance pushed by government incentives. After 1985 capital-intensive mechanized agriculture became the main deforestation driver, and in some cases, urbanization also played a key role in the expansion of agriculture. These results support the idea that the role of mechanization and urbanization, pointed out as two main driving forces in AA and FT hypotheses, depend on the interaction with other broader-scale explanatory variables in land use change pathway.

In agricultural frontiers, better access to international markets, increased mobility of workforce, finance support and mechanization may locally spur expansion of farmland and consequently increase deforestation (Fearnside, 2008). On the other hand, in developed regions the same variables may contribute to FT previsions, since migration to cities increases rural labor costs, and then local produced goods become more expensive, enduring the concurrence with foreign lower priced agricultural goods.

3.6 Final Considerations

This chapter presented a simplified Conceptual Framework that aims to help understanding the economic and social driving forces behind land use change. The Conceptual Framework incorporated particular Brazilian variables as the Forest Code and the Rural Credit law, which might influence the land use decisions in private farms. These laws could foster land sparing inside farms and the adoption of land-saving technologies, what represent a promising scenario to investigate the predictions of Agricultural Adjustment and Forest Transition hypotheses.

The results of the present study did not support AA and FT predictions regarding the causality effect of urbanization and agricultural technology on rural land use change. The predictions of both hypotheses were locally confirmed in some states, however, the analysis of local results must take into account broad-scale connections and fluxes that could mislead the conclusions. The occurrence of FT in one place can be linked to larger deforestation in others, through migration of workforce and trade of agricultural goods. This study highlighted the importance of designing future approaches on land use change analysis that must explicitly deal with explanatory variables as: relative agricultural costs, migratory fluxes and trade of agricultural goods. These variables would capture the comparative aspects that could lead to expansion or retraction of cultivated lands in different regions and then help to understand the connections between FT and deforestation in each of those regions. However, this economic-driven land use change leads to an "ecological adjustment", where ecosystems are displaced or conserved according to their potential agricultural profitability. Nature conservation would mainly take place in the less suitable lands for agricultural use. This approach is unlikely to converge to a reasonable land use pattern capable to support biological diversity and ecosystem services. Public policies on land use planning, economic incentives and subsides must be capable to integrate agricultural activities and environmental conservation objectives.

Bibliography

Angelsen, A. (1999). Agricultural expansion and deforestation: modeling the impact of population, market forces and property rights. Journal of Development Economics, 58, 185-218.

Angelsen, A., & Kaimowitz, D. (2001a). Introduction: the role of agricultural technologies in tropical deforestation. In A. Angelsen & D. Kaimowitz (Eds.), Agricultural technologies and tropical deforestation (p.1-18). Oxon, UK: CABI Publishing Wallingford.

Angelsen, A., & Kaimowitz, D. (2001b). Agricultural technology and forests: a recapitulation. In A. Angelsen & D. Kaimowitz (Eds.), Agricultural technologies and tropical deforestation (p.383-402). Oxon, UK: CABI Publishing Wallingford.

Angelsen, A., Van Soest, D., Kaimowitz, D., & Bulte, E. (2001). Technological change and deforestation: a theoretical review. In A. Angelsen & D. Kaimowitz (Eds.), Agricultural technologies and tropical deforestation (p.19-34). Oxon, UK: CABI Publishing Wallingford.

Balmford, A., Green, R.E., & Scharlemann, J.P.W. (2005). Sparing land for nature: Exploring the potential impact of changes in agricultural yield on the area needed for crop production. Global Change Biology, 11, 1594-1605.

Baptista, S.R., & Rudel, T.K. (2006). A re-emerging atlantic forest? Urbanization, industrialization and the forest transition in Santa Catarina, southern Brazil. Environmental Conservation, 33, 195-202.

Barbier, E.B. (2004). Explaining agricultural land expansion and deforestation in developing countries. American Journal of Agricultural Economics, 86, 1347-1353.

Barbier, E.B., Burgess, J.C., & Grainger, A. (2010). The forest transition: towards a more comprehensive theoretical framework. Land Use Policy, 27, 98-107.

BCB - Banco Central do Brasil (1985). Crédito rural: dados estatísticos 1969-1985. Rio de Janeiro, Brazil: BCB.

BCB - Banco Central do Brasil (1999). Anuário estatístico de crédito rural 1986-1999. Brasília, Brazil: BCB.

BCB - Banco Central do Brasil (2008). Resolução 3545/2008. Brasília, Brazil: BCB. Retrieved March 28, 2014, from https://www3.bcb.gov.br/normativo/detalharNormativo.do?N=108019002&method=detalharNormativo

BCB - Banco Central do Brasil (2012). Anuário estatístico do crédito rural 1999-2012. Brasília, Brazil: BCB. Retrieved March 28, 2014, from http://www.bcb.gov.br/?RELRURAL.

Clay, J. (2004). World agriculture and the environment: A commodity-by-commodity guide to impacts and practices. Washington DC, USA: Island Press.

Fearnside, P.M. (2008). The roles and movements of actors in the deforestation of Brazilian Amazonia. Ecology and Society, 13, 23.

FED – Federal Reserve Statistical Release (2012). Foreign Exchange Rates - Historical Rates for Brazil. Retrieved March 28, 2014, from http://www.federalreserve.gov/releases/h10/Hist/dat00_bz.htm

Grigg, D. (1982). The dynamics of agricultural change: the historical experience. New York, USA: St. Martin´s Press.

Hansen, M.C., Stehman, S.V., & Potapov, P.V. (2010). Quantification of global gross forest cover loss. Proceedings of the National Academy of Sciences, 7, 8650-8655.

IBGE - Instituto Brasileiro de Geografia e Estatística (1960a). Censo agrícola de 1960 – Brasil - VII Recenseamento geral do Brasil – Volume II. Rio de Janeiro, Brazil: IBGE.

IBGE - Instituto Brasileiro de Geografia e Estatística (1960b). Censo demográfico de 1960 – Brasil - VII Recenseamento geral do Brasil – Volume I. Rio de Janeiro, Brazil: IBGE.

IBGE - Instituto Brasileiro de Geografia e Estatística (2006). Censo agropecuário 2006 – Resultados preliminares. Rio de Janeiro, Brazil: IBGE.

IBGE - Instituto Brasileiro de Geografia e Estatística (2011). Censo demográfico 2010 – Características da população e dos domicílios – Resultados do universo. Rio de Janeiro, Brazil: IBGE.

IBGE - Instituto Brasileiro de Geografia e Estatística (2012). Área territorial oficial brasileira. Retrieved March 28, 2014, from http://www.ibge.gov.br/home/geociencias/areaterritorial/principal.shtm

Igari, A.T., Tambosi, L.R., & Pivello, V.R. (2009). Agribusiness opportunity costs and environmental legal protection: Investigating trade-off on hotspot preservation in the state of São Paulo, Brazil. Environmental Management, 44, 346-355.

Igari, A.T., & Pivello V.R. (2011). Crédito Rural e Código Florestal: Irmãos Como Caim e Abel? Ambiente e Sociedade, 14, 133-150.

Kaimowitz, D., & Smith, J. (2001). Soybean technology and the loss of natural vegetation in Brazil and Bolivia. In A. Angelsen & D. Kaimowitz (Eds.), Agricultural Technologies and Tropical Deforestation (p. 195-211). Oxon, UK: CABI Publishing Wallingford.

Lira, P.K., Tambosi, L.R., Ewers, R.M., & Metzger, J.P. (2012). Land-use and land-cover change in Atlantic Forest landscapes. Forest Ecology and Management, 278, 80-89.

Livanis, G., Moss, C.B., Breneman, V.E., & Nehring, R.F. (2006). Urban sprawl and farmland prices. American Journal of Agricultural Economics, 88, 915-929.

Malthus, T. (1798). An essay on the principle of population. London, UK: J. Johnson in St. Paul´s Church-Yard. Reprinted in 1998 by Electronic Scholarly Publishing Project.

Mather, A.S., & Needle, C.L. (1998). The forest transition: A theoretical basis. Area, 30, 117-124.

MEA – Millennium Ecosystem Assessment (2005). Ecosystems and human well-being: current state and trends. Washington DC, USA: Island Press.

Meadows, D.H., Meadows, D.L., Randers, J., & Behrens III, W.W. (1972). The limits to growth: a report for the Club of Rome's project on the predicament of mankind. New York, USA: Universe Books.

Meyfroidt, P., Rudel, T.K., & Lambin, E.F. (2010). Forest transitions, trade, and the global displacement of land use. Proceedings of the National Academy of Sciences of the United States of America, 107, 20917-20922.

Nepstad, D.C., Stickler, C.M., & Almeida, O.T. (2006). Globalization of the Amazon soy and beef industries: Opportunities for conservation. Conservation Biology, 20, 1595-1603.

Nepstad, D.C., & Stickler, C.M. (2008). Managing the tropical agriculture revolution. Journal of Sustainable Forestry, 27, 43-56.

Pfaff, A., & Walker, R. (2010). Regional interdependence and forest „transitions": Substitute deforestation limits the relevance of local reversals. Land use Policy, 27, 119-129.

Rezende, G.C. (2006). Políticas trabalhista, fundiária e de crédito agrícola no Brasil: uma avaliação crítica. Revista de Economia e Sociologia Rural, 44, 47-78.

Rudel, T.K. (1998). Is there a forest transition? Deforestation, reforestation, and development. Rural Sociology, 63, 533-552.

Rudel, T.K. (2001). Did a Green Revolution restore the forests of the American south? In A. Angelsen & D. Kaimowitz (Eds.), Agricultural technologies and tropical deforestation (p. 53-68). Oxon, UK: CABI Publishing Wallingford.

Rudel, T.K., Coomes, O.T., Moran, E., Achard, F., Angelsen, A., Xu, J., & Lambin, E. (2005). Forest transitions: Towards a global understanding of land use change. Global Environmental Change, 15, 23-31.

Rudel, T.K., Defries, R., Asner, G.P., & Laurance, W.F. (2009). Changing drivers of deforestation and new opportunities for conservation. Conservation Biology, 23, 1396-1405.

UNSD - United Nations Statistics Division (2013). Environment statistics country snapshot: Brazil. Retrieved March 28, 2014, from http://unstats.un.org/unsd/environment/envpdf/Country_Snapshots_Aug%202013/Brazil.pdf

Vegro, C.L.R., Ferreira, C.R.R.P.T., & Carvalho, F.C. (1997). Indústria brasileira de máquinas agrícolas: evolução e mercado: 1985-95. Informações Econômicas, 27, 1-26.

Walker, R. (2012). The scale of forest transition: Amazonia and the Atlantic Forests of Brazil. Applied Geography, 32, 12-20.

WTO - World Trade Organization (2013). International trade statistics 2013. Retrieved March 28, 2014, from http://www.wto.org/english/res_e/statis_e/its2013_e/its2013_e.pdf

Milton Cezar Ribeiro, Renata de Lara Muylaert, Pavel Dodonov, Giordano Ciocheti, Marcelo Magioli, Felipe Martello, Alessandro Rocha, Bruno Defane Borges, Carolina Carvalho, Claudia Zukeran Kanda, Karen G. Rodriguez-Castro

4 Dealing with Fragmentation and Road Effects in Highly Degraded and Heterogeneous Landscapes

4.1 Introduction

Over the past 100,000 years human activities have caused continuous and severe disturbances in natural ecosystems, resulting in the extinction of many species (Enrlich & Enrlich, 1988). According to Wilson (1988), these activities continue to sharply rise, leading to the local and regional extinction of hundreds of plant and animal species, many of which are yet unknown. Issues involving endangered species and conservation of natural resources are critical, especially in tropical and developing countries. Although great effort has been made in order to reduce biodiversity loss worldwide, in many regions habitat loss and fragmentation continue to negatively influence species and key ecological processes, particularly in tropical regions (Butchart et al., 2010).

The spatial arrangement of natural remnants in the landscape is an essential issue that could favor or hamper ecological processes such as dispersion and movement of individuals to succeed in maintaining the population of other animals and plants dispersed by them. In addition, the effects of reduced vegetation cover and the spatial arrangement of vegetation remnants, urban-agroecosystems and roads in the landscape have been shown to be important elements of different ecological aspects. Roads are vital to the economic growth of a nation and permit the creation of new service opportunities, jobs and the installation of new residential and industrial areas (Wilkie et al., 2000). Therefore, we should consider that traffic, the size of opening roads and the effect of barriers created by road construction are components that should be considered at the landscapes scale.

The growth of road networks has been causing deleterious effects on wildlife since it increases the risk of mortality by vehicles (Gunson et al., 2011) and profoundly changes the amount and arrangement of natural remnants (Frair et al., 2008). Studies indicate that the unwillingness or inability of vertebrates to cross roads can lead to isolation of animal populations, with negative effects on local and regional species maintenance (Jaeger et al., 2005; Taylor & Goldingay, 2010). The reduction in abundance of a great number of species is therefore likely to be influenced by the increase of road density (Frair et al., 2008).

According to Laurance et al. (2009), when a road leads to the fragmentation process in a natural landscape, there is an inhibition of dispersal and migration of species and a growth in processes that increase the impacts on wildlife (e.g. fire, pollution, biological invasion and hunting). Consequently, fragmentation of natural areas by roads negatively affects species that: i) do not survive well in areas under the influence of edge habitats, ii) are sensitive to human presence, iii) occur in low densities, iv), tend to avoid or are unable to cross roads (Van der Ree et al., 2011). Thus, it is important to identify which animal species use roads in some way, and stimulate the development of new studies about the use and behavior of these species along the roads. This would allow stakeholders, government and other sectors of society to predict the effects of roads on wildlife and on key ecological processes, and to develop actions to mitigate the impacts.

The present text overviews the advances in studies of cover and spatial structure of habitats and road effects on different groups within highly fragmented and heterogeneous landscapes. Using a spatial ecology approach, we also propose a potential research agenda for these fields, in order to stimulate new studies focusing on key issues to maintain and recovery biodiversity within agroecosystem-dominated landscapes. We focus on these issues with examples of studies conducted mostly in the Neotropical region due to its importance in terms of biodiversity and presence of agroecosystems and also because of the growing number of studies carried out in this region, permitting the detection of gaps in knowledge to which research efforts should be directed. We start by discussing general aspects of agroecosystems, roads and edge effects. Then we outline studies using genetics and stable isotope analysis for functional groups, which are useful tools for understanding landscape effects. Finally, we provide an overview of road and edge effects on two example animal groups, bats and ants, and we finish with highlights on landscape ecology and agroecosystems.

4.1.1 Urban-agroecosystem Gradients and Biodiversity

The urbanization process leads to large variation in the composition of natural communities and, therefore, in ecological dynamics. While many species are lost due to the suppression of their habitats, the development of the anthropic landscape also brings forth new micro-climatic spaces favorable to the settlement and development of species that cannot be found in other ecosystems in the surrounding area. This process may be seen as positive when considering the possibility of these species settling in the area, or negative, as many native species may lose their capacity to develop in this new environment, permitting the entry of exotic, more competitive, groups.

In some cases, even when exotic species are excluded from measurements of biodiversity the sub- and peri-urban systems still have more species than the environment that existed before the urbanization process (McKinney, 2008). This may be due to many reasons, such as the unavoidable creation of new micro-habitats that may harbor new animal and plant groups and the intentional introduction of species, as is the case of cultivated areas in city outskirts, which can increase the amount of resources available for other species. These peri-urban environments are the natural entryways and outlets of the material and energy flow between the predominantly human ecosystems formed by cities and the rural and natural ecosystems.

Moreover, the variation that exists in these sites cannot be viewed as a linear gradient between the urban and rural areas. Environmental and ecological conditions depend not on a point located along a straight line, but on characteristics that are also determined by the adjacent points (Ramalho & Hobbs, 2012). Thus, an area with high biodiversity may act both as a shelter for many species and as a source of propagules into neighboring sites, increasing the complexity and heterogeneity of these gradients.

There are many ways to achieve high biodiversity in urban areas, and one method that has been shown to be highly efficient in many environmental, social and economic aspects is the development of more sustainable landscapes though agroecosystems, particularly in urban-rural gradients (Alvey, 2006; Grimm et al., 2008). This combined cultivation technique that often takes place in peripheral environments may ensure not only better soil quality, but also the possibility to maintain the ecosystem services found in the area, such as water source preservation, pollination services and greater carbon sequestration (Alberti et al., 2003).

In addition, the development of agroecosystems permits a complete association between environmental preservation, improvement of urban environments, and increases in the income of the less favored population. All this makes Agroecosystems one of the main agents of biodiversity conservation projects in cities. No less important is the approximation between society and nature, which has been one of the big key points in the success of urban ecology-related actions, as decision making by public agencies is difficult in places without a good relationship between humans and their ecosystem.

4.1.2 Road Ecology, Landscape Structure and Biodiversity

Together with the urbanization process comes the transportation system, which is an increasing issue to biodiversity - although important to humans, roads promote several negative effects to species and ecological processes. One of the main effects of roads and highways is the interruption of the dispersal of animals and plants, breaking connectivity and leading to a change in habitat quality due to fragmentation. In turn, disturbance of between-fragment movement interferes with between-population migrations and, consequently, with the populations' structure and viability (Forman & Alexander, 1998), making the impacts of anthropic interventions on species more evident. The isolation generated by fragmentation also decreases the populations' genetic diversity, interfering with the adaptive plasticity of species in face of environmental changes that take place through time (Frankham et al., 2002).

The local and global variation generated by these interventions may influence evolutionary processes, and responses to these changes require high genetic variability to ensure a species' reproductive success and ongoing existence (Sork & Smouse, 2006). High levels of genetic variation are maintained by a constant flow of individuals between populations; however, the construction and presence of roads and the lack of ecological corridors in natural environments interrupt these flows, and species' spatial patterns come to be defined by habitat removal and by the subdivision of continuous areas (Miller et al., 1996; Forman & Deblinger, 2000).

The flow of individuals between populations is not constant for all the species that exist in a given fragment – for instance, Hawbaker et al. (2006) have shown that some species avoid crossing roads, whereas other species cross them. However, the individuals that take their chances in crossing the roads end up generating high levels of roadkill along these pathways (Forman & Alexander, 1998). These individuals are usually reproductive-age adults, and their mortality contributes to the fast decreases in population size of each of the species found in these fragments (Ashley & Robinson, 1996).

Animal roadkill is the most evident direct cause of mortality of wild vertebrates that is related to fragmentation by roads (Forman & Alexander, 1998). The yearly roadkill rates estimated by Forman (1995) and by Van der Zande et al. (1980) are disturbing: 159 thousand mammals and 653 thousand birds in the Netherlands; seven million birds in Bulgaria; five million frogs and reptiles in Australia and one million vertebrates in the United States. Reports of roadkill of vertebrate species are scarce in many areas; this situation is even more worrisome when considering the wide road and highway system and the country's biodiversity.

These impacts become even greater when considering that roads encourage the building of houses (Hawbaker et al., 2006) which, in turn, encourage the construction of new roads, leading to a vicious circle. It is also important to highlight that the species' responses to change take time, increasing the conflict due to the high speed with which construction-related changes occur in the environment. Another factor to be considered is the increase in traffic volume, which may lead to avoidance effects. Thus, maintenance of habitat quality demands the reduction of construction-related impacts; however, in order to mitigate the barrier effect produced, it is fundamental to assess the impacts and to identify the species most affected. Estimating the quantity of roadkill of small vertebrates is hard, especially due to difficulties in identifying the affected animals. Therefore, despite being hard to mitigate this impact, actions are urgently needed in order to reduce the negative effects of roads to biodiversity maintenance.

4.1.3 Overview of Edge Effects in Fragmented Landscapes

In addition to the direct effects of roads and other anthropic land uses on animal mortality and connectivity, it is important to consider how these land uses affect the structure and function of the adjacent natural ecosystems. These edge effects[1] may be defined as detectable differences between forest edge and the ecosystem on either side of the edge with regard to the ecosystem's structure, composition and function (Harper et al., 2005). Edge effects may be described by two parameters, the depth or distance of edge influence (also known as "edge width") and the magnitude of edge influence (Harper et al., 2005). Depth of edge influence is a measure of how far into the fragment statistically significant edge effects may be detected, whereas magnitude of edge influence is a measure of how different the edge and interior environments are in relation to each other. Edges may be characterized by different combinations of high and low depth and magnitude of edge influence (Harper et al., 2005). A variable may be influenced positively or negatively by the edge (Ries et al., 2004), and non-monotonic responses may also be observed (e.g. Dodonov et al., 2013).

Edge effects often result from changes in microclimate, such as increased light and wind incidence, which may result in greater temperature, lower moisture and increased wind throw at the edges (Saunders, 1991; Laurance & Curran, 2008). These changes in microclimate and the resultant modifications in vegetation structure and composition may, in turn, affect the resident fauna (Meyer et al., 2009) and vegetation (Cadenasso & Picket, 2000). In addition, as the edge vegetation regenerates or undergoes changes in structure and/or composition, the extent of microclimatic edge effects will also change (Didham & Lawton, 1999).

It is important, however, to note that edge effects are not caused solely by microclimatic changes, nor are they restricted to closed forests. For example, edge-related changes in vegetation composition have been observed at road edges of grasslands in Australia and Africa (Cilliers et al., 2008) and at different types of edges in the Brazilian savanna (Dodonov et al., 2013). In these areas, edge vegetation was largely dominated by exotic species, possibly because the establishment of these species was favored by disturbances or changes in soil chemistry. Another example of edge effects resulting from linear openings was observed by Smit & Asner (2012) in an African savanna, where there was a greater abundance of woody plants close

1 Some authors (e.g. Harper et al., 2005) recommend to use the term "edge influence" instead of "edge effects" for historical reasons: whereas the term "edge effects" was initially used to describe the increase in abundance of game species at forest edges (Leopold, 1933, cited by Harper et al., 2005), "edge influence" is more general and refers to all edge-related modifications in the ecosystem. However, we preferred to use "edge effects" in this chapter because this term is more commonly found in the literature (e.g. Fahrig, 2003; Ries et al., 2004).

to firebreak roads. This edge effect could have resulted from increased water run-off from the roads, thus increasing soil moisture at the edges (Smit & Asner, 2012).

These examples show the complexity of edge effects in different environments, and a number of factors have to be considered in order to successfully account for edge effects in a landscape study. The extent of edge effects may be affected by factors such as vegetation type (Delgado et al., 2007), adjacent land use (Pohlman et al., 2007), fragment size (Mascarúa-Lopez et al., 2006) and edge age (Chabrerie et al., 2013). Areas located close to more than one edge may also suffer greater edge effects than areas close to a single edge (Fletcher, 2005). In addition, it is important to note that different variables respond to edges in different ways, and there may be marked differences in the depth and magnitude of edge influence observed for variables such as seedling mortality and species diversity (Harper et al., 2005). Finally, different statistical analyses may give different results for the same data set (Harper & Macdonald, 2011). Therefore, detailed studies are needed before the extension and characteristics of edge effects in a given region may be predicted.

4.1.4 Understanding Landscape Effects on Genetic Diversity and Gene Flow

On a larger scale, roads and other sources of habitat fragmentation may lead to important changes in the genetics of different populations. A fragmented landscape is characterized essentially by the reduction of favorable habitats and by changes in its spatial configuration (Fahrig, 2003). Habitat loss and fragmentation may affect species richness (Martensen et al., 2012) and interaction processes such as seed dispersal, as well as decrease in population abundance (Fahrig, 2003; Uriarte et al., 2011). These are also considered the main factors that lead to decreases in the genetic variability of different species (Lowe et al., 2005; Aguilar et al., 2008), mostly due to a drastic decrease in population size (e.g. Andersen et al., 2004). In addition, in populations where size has been reduced there is a greater probability of mating between related individuals which, combined with genetic drift, also decreases genetic viability (e.g. Breed et al., 2012). Low genetic variability may in turn decrease the species' adaptive potential in face of environmental changes, which may lead to extinction (Hoffmann & Willi, 2008).

Movement and dispersal capacity of individuals is also affected by landscape fragmentation (Awade & Metzger, 2008). The movement of individuals between populations maintains species' connectivity; in plants, this connectivity is preserved by pollen and seed dispersal (Sork & Smouse, 2006). Animal and plant dispersal depends on natural history and landscape characteristics, such as matrix type (Umetsu et al., 2008), and distance between favorable habitats (Boscolo & Metzger, 2011). For example, population isolation and matrix resistance may limit the dispersal of individuals between populations, reducing genetic flow and consequently population connectivity (e.g. Lange et al., 2012). This may lead to a disturbance in the migration-drift balance, as the alleles lost due to genetic drift will not be replaced by the gene flow (Hamrick, 2004).

Even though the population genetics theory predicts the effects of habitat loss and fragmentation in reducing genetic variability (Young et al., 1996), many studies do not support these predictions (e.g. Collevatti et al., 2001; Winkler et al., 2011). This may be due mostly to the recent fragmentation of the area, considering the life cycle of the species studied (Collevatti et al., 2001; see also Lowe et al., 2005 for a review). In addition, the non-detection of the effects of habitat loss and fragmentation of genetic variability and gene flow may be due to the sampling designs often used in studies of this kind, as they do not compare different landscapes and often do not explicitly test the influence of landscape characteristics on the genetic structure of populations (e.g. Chambers & Garant, 2010 and see Storfer et al., 2010 for review).

Alternatively, there may be a reduction of genetic variability due to intense population reduction, as in the case of large mammals (Roelke et al., 1993). Combining studies to understand landscape effects on genetic variability, persistence of species and ecological processes is something needed in conservation biology. We live in a defaunated world (Dirzo & Galetti, 2013) and the use of genetics and other tools, such as the one we present below, may help provide guidelines for maintaining ecological processes.

4.2 Case Studies

4.2.1 Stable Isotope Analysis and Functional Diversity: Potential Tools for Subsidize Mammal Conservation Strategies in a Landscape Perspective

Mammals are one of the most vulnerable groups to threats such as habitat loss and fragmentation (Chiarello, 1999; Galetti et al., 2009; Dotta & Verdade, 2011), hunting pressure (Cullen et al., 2000; Peres, 2000) and roadkilling (Forman & Alexander, 1998; Scoss et al., 2004; Huijser et al., 2013). Medium- and large-sized mammals are key organisms for the structuring of biological communities (Dotta & Verdade, 2007), and are involved in fundamental ecological processes, such as herbivory, predation (Cuarón, 2000) and seed dispersal (Fragoso & Huffmann, 2000; Galetti et al., 2001). Therefore, they play a unique role in the maintenance, regeneration and dynamics of tropical forests (Jorge et al., 2013), even in modified and defaunated landscapes (Galetti et al., 2013).

In studies linking species, ecosystem fragmentation, and conservation, landscape ecology has been providing new study models (Metzger, 2001). Several studies indicate that mammals use the landscape matrix (Dotta & Verdade, 2007; Ferraz et al., 2010; Ferraz et al., 2012; Magioli et al., 2014a; Magioli et al., 2014b, Chapter 11 of this book), which is currently a dominant component in most Atlantic Forest domains, occupying approximately 84% of the biome original extension (Ribeiro et al., 2009).

Thus, in order to understand ecological processes, it is important to consider these modified habitats, since studies that included the effect of matrices in heterogeneous landscapes increased the explanatory power of ecological models (Umetsu et al., 2008; Prevedello & Vieira, 2010; Watling et al., 2011).

Several studies in landscape ecology have taken into account the influence of the landscape matrices over wildlife communities (Baum et al., 2004; Murphy & Lovett-Doust, 2004; Antongiovanni & Metzger, 2005; Dotta & Verdade, 2007; Umetsu et al., 2008; Pardini et al., 2009). However, knowledge on the effects of landscape structure on medium- and large-sized mammal species is scarce (Grelle, 2010; Prevedello & Vieira, 2010), raising some questions: 1) how do mammal species use these modified habitats? 2) which food resources are available and consumed? and 3) what is the contribution of persistent species for ecosystem functioning? Seeking new ecological approaches to answer these questions requires the combination of studies in landscape ecology. Two methods present potential for the acquisition of new information on modified landscapes: stable isotope analysis and functional diversity measurements.

Stable isotope analysis (SIA) generates information on trophic ecology, feeding patterns, origin of food resources, and has been used with considerable success in mammal studies (Kelly, 2000; Crawford et al., 2008; Ben-David & Flaherty, 2012). Carbon ($^{13}C/^{12}C$) and nitrogen ($^{15}N/^{14}N$) stable isotopes are used to assess animal diet and resource use, which also provides insight into trophic chain processes (De Niro & Epstein, 1978; Boecklen et al., 2011), and represents a useful tool for obtaining information on highly modified landscapes (Magioli et al., 2014b). SIA requires a tissue sample from the species of interest, with muscles and bone collagen being the most often used (Ben-David & Flaherty, 2012). Collecting these tissues requires the capture and sacrifice of the specimens, a strategy that is not particularly easy or ideal for studying rare or threatened species. Alternatively, the analysis of hair or feces offers a non-invasive method (Magioli et al., 2014b).

Thus, SIA can provide important information on trophic ecology, being complementary to diet studies and can help infer changes in habitat and resource use over both short and long time periods, which is of great value for species conservation (Kelly, 2000; Sponheimer et al., 2003; Codron et al., 2005; Crawford et al., 2008; Magioli et al., 2014b). However, the proper use of this tool requires i) a carefully elaborated sampling design, ii) questions and objectives that SIA may be able to answer, and iii) comparison of isotopic results with ecological and behavioral data of the studied species to avoid misinterpretations.

Functional diversity measures are presented as tools that may offer insights on the relationships between species and ecosystem functioning (Cadotte et al., 2011), and are being applied in various ecology fields and with different taxonomic groups (Cianciaruso et al., 2009). Functional diversity assesses the amount of interspecific variation in functional characteristics of a community, treating each species as unique (Poos et al., 2009).

There are several distance-based functional diversity measures (Mouchet et al., 2010), which can use species presence/absence data, such as the functional diversity index (FD) (Petchey & Gaston, 2002, 2006) and functional richness (FRic) (Villéger et al., 2008). Also, metric on abundance data can be used, such as Rao's quadratic entropy (Q) (Rao, 1982), functional evenness (FEve), and functional divergence (FDiv) (Villéger et al., 2008). When related to landscape characteristics, ecosystem processes and other metrics, each offer distinct insights on the relationship between species and ecosystem functioning. One of the most important things to consider when calculating functional diversity measures is to select an ecologically meaningful set of species traits. These traits need to be directly linked to the questions and objectives of the study (Cianciaruso et al., 2009), and preferably uncorrelated.

There is a large amount of information on the ecology and behavior of terrestrial vertebrates, particularly for mammals, which is widely used in conservation planning strategies (Jenkins et al., 2013). Over the last decade, functional diversity has been used to address several issues in mammal studies, providing insights for species conservation. Specifically, studies address the loss of functional diversity due to the intensification of land use by human activity, and the identification of thresholds between functional diversity and landscape metrics (Stevens et al., 2003; Blackburn et al., 2005; Flynn et al., 2009; Chillo & Ajeda, 2012; Magioli et al., 2015).

4.2.2 Anthropogenic Effects on Bats

Bats are a very large order of mammals in terms of diversity and abundance. They provide crucial environmental services such as pest control in agroforest ecosystems (Willians-Guillén et al., 2008) and function in ecological succession (Lobova et al., 2009). Therefore it is important to quantify the various ecological roles of bats and assess how intense human ecosystem modification affects these services. For example, bats play an important role in seed dispersal (Lobova et al., 2009), and contribute to the landscape-scale habitat regeneration of native species in agroecosystems.

Despite their abundance, bat populations are in decline in many areas (Altringham, 2011) with almost 25% of populations threatened by extinction (Mickleburgh et al., 2002). With advances in technology, new studies are able to monitor bat populations with higher accuracy (O´Mara et al., 2014), providing evidence for this population decline over decades (Betke et al., 2008).

Much has been done to understand the impact of human activities on bats, mainly in Poland and the USA. For example, great efforts have been made to minimize the effects of white nose syndrome in temperate regions (Blehert et al., 2009), which has killed thousands of bats over a decade. In the following section, we discuss two aspects of anthropogenic impacts: the presence of roads and a disease that affects both bats and humans.

4.2.3 Roads and Bats in Heterogeneous Landscapes

As optimal foragers (Pyke, 1984), bats always look for minimum cost paths in their daily routes (in migrating species this idea is more complex, see Fleming et al., 2008). Therefore, bats commonly use linear landscape features, such as open trails inside fragments, rivers and roads. Russel et al. (2009), investigated these aspects, demonstrating that bats cross roads in low flights (< 2 m high), with an annual mortality of 5%. Several papers (Shaub & Siemers, 2008; Siemers & Shaub, 2010, Berthinussen & Altringham, 2012a,b) showed an intense negative impact of roads on bat populations, but these effects remain unknown in the Neotropics, where few studies investigate this issue (Bernard et al., 2012).

Bats are common and abundant near cities and agroecosystems, being able to feed and roost in these environments (Bredt et al., 2012). Bats are also a known vector for many types of viruses (Calisher et al., 2006; Sabino-Santos Jr et al., 2015). An important issue affecting both bat and human health is rabies, a viral disease. In very altered ecosystems, such as those arising from deforestation, virus outbreaks can become more frequent due to increased contact of vampire bats with humans (Schneider et al., 2001) or alteration of viral rates on bat populations with unknown causes. Viral rates vary between different regions (Ruppretch et al., 2002) but in general the main rabies vectors are dogs and bats (Paéz et al., 2003). Several species of bats are well adapted to live in cities and houses (see Bredt et al., 2012), which increases contact risk to humans and also to herbivores in pastures (Gomes et al., 2010), such as cattle and horses.

Although instances of rabies are less common than other diseases, survival is rare once symptoms occur (Willoughby Jr et al., 2005). Therefore, strategies on how to prevent rabies are needed. This issue should be addressed mainly in cities with nearby extensive natural environments, such as in the "arc of deforestation" around the Amazon forest. In these areas, rapid infra-structural changes due to increases in agricultural crops and cattle fields can cause the rapid, opportunistic population growth of vampire bats.

In general, responses to fragmentation are still unclear for many bat species in the Neotropics (see Cunto & Bernard, 2012; but see also Ripperberger et al., 2015) and this remains especially true for agroecosystems, where more studies are still needed (Faria et al., 2007; Heer et al., 2015; Muylaert, 2015) in order to elucidate effective management strategies (see Chapter 8, this book). Moreover, isotopic analysis studies and functional diversity approaches could help determine how bats respond to heterogeneous landscape features.

4.2.4 Ants as Bioindicators of Fragmentation and Edge Effect in Neotropical Environments

Ants are among the most diverse organisms on the planet: they can inhabit any terrestrial environment, with the exception of the polar caps, and, due to their high richness and abundance, tend to be a dominant group in many habitats (Folgarait, 1988). Extreme ant biodiversity is observed in the Neotropics – for example, in the Amazon rainforest ants may represent more than 25% of the faunal biomass and over 40 species may be found in a single tree (Schultz, 2000; Wilson, 1987). This high diversity demonstrates the importance ants have in ecological processes, acting as disperses of seeds, decomposers, and agents of soil cycling and aeration (Lobry de Bruyn, 1999).

Due to their importance to ecosystems, high diversity, and the facility of sampling and identification, ants are considered good indicators of the impact of human activities on the environment (Majer, 1983; Andersen, 1993; Andersen & Majer, 2004). In addition, ant assemblage composition reflects trends in the composition of others groups more so than plants, terrestrial invertebrates and birds (Majer et al., 2007). Ant species can be grouped in functional groups (based on taxonomic, morphological, trophic and nesting behaviors) in order to reduce ecological complexity, enabling comparative analyses among environments that have few or no species in common (Philpott et al., 2010).

Compared to other regions and organism groups, there are few studies on the effect of fragmentation on ant assemblages in Neotropical natural habitats (Leal et al., 2012). Fragmentation characteristics, such as fragment size and isolation, are considered important factors driving the richness of ant assembly. On the contrary, the main factors affecting the richness of ant assemblages are the structural characteristics of vegetation and landscape, such as tree density and richness, as well as landscape heterogeneity (Carvalho & Vasconcelos, 1999; Pacheco & Vasconcelos, 2012; Ribas & Schoereder, 2007; Santos et al., 2006).

This relationship of ants with vegetation structure shows a relevant indirect effect of fragmentation on the assembly of ants. The hyperproliferation of pioneer tree species, as consequence of fragmentation, simplifies the forest structure, consequently reducing the heterogeneity and diversity of resources used by ants (nesting sites and food) (Leal et al., 2012). Studies that partitioned ant assemblages into functional groups show direct effects caused by fragmentation: functional groups such as army ants, specialist predators and cryptic species had their richness reduced when the fragment area was reduced (Leal et al., 2012; Petters et al., 2011). On the contrary, fragmentation usually benefits generalists, dominant and opportunist species of the Myrmicinae and Dolichoderinae subfamilies, as well as the leaf-cutting Attinae tribe (Dohm et al., 2011).

Edge effects influence ant assemblages in ways similar to fragmentation, reducing the diversity of some functional groups while benefiting generalist and opportunist species (Carvalho & Vasconcelos, 1999; Wirth et al., 2007; Dohm et al., 2011). Areas under edge influence usually have exotic species of ants (which generally belong to generalists, opportunistic and dominants functional groups), that directly compete with and reduce the diversity of native ants and others arthropods (Suarez et al., 1998; Holway, 2005). Some of these exotic species are among the most important invasive alien species (Lowe et al., 2000). These species often have a close relationship with human activity, which can cause accidental dispersal and offers a resource-rich environment free of natural predators, such as in urban areas, roads and powerlines (Lodge, 1993; Stiles & Jones, 1998; Suarez et al., 1998).

Therefore, studies that focus on the assembly of ants in fragmented environments are essential for Neotropical forest conservation on three different levels. First, the total diversity of ants could provide a good diagnostic measure regarding the vegetation structure and heterogeneity of the landscape. Second, by looking at functional groups of ants it is possible to see direct effects of fragmentation on ant assemblage, as well as the consequences of these changing patterns of ant assembly on ecological processes. Third, such studies could identify and delineate strategies to reduce the consequences of biological invasions caused by species of exotic ants.

4.3 Where Should We Go with Landscape Ecology within Agroecosystems?

The demand for information on how to manage landscapes in order to archive a better balance between human needs/activity and biodiversity conservation is by far higher than the information available to date. Government agencies, landowners, NGOs and a variety of industries ask for more precise information to be used in conservation planning, and prioritizing areas for conservation and restoration. But we still know very little about how habitat cover, configuration, matrix, edges and roads influences different species, functional groups and ecological processes. These are gaps that need to be urgently addressed in order to better understand the importance of landscape structure and the effects of roads on biodiversity, particularly within agriculture-dominated ecosystems:

- More than a habitat-matrix dichotomy: first we need to improve our concept of habitat, because many species use a variety of landscape features with different intensities and for different ecological functions (feeding, shelter, breeding, dispersal, escape from fire or pests). Thus, we need to avoid the idea that species interact with landscape features as habitat/matrix; however, information about how species uses natural and modified land cover types is scarce. Developing new research to effectively deal with the landscape as a complex mosaic is a great avenue for researchers during the next decade.
- Matrix types matter: we consider that it is important to broaden our vision of habitat types when classifying landscape elements, especially in diversified

landscapes such as agroecosystems. Within agroecosystems-dominated landscapes it is possible to have different types of matrices. This heterogeneity can be of good use for wildlife and also for economy, in terms of local economy and the appropriate use of soil, We need to consider how different matrices influence species persistence and movement at the landscape scale, as well as the effects of "matrix depth" (or matrix extent) on biodiversity maintenance. In this sense, we need to sample gradients of distance within the matrix in relation to habitats of reference within landscapes.

- Variation in edge effects: it is also important to consider both the vegetation type and the surrounding matrix when modelling edge effects in fragmented landscapes. Considering habitat and matrix as homogeneous, without accounting for differences in their structure, composition and function is likely to lead to an oversimplification of the proposed models.
- From patch-based to landscape-based studies: although landscape ecology advanced in different important directions during the last two decades, we need to move on toward a more landscape-based sampling design instead of patch-based surveys. Many of the studies with a landscape ecology approach still focus on patches as sampling units, often surveying only one patch or a particular type of habitat (e.g. forest). To effectively understand the full influence of landscape structure on biodiversity (see part II) we need to use landscapes as sampling unities and sample a variety of habitats and matrices within each landscape.
- Road ecology - beyond roadkill: road ecology studies became more common, but many of them still focus mainly on road kill-based research. There are plenty of ecological processes, for a great variety of taxa and regions, which need to be investigated in the light of road ecology. For example, what is the influence of road structure on facilitating or impeding road cross? What are the effects of roads on ecological processes such as frugivory, herbivory, hydrological regime, landscape regeneration etc?
- Movement ecology as a research agenda: movement is a key aspect of ecological processes because it is central to determine gene flow, dispersion, landscape regeneration etc. Understanding how landscape influences spatial-temporal movements at different scales will allow us to better define conservation and restoration priorities and strategies, especially for species or ecological processes that are more influenced by landscape spatial heterogeneity and habitat degradation. More than just knowing the home range of species, we need to effectively monitor and understand the movement patterns of keystone species, and consequently to better comprehend ecological processes that are more influenced by movements of different organisms.

Bibliography

Aguilar, R., Quesada, M., Ashworth, L., et al. (2008). Genetic consequences of habitat fragmentation in plant populations: susceptible signals in plant traits and methodological approaches. Molecular Ecology, 17, 5177-5188.

Alberti, M., Marzluff, J.M., Shulenberger, E., et al. (2003). Integrating humans into ecology: opportunities and challenges for studying urban ecosystems. BioScience, 53, 1169-1179.

Almeida, W.R., Meyer, S.T., Wirth, R., et al. (2007). Increasing densities of leaf cutting ants (*Atta* spp.) with proximity to the edge in a Brazilian Atlantic forest. Journal of Tropical Ecology, 23, 501--507.

Altringham, J.D. (2011). Bats: From Evolution to Conservation. 368 p. Oxford: Oxford University Press.

Alvey, A.A. (2006). Promoting and preserving biodiversity in the urban forest. Urban Forestry & Urban Greening, 5, 195--201.

Andersen, A.N. (1993). Ants as indicators of restoration success at a uranium mine in tropical Australia. Restoration Ecology, 1, 156-167.

Andersen, A.N., & Majer, J.D. (2004).Ants shows the Way Down Under: Invertebrates as bioindicators in Land Management. Frontiers in Ecology and Environment, 2 (6) 291-298.

Andersen, L.W., Fog, K., & Damgaard, C. (2004). Habitat fragmentation causes bottlenecks and inbreeding in the European tree frog (Hylaarborea). Proceedings of the Royal Society, 271, 1293-1302.

Antongiovanni, M., & Metzger, J.P. (2005). Influence of matrix habitats on the occurrence of insectivorous bird species in Amazonian forest fragments. Biological Conservation, 122, 441-451.

Aparicio, A., Hampe, A., Fernadez-Carrillo, L., & Albaladejo, R.G. (2012). Fragmentation and comparative genetic structure of four mediterranean woody species: complex interactions between life history traits and the landscape context. Diversity and Distributions, 18, 226-235.

Armbrecht, I., Perfecto, I., Philpott, S.M., et al. (2010). Ant Diversity and Function in Disturbed and Changing Habitats. In L. Lach, C.L. Parr, & K.L. Abbott (Eds.), Ant Ecology (p.137-156). New York: Oxford University Press.

Ashley, E.P., & Robinson, J.T. (1996). Road mortality of amphibians, reptiles and other wildlife on the Long Point Causeway, Lake Erie, Ontario. Canadian Field Naturalist, 110, 403-12.

Awade, M., & Metzger, J.P. (2008). Using gap-crossing capacity to evaluate functional connectivity of two Atlantic rainforest birds and their response to fragmentation. Austral Ecology, 33, 863-871.

Bacles, C.F.E., Burczyk, J., Lowe, A.J., & Ennos, R.A. (2005). Historical and contemporary mating patterns in remnant populations of the forest tree *Fraxinus excelsior*. Evolution, 59, 979-990.

Baum, K.A., Haynes, K.J., Dillemuth, F.P., & Cronin, J.T. (2004). The matrix enhances the effectiveness of corridors and stepping stones. Ecology, 85, 2671-2676.

Ben-David, M., & Flaherty, E.A. (2012).Stable isotopes in mammalian research a beginner's guide. Journal of Mammalogy, 93, 312-328.

Bennett, A.F., Radford, J.K., & Haslem, A. (2006).Properties of land mosaics, implications for nature conservation in agricultural environments. Biological Conservation, 133, 250-264.

Bernard, E., Aguiar, L.M.S. Brito, D., et al. (2012). Uma análise de horizontes sobre a conservação de morcegos no Brasil (2012) 65-77. In T.R.O Freitas & E.M. Vieira (Eds.) Mamíferos do Brasil: Genética, Sistemática, Ecologia e Conservação (p. 1-17). Rio de Janeiro: Sociedade Brasileira de Mastozoologia.

Berthinussen, A., & Altringham J. (2012a). The effect of a major road on bat activity and diversity. Journal of Applied Ecology, 49, 82-89, DOI:10.1111/j.1365-2664.2011.02068.x.

Berthinussen, A, & Altringham, J. (2012b). Do bat gantries and underpasses help bats cross roads safely? PLoS One, 7, e38775, DOI:10.1371/journal.pone.0038775,

Betke, M., Hirsh, D.E., Makris, N.C., McCracken, G.F., et al. (2008). Thermal imaging reveals significantly smaller Brazilian free-tailed bat colonies than previously estimated. Journal of Mammalogy, 89, 18-24.

Blackburn, T.M., Petchey, O.L., Cassey, P., Gaston, K.J., et al. (2005). Functional diversity of mammalian predators and extinction in island birds. Ecology, 86, 2916-2923.

Boecklen, W.J., Yarnes, C.T., Cook, B.A., & James, A.C., et al. (2011). On the use of stable isotopes in trophic ecology. Annual Review of Ecology and Systematics, 42, 411-440.

Boscolo, D., & Metzger, J.P. (2011). Isolation determines patterns of species presence in highly fragmented landscapes. Ecography, 34, 1018-1029.

Bredt, A., Uieda, W., & Pedro, W.A. (2012). Plantas e Morcegos na recuperação de áreas degradadas e na paisagem urbana. 1ª. ed. Brasília: Rede de Sementes do Cerrado.

Breed, M.F., Gardner, M.G., Ottewell, K.M., Navarro, C.M., Lowe, A.J., et al. (2012). Shifts in reproductive assurance strategies and inbreeding costs associated with habitat fragmentation in Central American mahogany. Ecology Letters, 15, 444-452.

Butchart, S.H.M., Walpole, M., Collen, B., et al. (2010). Global Biodiversity: Indicators of recent declines. Science, 328, 1164-1168.

Cadenasso, M.L., & Picket, S.T.A. (2000). Linking forest edge structure to edge function: mediation of herbivore damage. Journal of Ecology, 88, 31-44.

Cadotte, M.W., Carscadden, K., & Mirotchnick, N. (2011). Beyond species: functional diversity and the maintenance of ecological processes and services. Journal of Applied Ecology, 48, 1079-1087.

Calisher, C.H., Childs, J.E., Field, H.E., Holmes, K.V., & Schountz, T. (2006). Bats: important reservoir hosts of emerging viruses. Clinical Microbiology Reviews, 19, 531-545.

Carvalho, K.S., & Vasconcelos, H.L. (1999). Forest fragmentation in central Amazonia and its effects on litter dwelling ants. Biological Conservation, 91, 151-157.

Chabrerie, O., Jamoneau, A., Gallet-Moron, E., & Decocq, G. (2013). Maturation of forest edges is constrained by neighbouring agricultural land management. Journal of Vegetation Science, 24, 58-69.

Chambers, J.L., & Garant, D. (2010). Determinants of population genetic structure in eastern chipmunks (*Tamias striatus*): the role of landscape barriers and sex-biased dispersal. Journal of Heredity, 101, 413-422.

Chiarello, A.G. (1999). Effects of fragmentation of the Atlantic forest on mammal communities in south-eastern Brazil. Biological Conservation, 89, 71-82.

Chillo, V., & Ajeda, R.A. (2012). Mammal functional diversity loss under human-induced disturbances in arid lands. Journal of Arid Environments, 87, 95-102.

Cianciaruso, M.V., Silva, I.A., & Batalha, M.A. (2009). Diversidades filogenética e funcional: novas abordagens para a ecologia de comunidades. Biota Neotropica, 9, 93-103.

Cilliers, S.S., Williams, N.S.G., & Barnard, F.J. (2008). Patterns of exotic plant invasions in fragmented urban and rural grasslands across continents. Landscape Ecology, 23, 1243-1256.

Cleveland, C.J., Betke, M., Federico, P., Frank, J., et al. (2006). Economic value of the pest control service provided by Brazilian free-tailed bats in south-central Texas. Frontiers in Ecology and the Environment, 4, 238-243.

Codron, D., Codron, J., Sponheimer, M., et al. (2005). Assessing diet in savanna herbivores using stable carbon isotope ratios of faeces. Koedoe, 48, 115-124.

Collevatti, R.G., Grattapaglia, D., & Hay, J.D. (2001).Population genetic structure of the endangered tropical tree species *Caryocar brasiliense*, based on variability at microsatellite *loci*. Molecular Ecology, 10, 349-356.

Costa, L. P., Leite, Y.L.R., Mendes, S.L, et al. (2005). Mammal Conservation in Brazil. Conservation Biology, 19, 672-679.

Coulon, A., Cosson, J.F., Angibault, M., et al. (2004). Landscape connectivity influences gene flow in a roe deer population inhabiting a fragmented landscape: an individual-based approach. Molecular Ecology, 13, 2841-2850.
Crawford, K., McDonald, R.A., & Bearhop, S. (2008). Applications of stable isotope techniques to the ecology of mammals. Mammal Review, 38, 87-107.
Cuarón, A.D. (2000). A global perspective on habitat disturbance and tropical rainforest mammals. Conservation Biology, 14, 1574-1579.
Cullen, L., Bodmer, R.E., & Pádua, C.V. (2000). Effects of hunting in habitat fragments of the Atlantic forest, Brazil. Biological Conservation, 95, 49-56.
Cunto, G.C., & Bernard, E. (2012). Neotropical bats as indicators of environmental disturbance: what is the emerging message? Acta Chiropterologica, 14, 143-151.
Delgado, J.D., Arroyo, N.L., Arevalo, J.R., & Fernandez-Palacios, J.M. (2007). Edge effects of roads on temperature, light, canopy cover, and canopy height in laurel and pine forests (Tenerife, Canary Islands). Landscape and Urban Planning, 81, 328-340.
DeNiro, M.J., & Epstein, S. (1978). Influence of diet on the distribution of carbon isotopes in animals. Geochimica et Cosmochimica Acta, 42, 495-506.
Dibattista, J.D. (2008). Patterns of genetic variation in anthropogenically impacted populations. Conservation Genetics, 9, 141-156.
Didham, R.K., & Lawton, J.H. (1999). Edge structure determines the magnitude of changes in microclimate and vegetation structure in tropical forest fragments. Biotropica, 31, 17-30.
Differdorfer, J.E., Gaines, M.S., & Holt, R.D. (1995). Habitat fragmentation and movements of three small mammals (Sigmodon, Microtus, and Peromyscus). Ecology, 76 (3), 827-839.
Dodonov, P., Harper, K.A., & Silva-Matos, D.M. (2013). The role of edge contrast and forest structure in edge influence: vegetation and microclimate at edges in the Brazilian Cerrado. Plant Ecology, 214, 1345-1359.
Dohm, C., Leal, I.R., Tabarelli, et al. (2011). Leaf-cutting ants proliferate in the Amazon: an expected response to forest fragmentation? Journal of Tropical Ecology, 27, 645-649.
Dotta, G., & Verdade, L.M. (2007). Trophic categories in a mammal assemblage: diversity in an agricultural landscape. Biota Neotropica, 7 (2), 287-292.
Dotta, G., & Verdade, L.M. (2011). Medium to large-sized mammals in agricultural landscapes of south-eastern Brazil. Mammalia, 75, 345-352.
Ehrlich, P.R. (1988). The loss of diversity. Biodiversity. Washington, DC: National Academy Press.
Fahrig, L. (2003). Effects of habitat fragmentation on biodiversity. Annual Review of Ecology Evolution Systematics, 34, 487-515.
Ferraz, K.M.P.M.B., Siqueira, M.F., Alexandrino, E.R., et al. (2012). Environmental suitability of a highly fragmented and heterogeneous landscape for forest bird species in south-eastern Brazil. Environmental Conservation, 39, 316-324.
Ferraz, K.M.P.M.B., Siqueira, M.F., Martin, P.S., et al. (2010). Assessment of *Cerdocyon thous* distribution in an agricultural mosaic, southeastern Brazil. Mammalia, 74, 275-280.
Filgueiras B.K.C., Iannuzzi L., Leal I.R. (2011). Habitat fragmentation alters the structure of dung beetle communities in the Atlantic forest. Biological Conservation, 144, 362-369.
Filgueiras, B.K.C., Gomes, J.P., Leal, I.R., et al. (2012). Effects of habitat fragmentation on ant richness and functional composition in Brazilian Atlantic forest. Biodiversity and Conservation, 21, 1687-1701.
Fletcher, R.J. (2005). Multiple edge effects and their implications in fragmented landscapes. Journal of Animal Ecology, 74, 342-352.
Flynn, D.F.B., Gogol-Prokurat, M., Nogeire, T., et al. (2009). Loss of functional diversity under land use intensification across multiple taxa. Ecology Letters, 12, 22-33.
Folgarait, P.J. (1998). Ant biodiversity and its relationship to ecosystem functioning: a review. Biodiversity and Conservation, 7, 1221-1244.

Forman, R.T. (1995) Land mosaics: the ecology of landscapes and regions. Cambridge: Cambridge University Press.

Forman, R.T.T, & Deblinger R.D. (2000). The Ecological Road-Effect Zone of a Massachusetts (USA) Suburban Highway. Conservation Biology, 14, 36-46.

Forman, R.T.T., & Alexander, L.E. (1998). Roads and Their Major Ecological Effects. Annual Review of Ecology and Systematics, 29, 207-231.

Fragoso, J.M.V., & Huffmann, J.M. (2000). Seed-dispersal and seedling recruitment patterns by the last Neotropical megafaunal element in Amazonia, the tapir. Journal of Tropical Ecology, 16, 369-385.

Frair, J.L., Merrill, E.H., Beyer, H.L., et al. (2008). Thresholds in landscape connectivity and mortality risks in response to growing road networks. Journal of Applied Ecology, 45, 1504-1513.

Frankham, R., Ballou J., & Briscoe, D. (2002). Introduction to Conservation Genetics. Cambridge: Cambridge University Press.

Galetti, M., & Dirzo, R. (2013). Ecological and evolutionary consequences of living in a defaunated world. Biological Conservation, 163, 1-6.

Galetti, M., Giacomini, H.C., Bueno, R.S., et al. (2009) Priority areas for the conservation of Atlantic forest large mammals. Biological Conservation, 142, 1229-1241.

Galetti, M., Guevara, R., Cortes, M., et al. (2013). Functional extinction of birds drives rapid evolutionary changes in seed size. Science, 340, 1086-1090.

Galetti, M., Keuroghlian, A., Hanada, L., et al. (2001). Frugivory and seed dispersal by the Lowland Tapir (*Tapirus terrestris*) in southeast Brasil. Biotropica, 33, 723-726.

Gomes, M.N., Monteiro, A.M., Nogueira, V., & Gonçalves, C.A. (2010). Landscape risk factors for attacks of vampire bats on cattle in Sao Paulo, Brazil. Preventive Veterinary Medicine, 93, 139-146.

Grimm, N.B., Faeth, S.H., Golubiewski, N.E., et al. (2008). Global Change and the Ecology of Cities. Science, 319, 756-760.

Gunson, K.E., Mountrakis, G. & Quackenbush, L.J. (2011) Spatial wildlife-vehicle collision models: A review of current work and its application to transportation mitigation projects. Journal of Environmental Management, 92, 1074-1082.

Hamrick, J.L. (2004). Response of forest trees to global environmental changes. Forest Ecology and Management, 197, 323-335.

Harper, K.A., & Macdonald, S.E. (2011) Quantifying distance of edge influence: a comparison of methods and a new randomization method. Ecosphere, 2, 94.

Harper, K.A., Macdonald, S.E., Burton, P.K., et al. (2005). Edge influence on forest structure and composition in fragmented landscapes. Conservation Biology, 19, 768-782.

Hawbaker T.J., Radeloff V.C., Clayton M.K., et al. (2006). Road development, housing growth, and landscape fragmentation in northern Wisconsin: 1937-1999. Ecological Applications, 16, 1222-37.

Heer, K., Helbig-Bonitz, M., Fernandes, R.G., et al. (2015). Effects of land use on bat diversity in a complex plantation-forest landscape in northeastern Brazil. Journal of Mammalogy, 96 (4), 720-731.

Hoffmann, A.A., & Willi, Y. (2008). Detecting genetic responses to environmental change. Genetics, 9, 421-432.

Holway, D.A. (2005). Edge effects of an invasive species across a natural ecological boundary. Biological Conservation, 121, 561-567.

Huijser, M.P., Abra, F.D., & Duffield, J.W. (2013). Mammal road mortality and cost-benefit analyses of mitigation measures aimed at reducing collisions with capybara (*Hydrochoerus hydrochaeris*) in São Paulo State, Brazil. Oecologia Australis, 17, 129-146.

Jaeger, J.A., Bowman, J., Brennan, J., et al. (2005). Predicting when animal populations are at risk from roads: an interactive model of road avoidance behavior. Ecological Modelling, 185, 329-348.
Jenkins, C.N., Pimm, S.L., & Joppa, L.N. (2013). Global patterns of terrestrial vertebrate diversity and conservation. Proceedings of the National Academy of Science, 110, 2602-2610.
Johansson, M., Primmer, C. R., Merila, J. (2007). Does habitat fragmentation reduce fitness and adaptability? A case study of common frog (*Rana temporaria*). Molecular Ecology, 16, 2693-2700.
Jorge, M.L.S.P., Galetti, M., Ribeiro, M.C., et al. (2013). Mammal defaunation as surrogate of trophic cascades in a biodiversity hotspot. Biological Conservation, 163, 49-57.
Kelly, J.F. (2000). Stable isotopes of carbon and nitrogen in the study of avian and mammalian trophic ecology. Canadian Journal of Zoology, 78, 1-27.
Lange, R., Diekotter, T., Schiffmann, L.A., Wolters, V., & Durka, W. (2012). Matrix quality and habitat configuration interactively determine functional connectivity in a widespread bush cricket at a small spatial scale. Landscape Ecology, 27, 381-392.
Laurance, W.F., & Curran, T.J. (2008). Impacts of wind disturbance on fragmented tropical forests: A review and synthesis. Austral Ecology, 33, 399-408.
Laurance, W., Goosem, M. & Laurance, S. (2009). Impacts of roads and linear clearings on tropical forests. Trends in Ecology and Evolution, 24, 659-669.
Lobova, T.A., Geiselman, C.K., & Mori, S.A. (2009). Seed dispersal by bats in the Neotropics. New York, NY: Botanical Garden Press.
Lobry de Bruyn, L.A. (1999) Ants as bionidicators of soil function in rural environments. Agriculture, Ecosystems and Environment, 74, 425-441.
Lodge, D.M. (1993). Biological Invasions: Lesson for Ecology. Trends in Ecology and Evolution, 8, 133-137.
Lowe, A.J., Boshier, D., Ward, M., Bacles, C.F.E., & Navarro, C. (2005). Genetic resource impacts of habitat loss and degradation; reconciling empirical evidence and predicted theory for Neotropical trees. Heredity, 95, 255-273.
Lung, T., Peters, M. K, Schaab, G., et al. (2011). Deforestation and the population decline of the army ant Doryluswilverthi in western Kenya over the last century. Journal of Applied Ecology, 48, 697-705.
Magioli, M., Ferraz, K.M.P.M.B., & Rodrigues, M.G. (2014a). Medium and large-sized mammals of an isolated Atlantic Forest remnant, southeast São Paulo State, Brazil. Check List, 10, 850-856.
Magioli, M., Moreira, M.Z., Ferraz, K.M.B., et al. (2014b). Stable isotope evidence of *Puma concolor* (Felidae) feeding patterns in agricultural landscapes in southeastern Brazil. Biotropica, 46, 451-460.
Magioli, M., Ribeiro, M.C., Ferraz, K.M.P.M.B., et al. (2015). Thresholds in the relationship between functional diversity and patch size for mammals in the Brazilian Atlantic Forest. Animal Conservation, 18 (6), 499-511, DOI 10.1111/acv.12201.
Majer, J.D. (1983). Ants: bio-indicators of mine site rehabilitation, land-use, and land conservation. Environmental Management, 7, 375-383.
Majer, J.D., Orabi, G., & Bisevac, L. (2007). Ants (Hymenoptera: Formicidae) pass the bioindicator scorecard. Myrmecological News, 10, 69-76.
Martensen, A.C., Ribeiro, M.C.; Banks-Leite, C.; Prado, P.I.; Metzger, J.P. (2012). Associations of forest cover, fragment area, and connectivity with neotropical understory bird species richness and abundance. Conservation Biology, 26, 1100-1111.
Mascarúa Lopez, L.E., Harper, K.A., & Drapeau, P. (2006).Edge influence on forest structure in large forest remnants, cutblock separators, eand riparian buffers in managed black spruce forests. Ecoscience, 13, 226-233.

McKinney, M.L. (2008). Effects of urbanization on species richness: A review of plants and animals. Urban Ecosystems, 11, 161-176.
Meyer, S.T., Leal, I.R., & Wirth, R. (2009). Persisting hyper-abundance of leaf-cutting ants (Atta spp.) at the edge of an old Atlantic forest fragment. Biotropica, 41, 711--716.
Mickleburgh, S.P., Hutson, A.M., & Racey, P.A. (2002). A review of the global conservation status of bats. Oryx, 36, 18-34.
Miller J.R., Joyce L.A., Knight R.L. & King R.M. (1996). Forest roads and landscape structure in the southern Rocky Mountains. Landscape Ecology, 11, 115-27.
Mouchet, M.A., Villéger, S., Mason, N.H.W., & Mouillot, D. (2010). Functional diversity measures: an overview of their redundancy and their ability to discriminate community assembly rules. Functional Ecology, 24, 867-876.
Murphy, H.T., & Lovett-Doust, J. (2004). Context and connectivity in plant metapopulations and landscape mosaics: does the matrix matter? Oikos, 105, 3-14.
Muylaert, R.L. 2014. Influências multi-escala da paisagem e limiar da fragmentação em morcegos no Cerrado, Universidade Estadual Paulista, Rio Claro, Brazil. URL <http://repositorio.unesp.br/bitstream/handle/11449/121888/000814936.pdf?sequence=1> Accessed September 2015
O´Mara, T., Wikelski, M. & Dechmann, D.K.N. (2014). 50 years of bat tracking: device attachment and future directions. Methods in Ecology and Evolution, 5 (4), 311-319, DOI: 10.1111/2041-210X.12172
Pacheco, R. & Vasconcelos, H.L. (2012). Habitat diversity enhances ant diversity in a naturally heterogeneous Brazilian landscape. Biodiversity and Conservation, 21, 797-809.
Páez, A., Nũnez, C., García, C., et al. (2003). Molecular epidemiology of rabies epizootics in Colombia: evidence for human and dog rabies associated with bats. Journal of Genetic Virology, 84, 795-802.
Pardini, R., Faria, D., Accacio, G.M., et al. (2009). The challenge of maintaining Atlantic forest biodiversity: a multi-taxa conservation assessment of specialist and generalist species in an agro-forestry mosaic in Southern Bahia. Biological Conservation, 142, 1178-1190.
Peres, C.A. (2000). Effects of subsistence hunting on vertebrate community structure in Amazonian forests. Conservation Biology, 14, 240-253.
Petchey, O.L., & Gaston, K.J. (2002). Functional diversity (FD), species richness and community composition. Ecology Letters, 5, 402-411.
Petchey, O.L., & Gaston, K.J. (2006). Functional diversity: back to basics and looking forward. Ecology Letters, 9, 741-758.
Pohlman, C.L., Turton, S.M., & Goosem, M. (2007). Edge effects of linear canopy openings on tropical rain forest understory microclimate. Biotropica, 39, 62-71.
Poos, M.S., Walker, S.C., & Jackson, D.A. (2009). Functional-diversity indices can be driven by methodological choices and species richness. Ecology, 90, 341-347.
Prevedello, J.A., & Vieira, M.V. (2010). Does the type of matrix matter? A quantitative review of the evidence. Biodiversity and Conservation, 19, 1205-1223.
Pyke, G.H. (1984). Optimal foraging theory: a critical review. Annual Review in Ecoogy and Systematics, 15, 523-575.
Ramalho, C.E., Hobbs, R.J. (2012). Time for a change: dynamic urban ecology. Trends in Ecology and Evolution, 27, 179-188.
Rao, C.R. (1982). Diversity and dissimilarity coefficients: a unified approach. Theoretical Population Biology, 21, 24-43.
Ribas, C.R. & Schoereder, J.H. (2007). Ant communities, environmental characteristics and their implications for conservation in the Brazilian Pantanal. Biodiversity and Conservation, 16, 1511-1520.

Ribeiro, M.C., Metzger, J.P., Martensen, A.C., et al. (2009). The Brazilian Atlantic Forest: how much is left, and how is the remaining forest distributed? Implications for conservation. Biological Conservation, 142, 1141-1153.

Ries, L., Fletcher, R.J., Battin, J., & Sisk, T.D. (2004). Ecological responses to habitat edges: Mechanisms, models, and variability explained. Annual Review of Ecology, Evolution and Systematics, 35, 491-522.

Ripperger, S.P. Kalko, E.K.V., Rodríguez-Herrera, B., Mayer, F., & Tschapka, M., (2015). Frugivorous Bats Maintain Functional Habitat Connectivity in Agricultural Landscapes but Rely Strongly on Natural Forest Fragments. PLoS One, 10, e0120535.

Roelke, M., Martenson, J., & O´Brien, S. (1993). The consequences of demographic reduction and genetic depletion in the endangered Florida panther. Current Biology, 3, 340-350.

Rupprecht, C.E., Hanlon, C.A & Hemachudha, T., (2002), Rabies re-examined, The Lancet Infectious Diseases, 2, 327-343.

Russell, A.L., Butchoski, C., Saidak, L., et al. (2009). Road-killed bats, highway design, and the commuting ecology of bats. Endangered Species Research, 8, 49-60.

Sabino-Santos Jr, G., F. G. M. Maia, Vieira, T.M., Muylaert, R.L. Lima, S.M, et al. (2015). Evidence of Hantavirus Infection among Bats in Brazil. The American Journal of Tropical Medicine and Hygiene 93 (2), 404-406, DOI:10.4269/ajtmh.15-0032.

Saunders, D.A., Hobbs, R.J., & Margules, C.R. (1991). Biological consequences of ecosystem fragmentation - a review. Conservation Biology, 5, 18-32.

Schaub, A., Ostwald, J., & Siemers, B. M. (2008). Foraging bats avoid noise. Journal of Experimental Biology, 211, 3174-3780.

Schneider, M.C., Aron, J., Santos-Burgoa C., et al. (2001). Common vampire bat attacks on humans in a village of the Amazon region of Brazil. Caderno Saúde Pública,17, 1531-36.

Scoss, L.M., Júnior, P.D.M., Silva, E., et al. (2004). Uso de parcelas de areia para o monitoramento de impacto de estradas sobre a riqueza de espécies de mamíferos. Revista Árvore, 28, 121-127.

Shutz, T.R. (2000). In search of ant ancestors. Proceedings of the National Academy of Sciences, 97, 14028-14029.

Siemers, B.M., & Schaub, A. (2011). Hunting at the highway: traffic noise reduces foraging efficiency in acoustic predators. Proceedings of the Royal Society B, 278, 1646-1652

Silva Jr, A.P., & Pontes, A.R.M. (2008).The effect of a mega-fragmentation process on large mammal assemblages in the highly-threatened Pernambuco Endemism Centre, north-eastern Brazil. Biodiversity and Conservation, 17, 1455-1464.

Smit, I.P.J., & Asner, G.P. (2012). Roads increase woody cover under varying geological, rainfall and fire regimes in African savanna. Journal of Arid Environments, 80, 74-80.

Sork, V.L., & Smouse, P.E. (2006).Genetic analysis of landscape connectivity in tree populations. Landscape Ecology, 21, 821-836.

Sponheimer, M., Grant, C.C., Ruiter, D.J., Lee-Thorp, J.A., Codron, D.M., & Codron, J. (2003). Diets of impala from Kruger National Park evidence from stable carbon isotopes. Koedoe, 46, 101-106.

Stevens, R.D., Cox, S.B., Strauss, R.E., et al. (2003). Patterns of functional diversity across an extensive environmental gradient: vertebrate consumers, hidden treatments and latitudinal trends. Ecology Letters, 6, 1099-1108.

Stiles, J.H. & Jones, R.H. (1998). Distribution of the Red Imported Ant, Solenopsisinvicta, in Road and Powerline. Landscape Ecology, 13, 335-346.

Storfer, A., Murphy, M.A., Spear, S.F., et al. (2010). Landscape genetics: where are we now? Molecular Ecology, 19, 3496-3514.

Suarez, A.V.; Bolger, D.T. & Case, T.J. (1998). Effects of fragmentation and invasion on native ant communities in coastal southern in California. Ecology, 79, 2041-2056.

Umetsu, F., & Pardini, R. (2007).Small mammals in a mosaic of forest remnants and anthropogenic habitats - evaluating matrix quality in an Atlantic forest landscape. Landscape Ecology, 22, 517-530.
Umetsu, F., Metzger, J.P., & Pardini, R. (2008). Importance of estimating matrix quality for modeling species distribution in complex tropical landscape: a test with Atlantic forest small mammals. Ecography, 31, 359-370.
Uriarte, M., Anciães, M., Silva, M.T.B., et al. (2011). Disentangling the drivers of reduce long-distance seed dispersal by birds in an experimentally fragmented landscape. Ecology, 92, 924-937.
van der Ree, R., Jaeger, J.A. van der Grift, E.A. et al. (2011). Effects of roads and traffic on wildlife populations and landscape function: road ecology is moving towards larger scales. Ecology and Society, 16, 48.
Villéger, S., Mason, N.W., & Mouillot, D. (2008). New multidimensional functional diversity indices for a multifaceted framework in functional ecology. Ecology, 89, 2290-2301.
Watling, J.I., Nowakowski, A.J., Donnelly, M.A., et al. (2011). Meta-analysis reveals the importance of matrix composition for animals in fragmented habitat. Global Ecology and Biogeography, 20, 209-217.
Wilkie, D., Shaw, E., Rotberg, F., et al. (2000). Roads, development, and conservation in the Congo Basin. Conservation Biology, 14, 1614-1622.
Willians-Guillén, K., Perfecto, I., & Vandermeer, J. (2008). Bats Limit Insects in a Neotropical Agroforestry System. Science, 320, 70.
Willoughby Jr, R.E., Tieves, K.S., Hoffman, G.M., et al. (2005). Survival after treatment of rabies with induction of coma. New England Journal of Medicine, 352, 2508-2514.
Wilson, E.O. (1987).The arboreal ant fauna of Peruvian Amazon forests: A first assessment. Biotropica, 19, 245-251.
Wilson, E.O. (1988). The current state of biological diversity. Biodiversity, 521.
Winkler, M., Koch, M., & Hietz, P. (2011). High gene flow in epiphytic ferns despite habitat loss and fragmentation. Conservation Genetics, 12, 1411-1420.
Young, A., Boyle, T., & Brown, T. (1996).The population genetic consequences of habitat fragmentation for plants. Trends in Ecology and Evolution, 11, 413-418.

Part II: **Biodiversity Dimensions**

Dennis Goss-Souza, Clovis Daniel Borges, Lucas William Mendes, Acacio Aparecido Navarrete, Dilmar Baretta, Tsai Siu Mui

5 Exploring Diversity of Soil Microorganisms: A Multidimensional Approach

5.1 Introduction

Microorganisms represent the great majority of biodiversity in the biosphere. They play key roles in ecosystems services that support life, such as biogeochemical cycling (Madsen, 2011). Despite knowing the importance of soil microorganisms for the maintenance of dynamics and equilibrium in ecosystems, there is a lack of information about their diversity and functionality. Microbial ecologists have strongly increased their effort to unravel the "black box" of soil microbial diversity and functionality, surveying their genetic diversity (e.g. through metagenomics) and measuring the expression of functional traits (e.g. through rates of biogeochemical cycling). What remains unclear is how to integrate these dimensions of biodiversity with the associated environmental factors in order to elucidate ecological patterns (Fig. 5.1).

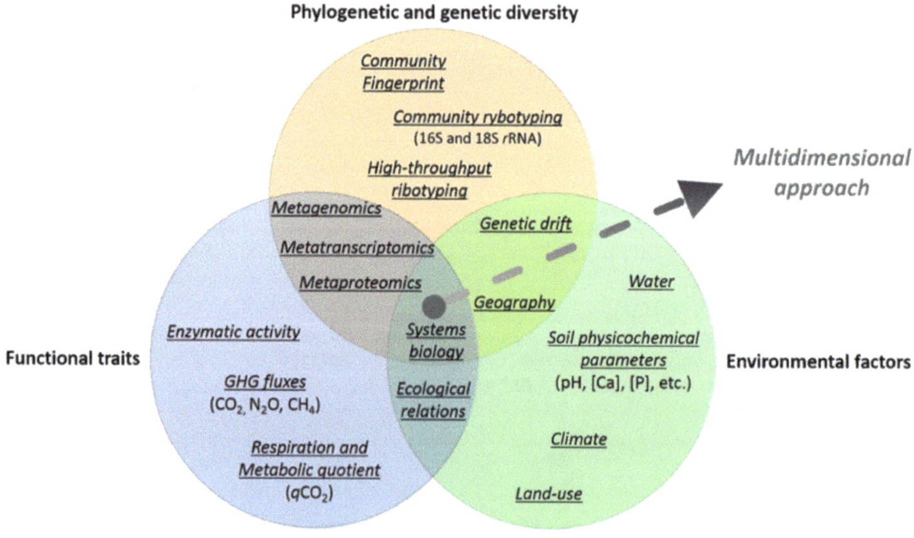

Fig. 5.1: Integrating microbial ecological data with environmental factors to create a comprehensive multidimensional approach.

Challenges concerning this issue have multiple aspects. First, there is the problem of scale, since it has been estimated that one gram of soil might contain billions of individual microbes and over 40,000 distinct lineages (Schloss & Handelsman, 2006). Several methods of sampling and survey fail to obtain a comprehensive picture of microbial diversity and its linkage with functional patterns. Moreover, interactions among these dimensions of microbial diversity and environmental factors that shape the variability are still scarce. Microbial ecologists agree that including environmental variables in a multiplex analysis system may elucidate microbial diversity, which will help the monitoring, management and conservation of agroecosystems.

Human activity on earth has altered the rates of environmental changes in a way that has never been seen before (Hegerl & Stott, 2014). Understanding how microbes can predict these changes requires looking forward on these dimensions. Recent advances have made possible the integration of microbial genetic, phylogenetic and functional diversity with environmental parameters. Fingerprint techniques, next-generation sequencing and related advances in bioinformatics are being massively applied to evaluate microbial diversity in soils (Myrold et al., 2014) and several other environments (Fierer & Lennon, 2011). Technological advances in ecosystem biogeochemistry (e.g. isotopic enrichment) now allow functional measurements at the necessary scale and resolution, linking microbial ecosystems to genetic and phylogenetic diversity.

In this chapter, we propose to combine advances in sequencing technologies, ecosystem biogeochemistry and systems science to determine the interactions and feedbacks among the genetic, phylogenetic and functional dimensions of microbial biodiversity. The environmental factors that modulate other ecological dimensions will also be considered, to enhance an emerging cornerstone, here named as the ecological multidimensional approach.

5.2 Sampling and Data Retrieval of Soil Microorganisms: From Field to Laboratory

One of the most important steps for obtaining results with quality and accuracy is to design a sampling that includes an appropriate number of samples in a well-defined space. Based on this, scientists have developed several strategies to obtain samples from the field, including soil sampling for evaluation of microbial diversity. An important factor to consider is the number of replicates that cover the sampling space. The number of replicates must guarantee faith in the data (Prosser, 2010), while also considering costs and logistics efficiency.

Biological and biochemical processes are of great importance in soils, since they mediate functions that support the terrestrial ecosystems. At the molecular level, characteristics linked to metabolism, such as organic compounds degradation and enzymatic-mediated reactions can be investigated (Klueglein et al., 2014). When focusing on the organism level, a broad range of methodologies are employed to survey taxonomic and phylogenetic description, functional genes analyses, regulation of gene transcription, enzymatic expression and cell kinetics (Mendes et al., 2014; Souza et al., 2015). Another level is the landscape scale, which includes all the prior levels in a macroview, in accordance with environmental characteristics that can affect biological attributes (e.g. soil acidity and fertility, temperature, geographic location) (Fierer et al., 2012).

Before data analysis, it is necessary and even more important to develop a workflow that includes an adequate experimental design and a sampling strategy that strives to guarantee the data integrity. The soil is a live, dynamic and open system. The heterogeneity of this complex matrix can be evidenced when we determined that there are hot spots of fertility and organism activity in soil. These hot spots usually constitute less than 10% of total soil volume, but corresponding up to 90% of soil total biological activity (Monard et al., 2008). Thus, sampling strategies should take this variability into account, avoiding over or underestimation of some attributes.

Studies of physicochemical and biological variables commonly use a strategy of sampling at the plot scale. Samples are collected from the study site and treated as individual or combined to form a composite sample. Samples for evaluating microbial patterns are often collected taking into account the scale of representation and can be grouped by sampling method (Fig. 5.2). The most commonly used sampling methods are described as follows:

Random – samples are collected in a simple random design, trying to cover the maximum possible area within the treatment. In areas in which topography and soil type vary by small distances, the suggestion is to divide the area into blocks with similar characteristics in order to decrease experimental error. This method is appropriate for small scales (areas less than 1 hectare).

Cross system – a cardinal-based method in which the samples are collected in all axes, in order to overcome the differences that may occur in small scales, such as slight differences in soil properties. Appropriate to small and medium scales (up to 10 hectares).

Transect – generally, the samples are taken via a geometrical exponential-fold distance scheme, designed in a straight line. Transects are widespread and largely used in microbial niche studies, due to the possibility to evaluate the occurrence of taxa according to distance (Rodrigues et al., 2013). Appropriate for covering large surfaces and areas. Transect sampling is a powerful tool in biogeography studies ranging from microcosm and mesocosm (Ramette & Tiedje, 2007), to landscape scale (Dequiedt et al., 2009), and from taxonomy to functional traits (Green et al., 2008).

Geogrid – samples collected from equidistantly designed points forming a grid. Recent advances in soil microbial studies are leading researchers to study soil microbial patterns of colonization and geographical distribution in a more systematized way (Bru et al., 2011). Geogrids can be used with advantage in biogeography studies, since they facilitate the local alignment and the construction of semivariograms for global alignment.

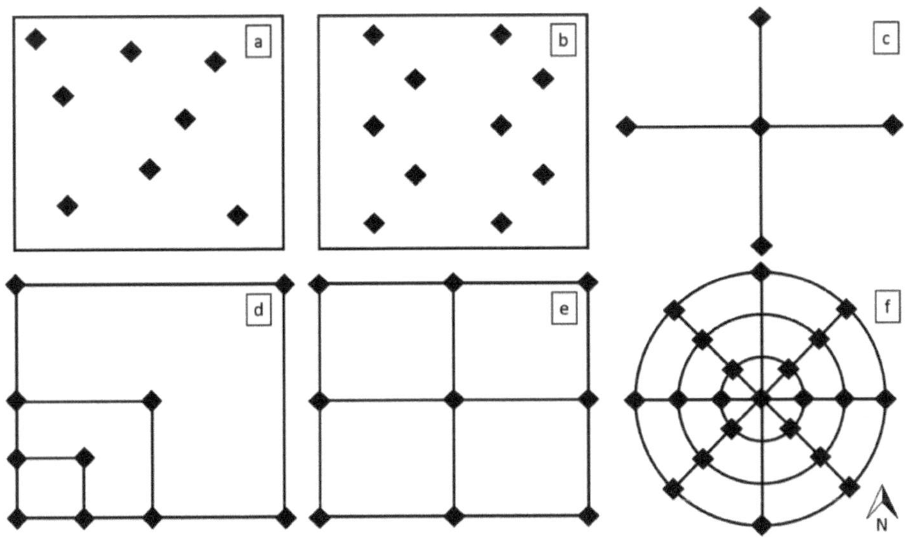

Fig. 5.2: Methods of sampling from field. Figures are ordinated by increasing level of complexity. (A) Random sampling (B) zig-zag (C) cross scheme (D) transect (E) geogrid and (F) radial scheme. Each diamond represents one sample or a composite sample.

Samples for analysis of microbial attributes need special care, since some environmental conditions may disturb biochemical aspects of the sample, leading to modifications of biological properties (e.g. degradation of DNA and RNA by nuclease enzymes from soil).

Samples for evaluation of biochemical properties, such as microbial biomass and enzyme activity, can be stored in moist form for up to 3 weeks at 4°C in cases where they cannot be processed immediately. For longer periods of storage, it is recommended to store the samples at -20°C. For evaluation of soil diversity through DNA and RNA analysis, additional care is required. It is largely recommended that the researcher have on hand a cooler with dry ice for the storage of soil immediately after sample collection. In cases that require keeping the samples frozen, some commercial products that preserve DNA and RNA integrity can be used (Saleh-Lakha et al., 2005).

After collection, samples should be transported as quickly as possible and stored at -20°C in a freezer, for samples to be treated immediately or at -80°C in an ultra-freezer, for samples to be stored for longer periods before treatment. It is recommended to sieve the soil through a 2-mm mesh screen to remove stones, roots and debris prior to analysis.

5.3 Assessing Diversity and Structure of Soil Microbial Communities by Molecular Tools

The adaptation of microbiological methods during the 19th and 20th centuries allowed for an increase in our comprehension of soil microorganisms, through techniques such as pure culturing, microscopy and chemical/biochemical assays. However, the rapid development of molecular biology towards the end of the 20th century and the successful application of these techniques to the study of microbial ecology has changed the view of microorganisms' life and function. For a long time the diversity of microbial communities was investigated by techniques based on culture and isolation, which are notoriously selective, and not able to cover and represent the microbial community in soils (Bent & Forney, 2008). These conventional methods used by soil microbiologists can retrieve only a tiny fraction of microbial species, which accounts for less than 1% by most estimates (Amann et al., 1995).

In the last few years, advances in the field of molecular microbial ecology, including DNA extraction, polymerase chain reaction (PCR), cloning and sequencing, have enabled the development of techniques not dependent on cultivation, which reduces the problems associated with conventional techniques (Handelsman & Smalla, 2003). These new molecular biology methodologies have revealed a major microbial diversity that was not detected previously with culture-dependent techniques. Additionally, these new tools have enabled the development of research in less time and cost with a high capacity of data production. Advances in molecular techniques applied to the study of the microbial ecology in complex systems have contributed significantly to the understanding of the presence and function of soil microbial communities. Several molecular methods are used to investigate the microbial diversity and changes in the microbial community structure in a wide range of environments. These techniques can be placed into two main groups: community fingerprinting, such as T-RFLP (Liu et al., 1997) and DGGE (Muyzer et al., 1993) and sequencing approaches, which are emphasized in this chapter.

5.3.1 Sequencing Technologies

Sequencing approaches are powerful tools for the study of microbial communities inhabiting soil and could be useful to predict changes in soil properties and quality. The assessment of microbial diversity is now in increasing advance, by the development of new technologies that answer some key questions about the "who, what, when, where, why and how" of microbial communities (Knight et al., 2012).

The first application of sequencing applied to microbial ecology was the construction of gene libraries, which is the amplification, cloning and sequencing (by Sanger technology) of genes present in organisms of the community, which are used for their identification.

Genomic libraries consist of a collection of DNA sequences constructed from genetic material obtained from environmental samples. The 16S rRNA gene has become the most used molecular marker for phylogenetic studies, providing an overview of community composition. The 16S rRNA gene is a fragment of approximately 1500 nucleotides, present in all prokaryotes and generates large amounts of useful information for phylogenetic inferences. The advantage of using this gene is the availability of a large number of sequences in databases (e.g. GenBank, NCBI), which allow for the comparison of new sequences obtained.

However, in recent years, technological advances have enabled the emergence of new sequencing platforms, increasing the quality, quantity, and size of the fragments sequenced. The next-generation sequencing (NGS) platform allowed microbial ecologists to advance from analyzing a few hundred sequences to millions per study. The increase in number of sequences per run from parallel pyrosequencing technologies such as the Roche 454 GS FLX™ (5×10^5) to Illumina GAIIx™ (1×10^8) is on the order of 1,000-fold and greater than the increase in the number of sequences per run from Sanger (1×10^3 through 1×10^4) to 454 (Caporaso et al., 2011). In addition, the use of barcode strategies allows for the analysis of thousands of samples in a single run. Advances in these technologies have allowed the read length to increase; however read lengths are still far shorter than the length obtained from traditional Sanger™ sequencing (~1000 bp) (Luo et al., 2012). The 454 pyrosequencing method was the first next-generation sequencing technology available as commercial product (Margulies et al., 2005) and can be considered the cornerstone of the sequencing revolution. The development of pyrosequencing allowed the advance of metagenomics studies by increasing the number of reads and decreasing the cost per sequence, enabling a deep phylogenetic community analysis.

Most recently, a variety of technologies have been created to improve sequencing capabilities, striving to increase number of reads and fragment length while reducing costs (Shendure & Ji, 2008) (Fig. 5.3). The Illumina HiSeq2000™ and MiSeq™ platforms decreased the costs for DNA sequencing dramatically, enabling high-throughput microbial genome analysis at the greatest coverage yet possible. The HiSeq produces >50 Gb per day and 1.6 billion of 100-base paired-end reads over a course of 10.8 days; on the other hand, the MiSeq generates 1,5 Gb per day from 5 million 150-base paired-end reads (Caporaso et al., 2012). The Illumina technology is still increasing the read length, with new chemical products generating fragments up to 800 bp.

The development of new technologies has allowed the emergence of new sequencing approaches, such as metagenomics. The metagenomics is the study of all genetic material recovered directly from environmental samples. The use of metagenomics in the studies of soil microbial communities has enable researchers to not only obtain an overview of diversity, but also the functional traits, which are also an important approach to defining microbiological parameters for monitoring soil quality. The analysis of functional diversity can also provide information on how adaptive microorganisms may influence soil fertility. The rapid advance of sequencing technologies allied with bioinformatics tools are increasing the possibility of massive studies on microbial ecology for a deep comprehension of the composition and function that soil microorganisms play in a wide range of ecosystems. The new information available will be useful for better understanding soil quality and improving the sustainable use of the ecosystem.

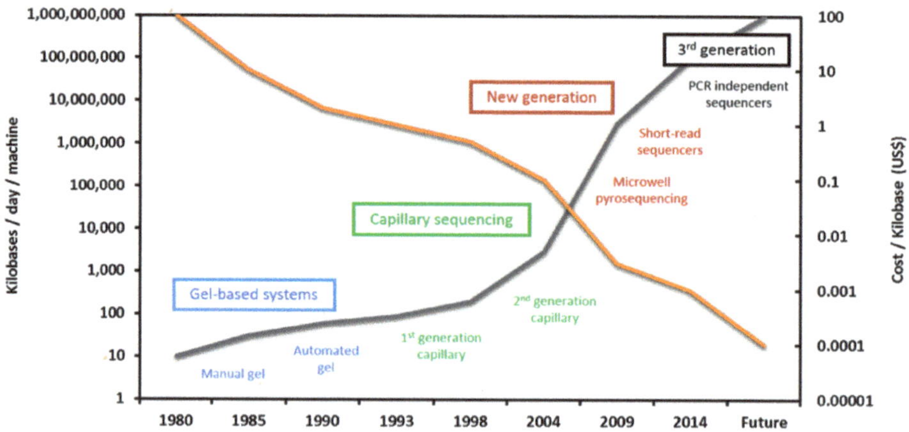

Fig. 5.3: Evolution of yield from sequencing technologies shown at the main axis (gray line). Reduction of costs along time shown at the secondary axis (brown line). Adapted from Metzker (2010)inexpensive and accurate genome information. This challenge has catalysed the development of next-generation sequencing (NGS.

High throughput sequencing technologies are very applicable to studies in soil microbial ecology. One of the major reasons for performing these studies is the fact that soil holds more unknown biodiversity than any other ecosystem on earth. Based on this, Delmont et al. (2012) used high throughput sequencing to assess the microbial phylogenetic and genetic metagenomics, in the famous Park Grass experiment at Rothamsted Research Station, the oldest soil science experiment in the world. Results revealed low seasonal effects on microbial phylogeny. In addition, gene functions related to carbohydrate metabolism were shown to be enriched. The experiment provided a better understanding of the microbial structures, through metagenomics, by linking natural and methodological fluctuations.

5.4 Biogeochemical Techniques for Evaluating Soil Quality and Fertility

Biogeochemical processes in soil are largely carried out by the microbial community. The microbial community structure, function, distribution and enzyme activity can be affected by soil factors such as soil type, plant cover species, management and crop rotation (Figuerola et al., 2012). Therefore, it is desirable to consider microbiological indicators when assessing the soil quality for ecosystems conservation.

5.4.1 Soil Microbial Biomass

Soil microbial biomass (SMB) has been defined by the portion of soil organic matter constituted by Archaea, Bacteria and eukaryotes, excluding roots and animals smaller than 5×10^3 μm^3 (Jenkinson & Ladd, 1981). The SMB has been often quantified through two methods: (I) fumigation-incubation (FI) (Jenkinson & Powlson, 1976), which is based on fumigation, re-inoculation with live microbial biomass, incubation under controlled conditions and measurement of differences in soil, and; (II) fumigation-extraction method (FE) (Vance et al., 1987), which extracts contents of microbial biomass with chemicals, immediately after fumigation.

Both methods have advantages and disadvantages: the simple determination of microbial respiration (CO_2 release) during the incubation process is the main advantage of FI method. However, the method cannot be used in areas with recent addition of organic material and pH below 5. On the other hand, the FE method is not directly dependent on the physiologic state of microbial soil population, which is an important advantage. The main disadvantage is the absence of a total carbon quantification, as the analytical procedures for determining the C extracted from samples are more complex and laborious, involving the use of toxic chemicals. However, the most important consideration point for application of these methods is the choice of the right conversion coefficient (Roscoe et al., 2006).

5.4.2 Metabolic Quotient (qCO_2)

The eco-physiological indices (metabolic quotients) can be generated by basal physiological performances (respiration, carbon uptake, growth / death ratio) on the total microbial biomass per unit of time (Anderson, 2003). A relationship between maintenance and growth of soil microbial communities is often observed, whereby the qCO_2 reflects the maintenance energy requirement. Some studies have attempted to relate the amount of C-CO_2 derived from respiration to microbial biomass in the same soil sample. This proposal is supported by the theory proposed by Odum (Odum, 1969), which states that the ratio (full breath) / (microbial biomass) decreases with time or succession in an ecosystem. This model can also be interpreted by the ratio (microbial respiration) / (microbial biomass) × time (Anderson & Domsch, 1990), in which the metabolic quotient (qCO_2) decreases the extent of soil maturity. Thus, this ratio can be used to clearly define and quantify microbial activity (Insam & Domsch, 1988), and to characterize the risk of soil degradation in terms of organic matter.

5.4.3 Microbial Enzymatic Activity

Soil microbial enzymes respond to soil land-use change and management long before other soil quality indicators (Pankhurst et al., 1995). These enzymes also play an important role in organic matter decomposition and nutrient cycling (Tab. 5.1). Some of them only facilitate the breakdown of organic matter (e.g. hydrolase, glycosidase), while others are involved in nutrient mineralization (e.g. amylase, urease, phosphatase, sulfatase) (Tate, 2000). With the exception of phosphatase activity, there is no strong evidence that directly links enzyme activity to nutrient availability or crop production (Finzi et al., 2006). The relationship may be indirect, considering that nutrient mineralization and plant uptaking are accomplished with the contribution of enzyme activity. Enzymes are generally measured indirectly by determining their activity in the laboratory, through biochemical assays. Enzyme assays reflect potential activity and do not represent true *in situ* activity levels, and therefore must be considered as indices.

Bacteria and fungi that break down insoluble nutrient sources in the soil produce extracellular enzymes. These are proteins that are produced inside the cell and exported out into the soil solution. The enzymes are active outside the cell where they catalyze reactions to break down the structure of the nutrient source to make it more accessible. The amount of an extracellular enzyme in the soil depends on the metabolic abilities of the soil organisms, the number of organisms present, the presence of substrate and the environment of the soil (pH, temp., ionic strength, etc.). Since enzymes are costly for the cells, they are tightly regulated and only metabolized when they are needed.

Tab. 5.1: Main soil enzymes. Predicted function and activity. Adapted from Tabatabai (1994).

Enzyme	Source product	End product	Significance	Predicted soil function
FDA Hydrolysis	Organic matter	Carbon and various nutrients	Energy and nutrients for microorganisms. Measure microbial biomass	Organic matter decomposition and nutrient cycling
Dehydrogenase	Carbon compounds	Glucose	Energy for microorganisms	Organic matter decomposition
Urease	Nitrogen (urea)	Ammonia (NH_3) and carbon dioxide (CO_2)	Plant available HN^{4+}	Nutrient cycling
Phosphatase	Phosphorus	Phosphate (PO^{4-})	Plant available P	Nutrient cycling

5.5 Microbial Biogeography and Diversity in Soil: A Top-down Approach in Landscape Ecology

In spite of the complexity of soil ecology, the effects of processes determining the dynamics of natural and agricultural areas on soil organisms can be assessed through a top-down approach in landscape ecology. This approach requires the combination of large-scale dynamics of land use change and local processes on vegetation and soil.

Because ecological systems are complex and composed of many interacting parts, it is useful to view their organization as a hierarchy, or a graded series with several levels of organization (Fig. 5.4).

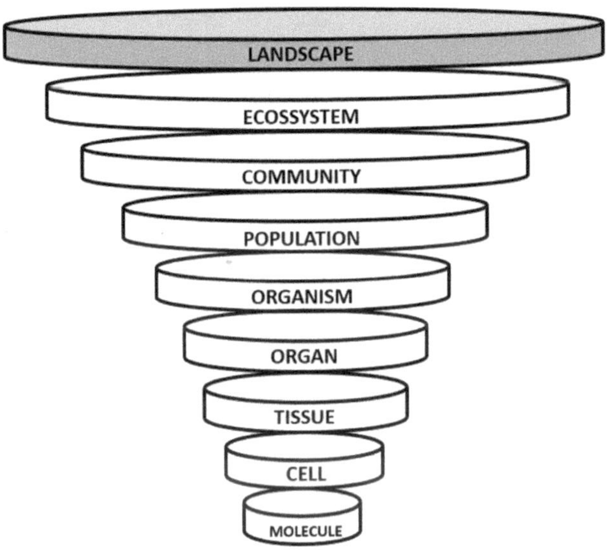

Fig. 5.4: Hierarchical organization of life. Adapted from Odum & Barret (2007).

On the basis of ecological hierarchy theory, it is presupposed that nature is working at multiple scales and has different levels of organization, which are part of a rate-structured, nested hierarchy. Specifically, it is claimed that, above the ecosystem level, a landscape level exists which is generated and identifiable by high interaction intensity between ecosystems, a specific interaction frequency and, typically, a corresponding spatial scale.

In general, species significantly vary across the landscape in association with environmental features. Global distributions of most of the world's flora and fauna are generally known. There has been much less emphasis on understanding and mapping the biogeography of microscopic organisms. Although molecular analysis has revealed that microbial diversity in soils far exceeds that of macroscopic organisms, the geographical patterns of this diversity and the factors controlling these patterns are only beginning to be examined. A common perception is that microorganisms are cosmopolitan in their distribution, being capable of growth in many different places worldwide. This idea goes back more than a century to Martinus Beijerinck, a Dutch soil microbiologist, who suggested that "everything is everywhere, the environment selects," meaning that microbial species can be found wherever their environmental requirements are met (de Wit & Bouvier, 2006). However, the concept of a common occurrence of soil microorganisms may be more theoretical than real due to a lack of information on microbial distributions. As more details of microbial populations are delineated with biochemical and molecular techniques, it may be found that at least some soil organisms are restricted to specific geographical areas (i.e. they are endemic).

Organism abundance and activity are not randomly distributed in soil, but vary both horizontally and vertically through the soil profile. Different groups of organisms exhibit different spatial patterns, because they each react to soil conditions in different ways (Klironomos, Rillig, & Allen, 1999). This spatial heterogeneity, which has been observed at the scale of millimeters to hundreds of meters, has been shown in some studies to correlate with gradients in site and soil properties, including bulk density, aggregation, texture, oxygen concentration, pH, moisture, soil organic matter content, inorganic N availability, precipitation levels, and vegetation dynamics. Some of these properties are important at the microscopic scale, whereas others act over larger distances. For example, acidobacterial subgroups 4, 6 and 7 responded to high content of soil Al, Ca, Mg, Mn, and B in agricultural management of soybean in Amazon forest soils (Navarrete et al., 2013). In other cases, soil characteristics have been found to explain a relatively minor amount (< 30%) of the spatial variation in organism abundance (Robertson & Freckman, 1995). Spatial heterogeneity can be high even in soils that appear relatively homogeneous at the plot or field scale (Franklin & Mills, 2003; Rodrigues et al., 2013).

Instead of asking what factors determine the distribution and abundance of organisms, a top-down approach to the study of landscape ecology seeks to explain how the distribution and abundance of organisms affects the entire collection of biodiversity and processes on the landscape (Sandersons & Harris, 2001). The understanding of keystone and indicator species to elucidate their roles in ecosystem functions and structuring processes is important. Understanding these roles in the context of heterogeneous landscapes is yet more important.

The ecology of landscapes we embrace extends Merriam's (1988) functional landscape ecology whereby organisms, as propagators of ecological functions, interact with, alter, and maintain their local environments. The functional pillars for a top-down approach are multi-level species evolution integration through time and space that is horizontally connected. These multi-level species can be better explored considering multiple dimensions of biodiversity (genetic, phylogenetic, taxonomic and functional).

Recent advances in sequencing technologies and bioinformatics have opened possibilities to explore these multiple dimensions of biodiversity for microbes. Technological advances in soil ecosystem biogeochemistry (for example in field-based isotopic measurements) now allow functional measurements at a scale and of a resolution necessary to link them to genetic and phylogenetic diversity. This information can be used to better understand how soil microbial biodiversity responds to environmental change including land use changes and local processes on vegetation and soil (i.e. top-down disturbances on soil).

5.6 Integrating Soil Microbial Molecular Data with Environmental Parameters

Ecological parameters such as diversity and structure do not follow normal distribution and are non-parametric data. The patterns of these complex communities often generate matrices with multiple zero (species that occur in one environment and disappear in others), few abundant species (dominant) and several low abundant species (rare). These observations lead the researchers to seek methods of analysis that reduce the complexity of ecological data to fair numerical and graphical representations.

Multivariate analysis emerges as the most accurate way to analyze ecological data (Ramette, 2007). The purpose of several multivariate methods is always to reduce one matrix of several ecological data to a simple number, vector or a coordinate that represents the structure of this data matrix. When analyzing multiple matrices together, the purpose is again to reduce both matrices, adding the fact that the matrix of explanatory variables must to be simplified to few variables that explain significantly the response variables in study (Blanchet et al., 2008). Several methods have been developed for multiple purposes. Here we summarize the most cited ones, with applications in microbial ecology.

5.6.1 Methods for Analysis in Soil Microbial Ecology

5.6.1.1 Linkage Methods

Linkage methods are often based on cluster analysis generated by distance, similarity and dissimilarity matrices. The choice of the type of cluster is always defined by the original dataset and objectives of the researcher. The most used types of clusters are the unrooted, the rooted and the network trees (Fig. 5.5).

In cases where you are looking for comparisons of ecological data from different environments (e.g. different land-uses) or have a matrix that angles different dataset, you can use a consensus unrooted tree (Widmer, FlieûBach, Laczkó, Schulze-Aurich, & Zeyer, 2001).

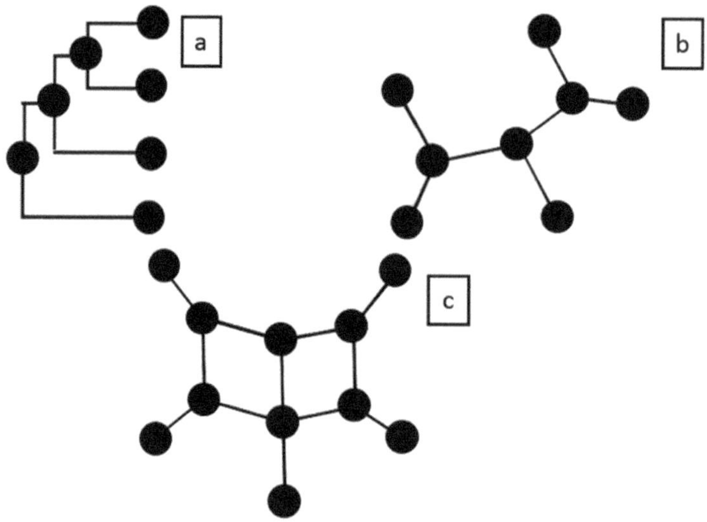

Fig. 5.5: Basic types of clusters linkage. (A) Represents an unrooted tree, (B) represents a rooted tree and (C) represents a network tree. Adapted from Kaufman & Roesseeuw (2005).

For studies in which the aim is to find a common linking point, like in studies of evolution when you have to define the last common ancestor, you can use a rooted tree. The permutation test (e.g. Monte-Carlo) could be used to define an out-group, which could represent, for example, a last common ancestor (Baldauf, 2003).

To evaluate ecological relations among species-species diversity data or species-environment correlation, you can use a network tree that shows the positive and negative interactions between the factors you have measured (Barberán et al., 2012). Elsewhere, for generating network matrices do you have to choose a correlation measure (e.g. Pearson, Spearman) or use a consensus matrix generated from permutations of this different correlation method. For studies with metagenomic data of species-species, function-function and species-function correlation, for example, only pairwise comparisons with R-value higher or equal to 0.7 and with p-value lower than 0.01 are often used as input for network graphs or correlation tables (Mendes et al., 2014).

The cluster analysis does not require a particular statistical method or measure, as we need for ordination methods, such as discriminant analysis. What you have to keep in mind is that the results are always based on the method you have chosen for measuring the distance among groups and what criteria you will use to consider the branches to be significantly different one from the others, defined by the separation algorithms (Everitt, 1979). Assumptions about the underlying distribution of dataset generally do not compromise the analysis. Groups of related variables (e.g. diversity indices resulting from different molecular techniques) can also be formed. The choice of a method depends on, among other things, the size of the data file.

5.6.1.2 Ordination Methods

Ordination methods are often based on a set gradient of multiple variables, whether ecological parameters (named response variables), or environmental factors (named explanatory variables). These methods are used in order to reduce the complexity of ecological dataset into a system of bidimensional or tridimensional axes that can explain the distribution and occurrence of groups of microbial communities in different environments or treatments (Meir & Whittaker, 1977). They also allow us to investigate environmental factors that may shape microbial diversity and functionality patterns. The use of multivariate ordination methods is very applicable in microbial ecology due the following characteristics related to microbial parameters and its explanatory variables:

Sparse data - diversity data generally have too many zeros, corresponding to the species or taxa that occurs in a determined environment and does not occur in other, often named as rare species.

Dominance - few species or taxa largely dominate the environment, manifesting great relative abundance in detriment of many species that have low or very low relative abundance (rare species).

Significant shaping factors - the number of factors that may potentially shape microbial diversity and functions is potentially very large, but in reality the number of factors that modulate ecological patterns is often very small.

Noise and redundancy - noise generally occurs when replicates vary substantially from each other. In addition, in most cases there is much redundant information. Redundancy tends to occur when we try to explain a phenomenon through explanatory variables that co-occur or are closely related to each other (e.g. soil organic carbon and soil organic matter). Application of redundant variables can increase the bias and reduce explainability of data.

Data transformation - most of the variables we use as explanatory variables are measured in different units and vary at different ranges. Thus, it is important to transform environmental explanatory variables prior to the analysis, (e.g. Box-Cox criterion) (Box & Cox, 1964), in order to avoid the under or overestimation of the effect of a single variable on the global data variability.

Ordination methods allow us to reduce the number of dimensions or gradients to be evaluated. The main goals of ordination are testing and exploration of data. Ordination methods have become very useful in ecology since scientists could better understand complex data in a resumed way. However, to link the goals of ordination with real results, it is necessary that a multi-testing validation approach is used to reach the best answers for an ecological problem, that includes both indirect and direct gradient analysis (ter Braak & Prentice, 1988).

Indirect gradient analysis – the dataset is organized in a single matrix. Any environmental factor or supplementary variable, if it exists, is often used as complementary information. This leads to the assumption that the values of explainability obtained from the axes by indirect gradient, result from the distribution of the response variables. Thus, what we look for when performing this kind of analysis is to find out the most important gradients that modulate variability of data in the samples.

Direct gradient analysis – the dataset is organized in at least two matrices. The first one contains the ecological data of the study (e.g. species richness/abundance matrix). The second matrix contains external environmental variables, (e.g. soil physicochemical parameters), which are expected to explain a certain part of the total variability of ecological data. In a simple way, the direct gradient analyzes the influence of the environmental variables on species composition, through regression.

Several ordination methods have been used in ecological studies. The most useful ones are listed in Table 5.2.

Tab. 5.2: Ordination methods applied in ecological studies. Adapted from Jongman, ter Braak, & van Tongeren (1995).

Ordination methods
Indirect gradient analysis
Distance-based approaches
- Bray-Curtis Polar Ordination (PO)
- Nonmetric Multidimensional Scaling (nMDS)
- Principal Coordinates Analysis (PCoA)
- Principal Coordinates of Neighbor Matrices (PCNM)
Eigen-analysis-based approaches
Linear model (*lengths of gradients ≤ 3)
- Principal Components Analysis (PCA)
Unimodal model (lengths of gradients > 4)
- Correspondence Analysis (CA)
- Detrended Correspondence Analysis (DCA)
Direct gradient analysis
Linear model (lengths of gradients ≤ 3)
- Redundancy Analysis (RDA)
Unimodal model (lengths of gradients > 4)
- Canonical Correspondence Analysis (CCA)
- Detrended Canonical Correspondence Analysis (DCCA)
Hybrid gradient analysis
- Hybrid-Redundancy Analysis (h-RDA)
- Hybrid-Canonical Correspondence Analysis (h-CCA)
- Hybrid-Detrended Canonical Correspondence Analysis (h-DCCA)

* When the lengths of the gradients from the first axis are between 3 and 4. you can opt for a linear or unimodal method to be performed.

5.6.2 Selective Approach on Molecular and Environmental Dataset Analysis

One of the central aims in Ecology is to unveil how and why species are structured (Blanchet et al., 2008). The advances in sequencing technologies and bioinformatics analysis, lead ecologists to an innovative world of possible inferences and data interpretation (see section 6.3.2 of this Chapter). Increased use and implementation of multivariate methods and algorithms allow us to explore ecological data and their explanatory variables by several tools and pipelines, as we never seen before.

Fierer et al. (2012) have applied a range of methodological approaches in order to analyze the structure and function of bacterial communities in an experiment with N-amended levels. For that, the authors used high throughput pyrosequencing of the universal marker 16S rRNA. Pyrosequencing revealed significant effects of the nitrogen levels on community composition. The authors found that copiotrophic taxa increased in relative abundance in plots with high N content while oligotrophic taxa, such as Acidobacteria, decreased. Whole genome shotgun (WGS) metagenomic approach also showed the differences between N-fertilization levels. The relative abundance of genes associated with DNA/RNA replication, electron transport and protein metabolism increased. Functional patterns were significantly correlated with phylogenetic and metagenomic results, leading to the assumption that theses dimensions could be linked with the environmental dimension (levels of N) that, ultimately, may modulate the shifts on microbial distribution and strategies predominance.

An emerging question from this placement is: what kind of analysis and algorithms should we use? Likewise, an even more challenging question: which criteria should be implemented to define the variables to be plotted? First, the analyses to be performed are often based on the type and distribution of the dataset, as defined in this chapter. Previous results using the same ecological data or methodology could also be helpful for the choice of an appropriate approach to analyze the data. To answer the second question, scientists have developed several statistical packages for selection of variables to be taken into account when linking them to ecological parameters. The principle is based on the concept of "forward selection" of explanatory variables. This selection comprises many steps with the objective of eliminating noise and non-significant variables and redundant information. Depending on the statistical package, the options for selection of explanatory variables can differ slightly, but are often based on the same assumptions:

a) Correct range of axes – as mentioned before, explanatory variables are measured by different methodologies and vary in different ranges. Thus, a global correction of the general explanatory matrix is needed prior the analysis. Several algorithms can be used to normalize the data, such as log transformation, square root transformation, normalized length and normalized percentage transformation. The matrix of ecological data should also be transformed as described above. Here we suggest the Box-Cox transformation for environmental variables, which corrects the data in order to reduce anomalies such as non-additivity, non-normality and heteroscedasticity. For a review of the characteristics of this transformation technique see Sakia (1992).

b) Three-step forward selection – after transformation of the explanatory variables matrix, a forward selection should be performed. The aim of the forward selection is to eliminate from the model the gradients or variables that increase the noise and reduce the explainability of data. Follow a workflow to select explanatory variables:

(I) *Co-variables* – eliminate those that are closely related to each other. Generally, they have a co-occurrence and describe the same phenomenon or discriminate for the same biological or ecological response. They may co-occur positively or negatively. A classic example of positive co-occurrence is that of soil organic matter (SOM) and soil organic carbon (SOC). Independent of the content and composition of the SOM it always has the same average content of SOC (about 60%). An example of negative co-occurrence is that of pH and aluminum concentration in the soil solution. The extent to which increases in the pH of soil (e.g. by liming), decreases the availability of aluminum in the soil solution. From pH = 5.2 all aluminum is precipitated as oxides and hydroxides. Thus, where the pH is low there is a lot of aluminum in solution, and, where the pH is high there is little or no aluminum in the soil solution. In both cases it is highly recommended to choose for one or the other. The co-occurrence can be verified by the inflation factor – variables with inflation factor higher than 25, co-occur with others in the model.

(II) *Tolerance* – variables with tolerance values higher than 0.6 are supposed to be independents in the model. Thus, they should be chosen in the model in detriment of that with lower values.

(III) *Wilks Lambda* – variables with high Wilks lambda values associated with significant values ($p \leq 0.05$) are selected for the model and in almost all cases explain most of the environmental influence on the ecological data variability. In ecological niche based or deterministic models, generally the percentage of explanation from environmental variables significantly increases the global explainability of species distribution. In neutral or stochastic models, the influence of explanatory variables is reduced, and most of the species distribution is explained in the ecological matrix itself.

Vasileiadis et al. (2013) identified the effects of land-use on microbial community patterns from soils. The authors have used all dimensions of biodiversity, and integrated them with the environmental dimension, in a multidimensional integrating approach. In addition, they searched for biodiversity drivers and tried to explain the patterns of microbial distribution through ecological theories and multivariate techniques. Microbial diversity was assessed through high throughput sequencing of partial 16S *r*RNA amplicons. The structure of the soil microbial community was significantly affected by land-use. Environmental factors that most shaped the variability were soil organic carbon and pH for bacterial community and pH for archaeal community. Higher microbial diversity was found in soils with lower enzymatic activity and less fertility. Based on this, niche-based theories, like resource heterogeneity hypothesis, are better fitted to the ecological diversity and functional traits patterns.

5.7 Final Considerations

In this chapter, we summarized tools and methods for studying microbial diversity in soils. Comprehensive integration of both genetic and phylogenetic dimension, functional traits dimension, and an emerging view of the environmental dimension, make it possible for us to elucidate some challenging questions about the "black box" of microbial ecological patterns in soil.

The advances in systems biology theories and methodologies, bioinformatics and biostatistics allow us to unravel the power of ecological multidimensional approach in deciphering the far unknown microbial world. Studies resulting from this integrative approach may create parameters for the establishment of guidelines and public polices for ecosystems conservation.

Bibliography

Amann, R.I., Ludwig, W., & Schleifer, K.H. (1995). Phylogenetic identification and in situ detection of individual microbial cells without cultivation. Phylogenetic Identification and In Situ Detection of Individual Microbial Cells without Cultivation. Microbiology and Molecular Biology Reviews, 59 (1), 143–169.

Anderson, T.-H. (2003). Microbial eco-physiological indicators to asses soil quality. Agriculture, Ecosystems & Environment, 98 (1-3), 285–293.

Anderson, T.-H., & Domsch, K.H. (1990). Application of eco-physiological quotients (qCO_2, and qD) on microbial biomasses from soils of different cropping histories. Soil Biology & Biochemistry, 22 (2), 251–255.

Baldauf, S.L. (2003). The deep roots of eukaryotes. Science, 300 (5626), 1703–1706.

Barberán, A., Bates, S.T., Casamayor, E.O., & Fierer, N. (2012). Using network analysis to explore co-occurrence patterns in soil microbial communities. The ISME Journal, 6 (2), 343–351.

Bent, S.J., & Forney, L.J. (2008). The tragedy of the uncommon : understanding limitations in the analysis of microbial diversity. The ISME Journal, 2, 689–695.

Blanchet, F.G., Legendre, P., & Borcard, D. (2008). Forward selection of explanatory variables. Ecology, 89(9), 2623–2632.

Box, G.E.P., & Cox, D.R. (1964). An Analysis of Transformations. Journal of the Royal Statistical Society, 26 (2), 211–252.

Bru, D., Ramette, A., Saby, N.P.A, Dequiedt, S., et al. (2011). Determinants of the distribution of nitrogen-cycling microbial communities at the landscape scale. The ISME Journal, 5 (3), 532–542.

Caporaso, J. G., Lauber, C. L., Walters, W. A., et al. (2011). Global patterns of 16S rRNA diversity at a depth of millions of sequences per sample. Proceedings of the National Academy of Sciences of the United States of America, 108, 4516–4522.

de Wit, R., & Bouvier, T. (2006). "Everything is everywhere, but, the environment selects"; what did Baas Becking and Beijerinck really say? Environmental Microbiology, 8 (4), 755–758.

Delmont, T.O., Prestat, E., Keegan, K.P., et al. (2012). Structure, fluctuation and magnitude of a natural grassland soil metagenome. The ISME Journal, 6 (9), 1677–1687.

Dequiedt, S., Thioulouse, J., Jolivet, C., Saby, N.P.A, et al. (2009). Biogeographical patterns of soil bacterial communities. Environmental Microbiology Reports, 1 (4), 251–5.

Everitt, B.S. (1979). Unresolved problems in cluster analysis. Biometrics, 35 (1), 169–181.

Fierer, N., Lauber, C.L., Ramirez, K.S., Zaneveld, J., Bradford, M.A., & Knight, R. (2012). Comparative metagenomic, phylogenetic and physiological analyses of soil microbial communities across nitrogen gradients. The ISME Journal, 6 (5), 1007–1017. doi:10.1038/ismej.2011.159

Fierer, N., & Lennon, J.T. (2011). The generation and maintenance of diversity in microbial communities. American Journal of Botany, 98 (3), 439–448.

Figuerola, E.L.M., Guerrero, L.D., Rosa, S.M., Simonetti, L., et al. (2012). Bacterial indicator of agricultural management for soil under no-till crop production. PloS One, 7 (11), 1–12.

Finzi, A. C., Sinsabaugh, R.L., Long, T.M., & Osgood, M.P. (2006). Microbial Community Responses to Atmospheric Carbon Dioxide Enrichment in a Warm-Temperate Forest. Ecosystems, 9 (2), 215–226.

Franklin, R.B., & Mills, A.L. (2003). Multi-scale variation in spatial heterogeneity for microbial community structure in an eastern Virginia agricultural field. FEMS Microbiology Ecology, 44 (3), 335–346.

Green, J.L., Bohannan, B.J.M., & Whitaker, R.J. (2008). Microbial biogeography: from taxonomy to traits. Science, 320 (5879), 1039–1043.

Handelsman, J., & Smalla, K. (2003). Techniques: conversations with silent majority. Current Opinion in Microbiology, 6 (3), 271–273.

Hegerl, G., & Stott, P. (2014). Atmospheric science. From past to future warming. Science, 343 (6173), 844–845.

Insam, H., & Domsch, K.H. (1988). Relationship between soil organic carbon and microbial biomass on chronosequences of reclamation sites. Microbial Ecology, 15 (2), 177–188.

Jenkinson, D.S., & Ladd, J.N. (1981). Microbial biomass in soil: measurement and turnover. In E.A. Paul & J.N. Ladd (Eds.), Soil Biochemistry (p.415–471). New York: Marcel Dekker.

Jenkinson, D.S., & Powlson, D.S. (1976). The effects of biocidal treatments on metabolism in soil-V. Soil Biology & Biochemistry, 8 (1), 209–213.

Jongman, R.H.G., ter Braak, C.J., & van Tongeren, O.F.R. (1995). Data analysis in community and landscape ecology. Cambridge: Cambridge University Press.

Kaufman, L., & Roesseeuw, P. J. (2005). Finding groups in data: An introduction to cluster analysis (Wiley seri.). Hoboken: Wiley.

Klironomos, J.N., Rillig, M.C., & Allen, M.F. (1999). Designing belowground field experiments with the help of semi-variance and power analyses. Applied Soil Ecology, 12 (1), 227–238.

Klueglein, N., Zeitvogel, F., Stierhof, Y.-D., et al. (2014). Potential role of nitrite for abiotic Fe(II) oxidation and cell encrustation during nitrate reduction by denitrifying bacteria. Applied and Environmental Microbiology, 80 (3), 1051–1061.

Knight, R., Jansson, J., Field, D., Fierer, N., et al. (2012). Unlocking the potential of metagenomics through replicated experimental design. Nature Biotechnology, 30 (6), 513–520.

Liu, W., Marsh, T.L., Cheng, H., & Forney, L.J. (1997). Characterization of Microbial Diversity by Determining Terminal Restriction Fragment Length Polymorphisms of Genes Encoding 16S rRNA, Applied and Environmental Microbiology, 63 (11), 4516–4522.

Luo, C., Tsementzi, D., Kyrpides, N., Read, T., & Konstantinidis, K.T. (2012). Direct comparisons of Illumina vs. Roche 454 sequencing technologies on the same microbial community DNA sample. PloS ONE, 7 (2), 1–12.
Madsen, E.L. (2011). Microorganisms and their roles in fundamental biogeochemical cycles. Current Opinion in Biotechnology, 22 (3), 456–464.
Margulies, M., Egholm, M., Altman, W. E., Attiya, S., et al. (2005). Genome sequencing in microfabricated high-density picolitre reactors. Nature, 437 (7057), 376–380.
Meir, N., & Whittaker, R.H. (1977). Continuous multivariate methods in community analysis: some problems and developments. Plant Ecology, 33 (2), 79–98.
Mendes, L.W., Kuramae, E.E., Navarrete, A.A., van Veen, J.A., & Tsai, S.M. (2014). Taxonomical and functional microbial community selection in soybean rhizosphere. The ISME Journal, 8, 1577–1587.
Merriam, G. (1988). Landscape dynamics in farmland. Trends in Ecology & Evolution, 3 (January), 16–20.
Metzker, M.L. (2010). Sequencing technologies - the next generation. Nature Reviews. Genetics, 11 (1), 31–46.
Monard, C., Binet, F., & Vandenkoornhuyse, P. (2008). Short-term response of soil bacteria to carbon enrichment in different soil microsites. Applied and Environmental Microbiology, 74 (17), 5589–5592.
Muyzer, G., de Waal, E.C., & Uitterlinden, A.G. (1993). Profiling of complex microbial populations by denaturing gradient gel electrophoresis analysis of polymerase chain reaction-amplified genes coding for 16S rRNA. Applied and Environmental Microbiology, 59 (3), 695–700.
Myrold, D.D., Zeglin, L.H., & Jansson, J.K. (2014). The Potential of Metagenomic Approaches for Understanding Soil Microbial Processes. Soil Science Society of America Journal, 78 (1), 3–10.
Navarrete, A.A., Kuramae, E.E., de Hollander, M., Pijl, A.S., van Veen, J.A., & Tsai, S.M. (2013). Acidobacterial community responses to agricultural management of soybean in Amazon forest soils. FEMS Microbiology Ecology, 83 (3), 607–621.
Odum, E.P. (1969). The Strategy of Ecosystem Development. Science, 164, 262–270.
Odum, E.P., & Barret, G.W. (2007). Fundamentals of Ecology. Farmington Hills: Thomson.
Pankhurst, C.E., Hawke, B.G., McDonald, H.J., Kirkby, C. A., et al. (1995). Evaluation of soil biological properties as potential bioindicators of soil health. Australian Journal of Experimental Agriculture, 35 (1), 1015–1028.
Prosser, J.I. (2010). Replicate or lie. Environmental Microbiology, 12 (7), 1806–1810, DOI:10.1111/j.1462-2920.2010.02201.x
Ramette, A. (2007). Multivariate analyses in microbial ecology. FEMS Microbiology Ecology, 62 (2), 142–160.
Ramette, A., & Tiedje, J.M. (2007). Multiscale responses of microbial life to spatial distance and environmental heterogeneity in a patchy ecosystem. Proceedings of the National Academy of Sciences of the United States of America, 104 (8), 2761–6.
Robertson, G.P., & Freckman, D.W. (1995). The spatial distribution of nematode trophic groups across a cultivated ecosystem. Ecology, 76 (5), 1425–1432.
Rodrigues, J.L.M., Pellizari, V.H., Mueller, R., et al. (2013). Conversion of the Amazon rainforest to agriculture results in biotic homogenization of soil bacterial communities. Proceedings of the National Academy of Sciences of the United States of America, 110 (3), 988–993.
Roscoe, R., Mercante, F.M., Mendes, I.C., Reis-Júnior, F.B., Santos, J.C.F., & Hungria, M. (2006). Biomassa microbiana no solo; fraçao mais ativa da matéria orgânica. In R. Roscoe, F. M. Mecante, & J. C. SAalton (Eds.), Dinâmica da matéria orgânica do solo em siste,as conservacionsistas - modelagem matemática e métodos auxiliares (p. 163–200). Dourados: Embrapa.
Sakia, R.M. (1992). The Box-Cox Transformation Technique: A Review. The Statistician, 41 (2), 169–178. DOI:10.2307/2348250

Saleh-Lakha, S., Miller, M., Campbell, R.G., Schneider, K., Elahimanesh, P., Hart, M.M., & Trevors, J.T. (2005). Microbial gene expression in soil: methods, applications and challenges. Journal of Microbiological Methods, 63 (1), 1–19.

Sandersons, J., & Harris, L.D. (2001). A Top-Down Approach to Landscape Ecology Landscape Ecology. Ecology, 82 (2), 604–606.

Schloss, P.D., & Handelsman, J. (2006). Toward a census of bacteria in soil. PLoS Computational Biology, 2 (7), 786–793.

Shendure, J., & Ji, H. (2008). Next-generation DNA sequencing. Nature Biotechnology, 26 (10), 1135–1145.

Souza, D.G., Mendes, L.W., Navarrete, A.A., & Tsai, S.M. (2015). Microbial assembly in agroecosystems: from the small arise the big. In J.A. Blanco & Y.-H. Lo (Eds.), Biodiversity in ecosystems -Linking structure and function (p.187–201). Rijeka: Intech.

Tabatabai, M.A. (1994). Soil enzymes. In R.W. Weaver, J.S. Angle, & P.S. Bottomley (Eds.), Methods of soil analysis (p.775–833). Madison: SSSA Book Series.

Tate, R.L. (2000). Soil Microbiology (2nd ed.). New York: John Wiley.

ter Braak, C.J., & Prentice, I.C. (1988). A Theory of Gradient Analysis. Advances in Ecological Research, 18 (1), 271–317.

Vance, E.D., Brookes, P.C., & Jenkinson, D.S. (1987). An extraction method for measuring soil microbial biomass. Soil Biology & Biochemistry, 19 (6), 703–707.

Vasileiadis, S., Puglisi, E., Arena, M., Cappa, F., van Veen, J.A., Cocconcelli, P.S., & Trevisan, M. (2013). Soil microbial diversity patterns of a lowland spring environment. FEMS Microbiology Ecology, 86 (2), 172–184.

Widmer, F., FlieûBach, A., Laczkó, E., Schulze-Aurich, J., & Zeyer, J. (2001). Assessing soil biological characteristics : a comparison of bulk soil community DNA-, PLFA-, and Biolog-analyses. Soil Biology & Biochemistry, 33 (1), 1029–1036.

Eduardo A. Athayde, Leonardo F. Cancian, Leonor Patricia C. Morellato

6 The Value of Agricultural Landscape for Tropical Trees

Abstract: Growing demands for food, fuel and fiber are driving the intensification and expansion of agricultural land use through a corresponding displacement of native forests, woodlands, savanna and shrublands in the Neotropics. In this modified landscape, it isn't clear what role farmland can play in supporting biodiversity through preservation of important ecosystem elements at a fine scale. Native species able to establish and survive in this modified landscape are considered generalists. However, many native species are restricted to remnant areas of native vegetation near plantations. Therefore, these agricultural landscapes constitute an environmental mosaic, with an anthropogenic matrix and associated areas with native vegetation at different successional stages, which influence the distribution of native species at the mesoscale. In that sense, we demonstrate that areas of native vegetation increase the diversity and richness of native tree species in the agricultural landscape, where the distribution of these species is related to environmental gradients, especially soil types and management. Therefore, the present work highlights the value of remnants of native vegetation for biodiversity conservation in agricultural landscapes. We also demonstrate that scattered trees are a prominent feature of agricultural landscapes and of key importance to the development of management and conservation plans.

6.1 Introduction

With increases in human population and affluence, demand for agricultural goods is projected to double by 2050 (based on year 2000 levels), while halting the loss of biodiversity and ecosystem services (Mendenhall et al., 2011). The conflict inherent in these goals has produced much debate, yielding two contrasting strategies – each at an extreme spatial scale (Green et al., 2005; Fischer et al., 2008). At the large scale, a so-called "land sparing" strategy pursues maximal yields through intense industrial farming in places with high potential productivity while setting aside separate reserves for biodiversity. Its opposite is a "wildlife-friendly" approach that integrates agricultural production and conservation on a fine scale on land managed for both (Mendenhall et al., 2011).

Despite a decline in growth rate by almost half during the past 40 years, estimates of the global human population place it at 8 billion in 2014, with more than 9 billion expected by 2050 (Roberts, 2011; Tilman et al., 2011). This latter figure represents an increase of the current world population by about 30% but the corresponding percentage change in food crop production to meet projected world demand will be much larger because it is driven by not only population growth but also by personal income growth (Kearney, 2010; Tilman et al., 2011). This large relative increase in food crop production will have to come mainly from increasing crop yield per hectare planted – crop intensification – not from converting more land to agricultural use. The challenge posed becomes even more daunting when considered the evident stagnation or even decline in food crop yield over the past decade along with the dramatically increasing competition for resources for nonfood crops, particularly biofuels (Gregory & George, 2011; Sposito, 2013) and the climatic change (Manning et al., 2009).

Brazil, the country with the largest area of tropical forest in the world (estimated at about 509 Mha), has lost a large proportion of its forested area through deforestation and land use change (FAO, 2006; Brockerhoff et al., 2013). On the other hand, planted forests, represented by few genera of plants such as *Pinus* spp., *Eucalyptus* spp. and *Acacia* spp., cover a significant portion of the Brazilian territory, reaching 6.8 Mha in 2010 (SFB, 2012). In addition, the key objective of the Brazilian National Forest Program (PFN, 2007) is the expansion of forests in degraded areas, with plans to cultivate 2.2 Mha of trees on private properties and abandoned farmland. Thus, the coverage of native and non-native forests will probably increase in the future, driven by private investment in carbon sequestration projects in the tropics (Yu, 2004), that meet the growing interest in biofuel and fibers (Pacala & Socolow, 2004; Barlow et al., 2007).

In this context, conservation ecologists are divided over the extent of land in natural regeneration and in forest plantations that will be needed to offset the loss of biodiversity by deforestation in the tropics (Barlow et al., 2007). A wide range of scientists have studied the role of forest plantations in biodiversity conservation and have demonstrated that these areas may harbour a high diversity of animal, plants and fungi species (Parrotta et al., 1997; Humphrey et al., 2000; Lindenmayer et al., 2003; Barbaro et al., 2005; Brockerhoff et al., 2008; Fonseca et al., 2009; Felton et al., 2010). The number and abundance of species that forest plantations can maintain depend on several factors such as the type of silviculture management of the plantation; the type of land use within and surrounding the areas upon which these plantations are established; the species of tree involved in the plantation; the land use alternatives found in the region (Brokerhoff et al., 2008); and the complexity of the landscape, i.e. the sum and spatial distribution of non-cultivated environments (Concepción et al., 2008).

Most planted forests are grown primarily for efficient wood production and are typically characterized by lower levels of biodiversity than mixed natural forests (Barlow et al., 2007; Brockerhoff et al., 2008). In addition, such forests are likely to be less capable of providing ecosystem services that are linked to biodiversity (Kelty, 2006; Thompson et al., 2009). There is evidence, for example, that forest "monocultures" are more prone to damage from insect herbivores than mixed stands (Jactel & Brockerhoff, 2007). Apart from more recent considerations of effects on ecosystem function, the proximate and ultimate effects of plantation forestry on biodiversity have been debated vigorously over the last two decades. There is growing recognition that the consideration of biodiversity issues is an important element of planning and management of planted forests as well as wider considerations of biodiversity conservation across landscapes (e.g. Hartley, 2002; Brockerhoff et al., 2008; Brockerhoff et al, 2013).

Local impacts on forest biodiversity can often be partially mitigated by retaining forest cover in plantations (Barlow et al., 2007; Fonseca et al., 2009) or agroforestry such as cocoa *Theobroma cacao* (Pardini et al., 2009). Nevertheless, efforts to retain biodiversity in modified landscapes are heavily dependent on the wider landscape context, ecological connectivity and the proximity of source populations (Ferreira et al., 2012).

To examine the impact of planted forests on biodiversity, in terms of variation of the composition and structure of native tree communities, we compared the diversity of plant species between areas planted with eucalyptus, pasture and areas of native vegetation. We took as a model of study two rural properties with eucalyptus plantations, seeking to answer the following questions: 1) what plant diversity and native species were retained in the area of eucalyptus plantation? 2) do areas of planted forests and pastures retain plant species common to areas of native vegetation, and what proportion of native species is retained in areas of planted eucalyptus forest and pasture in regeneration in relation to native vegetation? and 3) what environmental factors determine the distribution of native species in the agricultural landscape?

6.2 Material and Methods

6.2.1 Study Area

The study was conducted in two adjoining farms: Três Lagoas farm (23°22'0" S and 48°28'0" W) and the Arca farm (23°20'0"S and 48°27'30"W), with areas of 3209.93 ha and 1122.77 ha, respectively, located in the central-south state of São Paulo in the municipality of Angatuba, southeastern Brazil (Fig. 6.1). In the year 2007, the Três Lagoas and Arca farms went through a process of landscape transformation through the implantation of eucalyptus forests in exotic grassland areas of low productivity, in an extensive territory of 2223.9 ha and 721.9 ha in the Três Lagoas and Arca farms, respectively. In Brazil, the eucalyptus forest plantations occur mainly in land deforested decades ago for large-scale agricultural development (Brockerhoff et al., 2008). The approximately 269.23 ha of native vegetation on the Três Lagoas farm were kept unchanged and formed the farm's permanent protection area (APP-art. 4, Law 12,651/12). The areas of pasture in regeneration surround the forest fragments with a total of 586.52 ha, composing the farm's area of Legal Reserve (RL-art. 12, Law 12,651/12). The same occurred with the Arca farm, resulting in 293.98 ha of pasture in regeneration (which form the Legal Reserve) and 66.6 ha of native vegetation that form the APP.

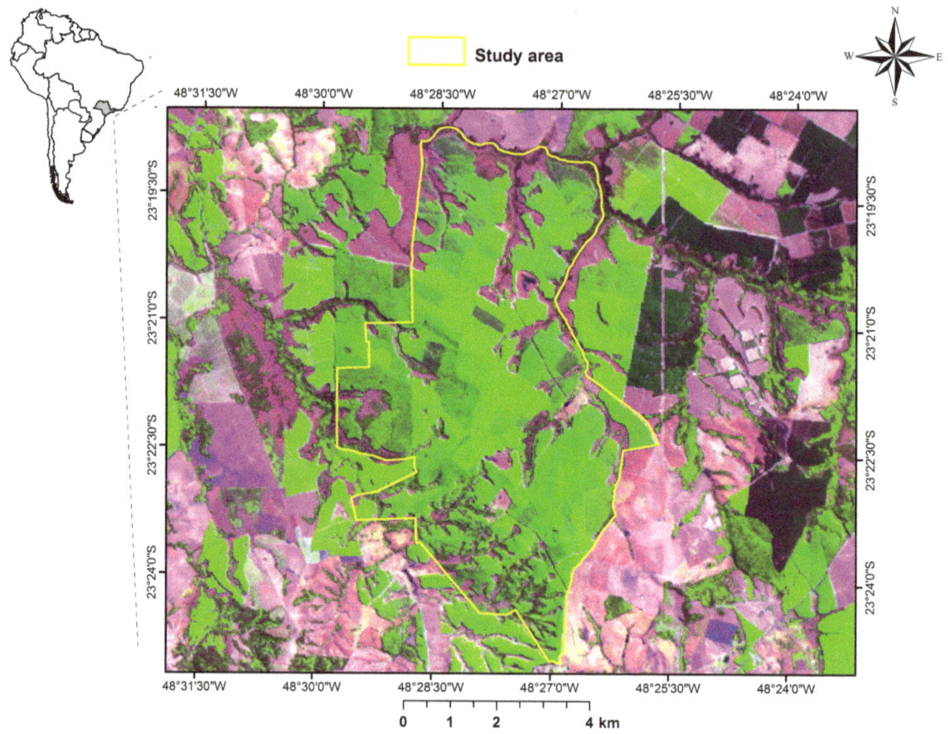

Fig. 6.1: Location of the study area in Angatuba, São Paulo, South-eastern Brasil and the general vision of the sampled area on the Três Lagoas and Arca farms.

The farms are formed by a complex landscape that comprise several types of cerrado and semi-deciduous seasonal forest phytophysiognomies, treated here as native vegetation, and anthropically altered environments, here represented by the eucalyptus plantation and pasture in regeneration, that make up the types of land use on the farms. The native vegetation went through a process of selective wood extraction in the 1970s (Martin, 2010).

The eucalyptus plantation occupies the largest territorial expanse of the farms, and is commercially managed; the understory vegetation is trimmed and supressed periodically with herbicide (Gliphosate e Isoxaflutole), especially in the initial phase of crop implantation. However, present in those areas are isolated native trees and small fragments of secondary forest not exceeding five hectares in size, which are characterized by tree species such as *Anadenanthera peregrina*, *Aegephylla sellowiana*, *Gochnatia polymorpha*, *Copaifera langsdorffii*, *Davilla elliptica* and annual species of the Piperaceae and Rubiaceae families.

6.2.2 Sampling Design

The study design is based on the methodology of the Biodiversity Research Program where sampling units form a grid with nodules distanced one kilometer from each other (Magnusson et al., 2005). This methodology is appropriate for long-term ecological research because it allows rapid biota inventories (Magnusson et al., 2005). In this case, 30 plots (located in nodule grid) were installed in the study area, taking into account the occurrence proportion of the three elements in landscape: *Eucalyptus* plantations (n = 18), native vegetation (n = 8) and abandoned pasture (n = 4) (Martin et al., 2010).

Plots measuring 250 m × 40 m were positioned following the isocline of the topography to minimize the variation in altitude and soil (Costa et al., 2005; Kinupp & Magnusson, 2005). The geographic location of plots installed on the farm was obtained through GPS (Global Positioning System). A professional surveyor measured the altitude quota in the 30 plots. The altitude value of each plot corresponds to the altitude quota compared to a previously known point of altitude on the farm (Costa et al., 2005). Slope was measured with a clinometer every 25 m, perpendicular to the main axis of the plot, and summarized as the plot average (Costa et al., 2005).

To sample the tree individuals with DBH (diameter at breast height) ≥ 30 cm, we used the total area of the plot, where all arboreal individuals that satisfied the inclusion criterion and fell within the demarcated area were sampled and identified, and the DBH, height were measured. For the sampling of arboreal individuals with 5 cm ≤ DBH < 30 cm, a range of 20 m wide and 250 m in length was established in the central part of the plot, where all individuals that satisfied the inclusion criteria and fell within the demarcated area were sampled, identified, the DBH, height and measured (Kinupp & Magnusson, 2005). Material from all the species was collected and deposited at the Herbário Rioclarense (HRCB) and identified through consultation of specialized literature and expert confirmation.

Soil samples were performed at a depth of 10 cm at the same points where the terrain slope was measured, and were blended to produce a composite sample for each plot. Before the analysis, the roots were removed from the samples, air-dried and sieved with a 2 mm sieve. Analyses included the texture (percentage of clay, sand and silt), pH, C, N, Ca^{2+}, Mg^{2+}, Na^{+}, K^{+}, P, Cu, Zn, Fe, Al^{3+}, (Al^{3+} + H^{+}) and Mn^{2+}; the cationic exchange capacity (CTC) was established based on the sum of K, Ca and Mg; base saturation (V) was calculated as a percentage of the total CTC; aluminum saturation (m) was calculated based on effective cationic exchange capacity, and the sum of the bases (SB) was represented by Ca + Mg + K. All analyses were performed in the Laboratory of Soil Science at the University of São Paulo (USP) and followed the protocol established by Embrapa (1997).

6.2.3 Statistical Analyses

Principal component analysis (PCA), using all the measured soil variables, was applied to summarize the main gradients in the soil structure (Costa et al., 2005). Soil variables were standardized by dividing the square root of the sum of squares of each variable, to give equal lengths to the variables in the analysis (Costa et al., 2005). After a preliminary analysis, we eliminate variables with high multicolinearity from the ordination analysis (Ruggiero et al., 2002).

The dimensionality in the composition of the assembly of plants was reduced by multivariate ordination through principal coordinate analysis (PCoA) (Kinupp and Magnusson, 2005). We used this technique to test for the existence of floristic gradients and to determine the relative importance of topographic, soil variables and land use types in the distribution of the species (Kinupp & Magnusson, 2005). Similar multivariate analyses have been used to describe plant communities, classify vegetation, and relate environmental variables to dissimilarities in plant communities (Oliveira & Nelson, 2001; Terborgh & Andresen, 1998; Tuomisto et al., 2003; Kinupp & Magnusson, 2005; Costa et al., 2005).

Two ordinations were performed with the plant data: one based on quantitative data and another based on qualitative data. Quantitative ordinations were performed with distance measurements through the Bray-Curtis method of standardized plot data (percentage of abundance of each species in the plot), generating a matrix of 30 plots × 199 species. The ordination of the quantitative data was used to capture the patterns exhibited by the most abundant species, as they have the largest quantitative contribution to the difference between the plots. The Sorensen index was used in the qualitative ordination. This ordination captures the pattern displayed by the rare species, because the most abundant species generally occur in many plots and so contribute little to the difference between the plots (Costa et al., 2005; Kinupp & Magnusson, 2005; Zuquim et al., 2009).

Principal coordinate analysis (PCoA) was used on the dependent variables to obtain linearity, i.e. orthogonal (axis) variables describing the composition of the community that satisfy the assumptions of multivariate orthogonal analyses (Anderson & Willis, 2003); inference tests were conducted using multivariate analysis of covariance (Kinupp & Magnusson, 2005). The first two axes derived from the PCoA describe the composition of the community, and were used as the dependent variable in the inference tests of the effects of land use type, soil fertility, and terrain slope. These axes explain 65.29% of the variance in the original variables for the quantitative data, and 58.25% of the variance for the qualitative data of presence-absence.

The variance explained by the regression of the independent variables on the multivariate axes of composition is less than the variance in the original variables explained by the regression, because the multivariate axes do not contain all the variation of the original data. To estimate the proportion of the variance in the original variables explained by multivariate regression of species composition data on the type of land use, soil fertility and terrain slope, we multiplied the proportion of variance on the multivariate axes explained by the regression by the proportion of variance in the original variables associated with these multivariate axes (Costa et al., 2005).

6.3 Results

6.3.1 Soil Analyses

The principal components of ordination of the soil variables produced three main gradients. The coefficients of correlation between topographic variables and multivariate ordination axes of the soil are shown in Table 6.1, where the most significant variables for the first three axes of soil ordination, according to Pearson's correlation, can be distinguished.

Tab. 6.1: Correlations of the textural and chemical properties of soil with the composite axes of soils derived from the PCA and with topographic characteristics of the 30 sampled plots.

	PC1	PC2	PC3	Altitude	Slope
Sand (%)	0.369	-0.086	-0.795***	-0.223	-0.376
Silt (%)	0.225	-0.42	-0.415	0.089	-0.282
Clay (%)	-0.385	0.134	0.822***	0.208	0.399
Water pH	-0.462	0.712***	-0.291	-0.306	0.018
C	-0.811***	-0.485	0.068	-0.047	0.631**
N	-0.939***	-0.235	-0.038	-0.105	0.663***
Cu	-0.423	-0.297	-0.579**	0.048	0.167
F	-0.029	-0.872***	0.005	0.306	0.117
Zn	-0.944***	-0.149	-0.113	0.015	0.512
P^+	-0.874***	0.116	0.154	0.088	0.381
Na^+	-0.592**	-0.098	0.238	-0.065	0.309
K^+	-0.971***	0.089	-0.011	-0.22	0.493
Ca^{2+}	-0.957***	0.177	-0.012	-0.174	0.452
Mg^{2+}	-0.95***	0.172	-0.117	-0.323	0.448
Mn^{2+}	-0.925***	-0.035	-0.302	-0.225	0.449
Al^{3+}	0.416	-0.773***	0.302	0.401	0.10
$H^+ + Al^{3+}$	-0.412	-0.854***	0.063	0.142	0.589**
Base δ	-0.967***	0.173	-0.038	-0.214	0.458
CTC δ	-0.986***	-0.0017	-0.022	-0.176	0.543**
V δ	-0.882***	0.346	-0.155	-0.235	0.34
M δ	0.698***	-0.488	0.229	0.242	-0.187
Altitude δ	0.128	-0.291	0.279		
Slope δ	-0.549**	-0.259	0.274	-0.031	
Explained variance (%)	48.44	19.36	12.95		

** $p < 0.001$; *** $p < 0.0001$; δ Not used in the ordination

The first axis (PCA1) accumulated 48% of the variation and presented significant correlation coefficient with C, N, Zn, P^{+}, Na^{+}, K^{+}, Ca^{2+}, Mg^{2+}, Mn^{2+} (Tab. 6.1). This axis is moderately correlated with slope (r^2 = -0.549; p = 0.001), but showed no significant correlation with altitude (r^2 = 0.128; p = 0.5) (Tab. 6.1). Soil variables associated with the first axis, such as C and N, were also significantly correlated with the slope (Tab. 6.1). There was also a strong correlation of this first axis with the sum of the bases (r^2 =-0.967; p < 0.0001), with the total exchangeable bases (r^2 = -0.986; p < 0.0001), with base saturation (r^2 =-0.882; p < 0.0001) and with aluminum saturation (r^2 = 0.698; p < 0.0001). Thus, the first axis reflects mainly the fertility of the soil, which partially correlated to the slope of the terrain.

The second axis (PCA2) accumulated 19% of the variation and presented significant correlation coefficient with pH, F, Al^{3+} and $H^{+} + Al^{3+}$ (Tab. 6.1). This second axis was not correlated with altitude (r^2 = -0.291; p = 0.12) or with slope (r^2 =-0.259; p = 0.17). None of the soil variables associated with the second axis were individually correlated with altitude (Tab. 6.1), but $H^{+} + Al^{3+}$ were individually correlated with slope (r^2 = 0.589; p = 0.0006). Thus, this second axis reflects mainly the acidity of the soil.

The third axis (PCA3), with 13% of variance, presented significant correlation coefficient with the principal texture variables (content of sand and clay) and with Cu (Tab. 6.1). This axis did not correlate with the altitude, slope or exchangeable bases (Tab. 6.1). This third axis reflects mainly the soil texture variables.

6.3.2 Species Composition and Relationship with Environmental Variables

The ordination of plots in two dimensions shows that the composition of species in eucalyptus plantation differs from plots sampled in native vegetation and pasture in regeneration, which are more similar to each other as seen in the first axis resulting from multivariate ordination (Fig. 6.2). The second axis of ordination separated the native vegetation areas in cerrado, cerradao and semi-deciduous seasonal forest as expected (Fig. 6.2). The species abundance data reduced to the two axes of ordination were significantly related to land use and soil fertility (Tab. 6.2).

The first ordination axis (PCoA1), as much for the quantitative ordination as for the qualitative ordination, separated the sampled plots according to the types of land use, as the plots on native vegetation and pasture in regeneration differed from the plots in eucalyptus plantation in terms of their species composition (Fig. 6.2 and Tab. 6.2). The second axis of ordination (PCoA2), for both quantitative and qualitative data, separated the sampled plots according to soil fertility, grouping the native vegetation in areas of cerrado *sensu stricto*, cerradão and semi-deciduous seasonal forest (Fig. 6.2 and Tab. 6.2). In the most nutrient-rich soils, we found areas of semi-deciduous seasonal forest, which change gradually into a sparse cerrado *sensu stricto* in areas where the amount of soil nutrients was low. The plots of pasture in regeneration resembled the areas of cerrado *sensu stricto* with more nutrient-poor soils, but were different from areas of semi-deciduous seasonal forest that resembled more the eucalyptus plantation areas as shown by the second axis resulting from the ordination (Tab. 6.2). The relative importance of environmental variables in explaining patterns of species composition in the community was tested for the whole community as a response variable with our model including the land use type, terrain slope, and soil fertility. This model explained a large proportion of the variance in the two dimensions of ordination and explained more than half of the variance in the original community data (Tab. 6.3).

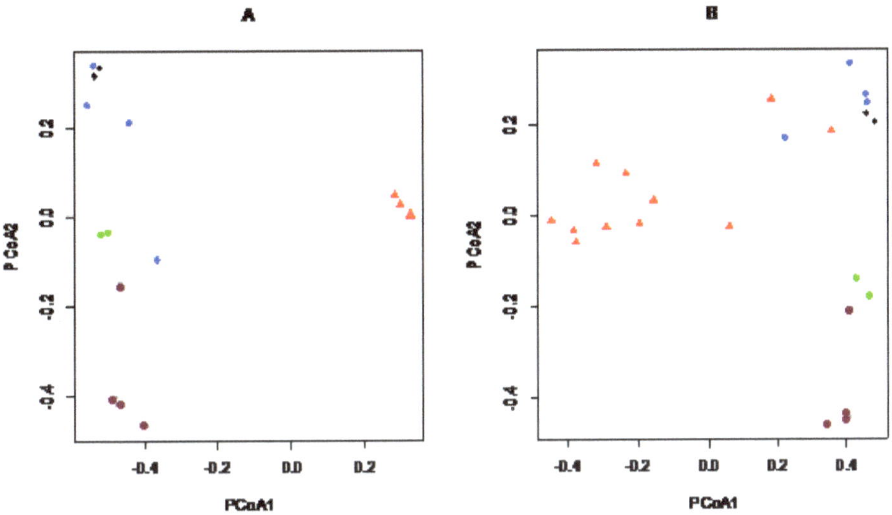

Fig. 6.2: Ordination through Analysis of Principal Coordinates (PCoA) in two dimensions of the arboreal plant community in relation to land use and soil fertility on the Três Lagoas and Arca farms, Angatuba, São Paulo, Brazil. Pink: semi-deciduous forest; Green: cerradão; Black: cerrado *sensu stricto*; Blue: abandoned pasture; Red: eucalyptus plantation. A) Quantitative data: number of individuals per species in each sampled plot, and B) qualitative data: presence-absence data per species for each sampled plot.

Tab. 6.2: Relationship between the axes of ordination of the Principal Coordinate Analysis (PCoA) and the land use type, fertility, and slope for the 30 sampled plots. The land use types (eucalyptus, forest and pasture) shown by different letters are significantly different.

		F values				Land use		
		Land use	**Fertility**	**Slope**	r^2	**Eucalyptus**	**Forest**	**Pasture**
Qualitative composition	PCoA 1	-0.6936***	0.118^{NS}	0.1036^{NS}	0.7735***	a	b	b
	PCoA 2	-0.4344**	0.6784***	0.0967^{NS}	0.7673***	a	a	b
Quantitative composition	PCoA 1	0.7964***	-0.0469^{NS}	-0.0413^{NS}	0.9932***	a	b	b
	PCoA 2	-0.1467^{NS}	0.8466***	0.0615^{NS}	0.6533***	a	a	b

^{NS}not significant; *$p < 0.05$; ** $p < 0.01$; *** $p < 0.001$

Tab. 6.3: Percentage of the explained variance (% VE) by environmental factors (land use, fertility, slope) for the two types of communities of woody plants, the principal gradient of the composition of the community, extracted by PCoA, and the original data of the community. The percentage of variance explained in the original data and the percentage of the variance captured by the axes of ordination times the percentage of the variance explained by the environmental variables on the two axes of ordination.

Type of data of the community	% variance captured by the axes of ordination	% variance explained by environmental variables on axes of ordination	% variance explained by environmental variables in the matrix of original community data
Qualitative	58.25	77.35	45.06
Quantitative	65.29	99.32	64.84

6.3.3 Vegetation Maps

According to the ordination of the variables of the soil, plots in semi-deciduous seasonal forest were separated from the plots in other environments and have relationship with the nutrients present in the soil (calcium, magnesium, phosphorus, potassium, manganese, sodium, zinc, copper, nitrogen, organic carbon) and with the percentage of clay in the soil. Plots in cerradão phytophysiognomy showed a relationship with the amount of iron, aluminium and $H^{+} + Al^{+3}$ and with the percentage of silt, showing intermediate levels of soil fertility. The plots in cerrado *sensu stricto* presented a relationship with the amount of sand present in the soil, and have nutrient-poor soils. The plots in eucalyptus plantation were distributed over the ordination axis, indicating great variability in soil fertility. The plots in pasture in regeneration were very similar to the plots in cerrado *sensu stricto,* possessing a greater amount of sand and nutrients-poor soils. Based on these results we produce a land use map of the Três Lagoas e Arca farms (Fig. 6.3).

6.4 Discussion

We observe changes in the richness and composition of plant species between plots managed with eucalyptus plantation and plots of pasture in regeneration and of native vegetation. Notably, the eucalyptus plantation and pasture in regeneration retained a subset of the species that are present in native vegetation. The loss of species in eucalyptus plantation occurred in a random manner dependent on the previous land use and farm management. In addition, soil fertility was a good predictor of the tree community structure between the different vegetation types found in the landscape, where nutrient-rich soils determine more exuberant vegetation, and soils with intermediate to poor nutrients determine vegetation with short stature

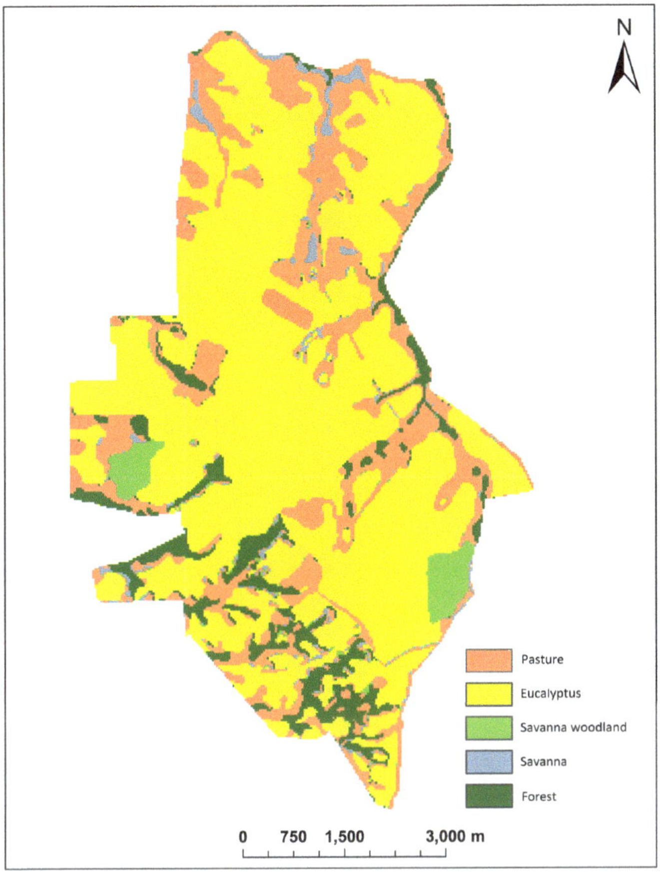

Fig. 6.3: Map of coverage of land use types and of the distribution of the different vegetation phytophysiognomies in the agroforestry landscape of the Três Lagoas and Arca farms, Angatuba, southeastern Brazil.

and greater openness of the canopy. The plots in abandoned pasture presented low soil fertility, perhaps a legacy of previous land use. However, the plots in eucalyptus plantation recovered the soil chemical and physical properties, probably due to the implementation of soil fertilizers in the initial stage of its cultivation and the gradual increase in the soil coverage resulting from the growth of trees.

Our analysis revealed that the dominant gradient in the composition of tree species in agricultural landscape is significantly related to soil fertility, differentiating the native vegetation in the cerrado *sensu stricto*, cerradão and semi-deciduous seasonal forest. Gradients in species composition between different cerrado phytophysiognomies were also found by other studies (Goodland & Pollard, 1973; Oliveira-Filho & Ratter, 2002; Ruggiero et al., 2002). Therefore, agricultural systems of eucalyptus plantation are characterized by a complex landscape mosaic, generally consisting of various land use types and phytophysiognomies. The environmental gradients of soil fertility determine the distribution of tree species in this system and increase the beta diversity of landscape as a whole (Barlow et al., 2007).

The strong gradient in the species distribution pattern between the different vegetation types we found was determined mainly by the characteristics of the soil fertility and farm management. Other studies have demonstrated the influence of soil fertility in the distribution pattern of plant species, both in natural environments and anthropogenic environments such as agricultural landscapes (Lugo, 1992; Michelsen et al., 1996; Chen et al., 1997; Clarck et al., 1999; Ruggiero et al., 2002; Oliveira-Filho & Ratter, 2002; Costa et al., 2005). Ruggiero et al. (2002) found a significant relationship between soil fertility and plant species composition of semi-deciduous forest and the cerrado phytophysiognomies. Accordingly, Michelsen et al. (1996) studied the understory vegetation and soil fertility in a silviculture system, pointing out the greater fertility of the soil in areas with native forest in comparison with eucalyptus plantation areas.

Plots of eucalyptus plantation and of pasture in regeneration retained a subset of species of the native vegetation, indicating that the species composition is severely impoverished compared with the neighboring areas of native vegetation (Appendix). The maintenance of structural complexity in eucalyptus plantation is critical for biodiversity conservation because it can allow the persistence of organisms in this landscape, which otherwise would be eliminated by the absence of their natural environment (Lindenmayer et al., 2006). Our study suggests that the presence of small patches of forest amid eucalyptus plantation, and of scattered trees as biological legacies, increase structural complexity and heterogeneity of the plantation landscape, minimizing the effects of habitat loss and, consequently, the loss of biodiversity in these conditions (Franklin et al., 2000, Lindenmayer et al., 2006). Therefore, if trees and small patches of native vegetation are maintained, the eucalyptus plantation should not be considered a „green desert" (Hartley, 2002), giving support to the prediction that human impact in managed landscapes can have a negative effect on alpha-diversity patterns, as it promotes beta diversity at the landscape scale (Barlow et al., 2007).

The different vegetation types in the studied agricultural landscape, which in this study include forest remnants in different stages of succession and cerrado areas, and their spatial arrangement among others attributes, increased spatial heterogeneity, as discussed by Lindenmayer et al. (2003). Additionally, landscapes are characterized by the presence of environmental gradients such as topography, climate, and soil type and depth. As such, the landscape heterogeneity will correspond to the remaining forest mosaic representing different compositions and age classes of vegetation when different environmental conditions occur (Tews et al., 2004). Therefore, the permanence of isolated trees amid the eucalyptus plantation as a biological legacy maintains moderate levels of species diversity in the matrix (Athayde et al., 2015).

Our results demonstrate the importance of preserving the remnants of native vegetation in the eucalyptus matrix, including patches of reduced area and composed of species of initial successional stages, as well as isolated trees amid the eucalyptus plantation. Our results show that agricultural landscapes with a predominance of eucalyptus may maintain rich assemblies of native trees as postulated by Lindenmayer and Cunningham (2012), which help to maintain high levels of structural complexity in the plantation and high spatial heterogeneity essential to the persistence of many taxa in these landscapes (Tews et al., 2004; Pardini et al., 2009). The structural complexity of the plantation and landscape heterogeneity incorporate not only particular types of attributes on the plantation, but also their arrangement (Lindenmayer et al., 2006).

Therefore, highly productive tropical landscapes can be managed to maintain an appreciable component of native biodiversity by preventing the creation of large tracts of very homogeneous land (Pardini et al., 2009). Approaching this goal in Brazil should be somewhat simple by conforming to the Código Florestal Federal (Forest Code), which defines that each property must maintain a proportion (ranging from 20% to 80%) of land coverage for areas of biodiversity conservation.

6.5 Final Considerations

We have found that sites in eucalyptus plantation and in abandoned pastures in regeneration of the same age can retain a significant amount of the native species pool of the region. However, these places alone are not sufficient to mitigate the erosion of biodiversity in tropical forests. Our results indicate that investments in conservation of tree diversity in agricultural systems should focus on primary forest sites (Barlow et al., 2007) as well as forest remnants that are still present in the landscape of different vegetation types and in different stages of succession (Duncan & Chapman, 1999; Franklin et al., 2000; Lindenmayer & Cunningham, 2012), as is common in southeastern Brazil. In addition, the forest remnants present in the landscape, along with the isolated trees amid the eucalyptus matrix, enhance the landscape heterogeneity and the complexity of the plantation that is of paramount importance for the maintenance of various taxa of plants and animals in agroforestry landscapes (Lindenmayer et al., 2006; Pardini et al., 2009; Lindenmayer & Cunningham, 2012) and should be considered for the management of these landscapes.

Acknowledgments: We thank the owners of the Três Lagoas and Arca farms for the permission and support for the development of this research. To the CNPq (National Council of Scientific and Technological development) for the doctoral scholarship granted by the EAA and for the research productivity scholarship to LPCM. The Laboratory of Phenology and the Plant Phenology and Seed Dispersal Group receives support from FAPESP. The present study is an integral part of the inventory survey called „Socio-environmental Changes in the State of São Paulo and Prospects for Conservation" financed by São Paulo Foundation for Science (FAPESP grant #. 06/60954-4) and a TTV fellowship to LFC (#2014/13354-8).

Bibliography

Anderson, M.J., & Willis, T.J. (2003). Canonical analyses of principal coordinates: a useful method of constrained ordination for ecology. Ecology, 84, 511-525.

Athayde, E.A., Cancian, L.F., Verdade, L.M., et al. (2015). Functional and phylogenetic diversity of scattered trees in an agricultural landscape: implications for conservation. Agriculture, Ecosystems and Environment, 199, 272-281.

Barbaro, L., Pontcharraud, L., Vetillard, F., et al. (2005). Comparative responses of bird, carabid and spiders assemblages to stand and landscape diversity in maritime pine plantation forests. Ecoscience, 12, 110-121.

Barlow, J., Mestre, L.A.M., Gardner, T.A., et al. (2007). The value of primary, secondary and plantation forests for Amazonian birds. Biological Conservation, 136, 212-231.

Brockerhoff, E.G., Jactel, H., Parrotta, J.A., et al. (2008). Plantation forests and biodiversity: oxymoron or opportunity? Biodiversity and Conservation, 17, 925-951.

Brockerhoff, E.G., Jactel, H., Parrotta, J.A., et al. (2013). Role of eucalypt and other planted forests in biodiversity conservation and the provision of biodiversity-related ecosystem services. Forest Ecology and Management, 301, 43-50.

Chen, Z., Hsieh, C., Jiang, F., et al. (1997). Relations of soil properties to topography and vegetation in a subtropical rain forest in southern Taiwan. Plant Ecology, 132, 229-241.

Clarck, D.B., Palmer, M.W., & Clarck, D.A. (1999). Edaphic factors and the landscape-scale distributions of tropical rain forest trees. Ecology, 80, 2662-2675.

Concepción, E.D., Díaz, M., & Baquero, R.A. (2008). Effects of landscape complexity on the ecological effectiveness of agri-environment schemes. Landscape Ecology, 23, 135-148.

Costa, F.R.C., Magnusson, W.E., & Luizão, R.C. (2005). Mesoscale distribution patterns of Amazonian understorey herbs in relation to topography, soil and watersheds. Journal of Ecology, 93, 863-878.

EMBRAPA (1997). Manual de métodos de amostragem de solo. Empresa Brasileira de Pesquisa Agropecuária e Centro Nacional de Pesquisas de Solos, Rio de Janeiro, Brazil.

FAO (2006). Global Forest Resources Assessment 2005 – Progress towards sustainable forest management. Food and Agriculture Organization of the United Nations, Rome, Italy.

Felton, A., Knight, E., Wood, J., et al. (2010). A meta-analysis of fauna and flora species richness and abundance in plantations and pasture lands. Biological Conservation, 143, 545-554.

Ferreira, J., Pardini, R., Metzger, J.P., et al. (2012). Towards environmentally sustainable agriculture in Brazil: challenges and opportunities for applied ecological research. Journal of Applied Ecology, 49, 535-541.

Fischer, J., Brosi, B., Daily, G.C., et al. (2008). Should agriculture policies encourage land sparing or wildlife-friendly farming? Frontiers in Ecology and the Environment, 6, 380-385.

Fonseca, C.R., Ganade, G., Baldissera, R., et al. (2009). Towards an ecologically-sustainable forestry in the Atlantic Forest. Biological Conservation, 142, 1209-1219.

Franklin, J.F., Lindenmayer, D.B., MacMahon, J.A., et al. (2000). Threads of continuity. Conservation in Practice, 1, 8-17.

Goodland, R., & Pollard, R. (1973). The brazilian cerrado vegetation: a fertility gradient. Journal of Ecology, 61, 219-224.

Green, R.E., Cornell, S.J., Scharlemann, J.P.W., et al. (2005). Farming and the fate of wild nature. Science, 307, 550-555.

Gregory, P.J., & George, T.S. (2011). Feeding nine billion: The challenge to sustainable crop production. Journal of Experimental Botany, 62, 5233-5239.

Hartley, M.J. (2002). Rationale and methods for conserving biodiversity in plantation forests. Forest Ecology and Management, 155, 81-95.

Humphrey, J.W., Newton, A.C., Peace, A.J., et al. (2000). The importance of conifer plantations in northern Britain as a habitat fungi. Biological Conservation, 96, 241-252.
Jactel, H., & Brockerhoff, E.G. (2007). Tree diversity reduces herbivory by forest insects. Ecology Letters, 10, 835-848.
Kearney, J. (2010). Food consumption trends and drivers. Philosophical Transactions Royal Society B, 365, 2793-2807.
Kelty, J.M. (2006). The role of species mixtures in plantation forestry. Forest Ecology and Management, 233, 195-204.
Kinupp, V.F., & Magnusson, W.E. (2005). Spatial patterns in the understorey shrub genus *Psychotria* in Central Amazonia: effects of distance and topography. Journal of Tropical Ecology, 21, 363-374.
Lindenmayer, D.B., Hobbs, R.J., & Salt, D. (2003). Plantation forests and biodiversity conservation. Austral Forestry, 66, 62-66.
Lindenmayer, D.B., Franklin, J.F., Fischer, J. (2006). General management principles and a checklist of strategies to guide forest biodiversity conservation. Biological Conservation, 131, 433-445.
Lindenmayer, D.B., & Cunningham, S.A. (2012). Six principles for managing forests as ecologically sustainable ecosystems. Landscape Ecology, 28, 1099-1110.
Lugo, A.E. (1992). Comparison of tropical tree plantations with secondary forests of similar age. Ecological Monographs, 62, 1-41.
Magnusson, W.E., Lima, A.P., Luizão, R., et al. (2005). RAPELD: a modification of the gentry method for biodiversity surveys in long-term ecological research sites. Biota Neotropica, 5, 1-6.
Manning, A.D., Gibbons, P., & Lindenmayer, D.B. (2009). Scattered trees: a complementary strategy facilitating adaptive responses to climate change in modified landscape? Journal of Applied Ecology, 46, 915-919.
Martin, P.S. (2010). Distribuição e abundância de mamíferos neotropicais não voadores de pequeno porte em paisagem silvicultural da bacia do Alto Paranapanema, São Paulo, Brasil. Dissertação de Mestrado. Escola Superior de Agricultura "Luiz de Queiroz", Piracicaba, Brazil.
Mendenhall, C.D., Sekercioglu, C.H., Brenes, F.O., et al. (2011). Predictive model for sustaining biodiversity in tropical countryside. Proceedings of the National Academy of Sciences of the United States of America, 108, 16313-16316.
Michelsen, A., Lisanework, N., Friis, I., et al. (1996). Comparisons of understorey vegetation and soil fertility in plantations and adjacent natural forests in the Ethiopian highlands. Journal of Applied Ecology, 33, 627-642.
Oliveira, A.A., & Nelson, B.W. (2001). Floristic relationships of terra firme forests in the Brazilian Amazon. Forest Ecology and Management, 146, 169-179.
Oliveira-Filho, A.T., & Ratter, J.A. (2002). Vegetation physiognomies and woody flora of the cerrado biome. In Oliveira, P.S., & Marquis, R.J. (Eds.) The cerrados of Brazil: Ecology and natural history of the Neotropical savanna (pp. 91-120). New York, Columbia University Press.
Pacala, S., & Socolow, R. (2004). Stabilization wedges: solving the climate problem for the next 50 years with current technologies. Science, 305, 968-972.
Pardini, R., Faria, D., Accacio, G.M., et al. (2009). The challenge of maintaining Atlantic forest biodiversity: A multi-taxa conservation assessment of specialist and generalist species in an agro-forestry mosaic in southern Bahia. Biological Conservation, 142, 1178-1190.
Parrotta, J.A., Turnbull, J., & Jones, N. (1997). Catalyzing native forest regeneration on degraded tropical lands. Forest Ecology and Management, 99, 1-8.
PFN (2000). Programa Nacional de Florestas – Decreto No 3.420, de 20 de abril de 2000. Ministério do Meio Ambiente, Brasil.
Roberts, L. (2011). Nine billion? Science, 333, 540-543.
Ruggiero, P.G.C., Batalha, M.A., Pivello, V.R., et al. (2002). Soil-vegetation relationships in cerrado (Brazilian savanna) and semideciduous forest, southeastern Brazil. Plant Ecology, 160, 1-16.

SFB (2012). Serviço Florestal Brasileiro. http://www.florestal.gov.br/ (access verified October 2013).
Sposito, G. (2013). Green water and global food security. Vadose Zone Journal, 1, 1-6.
Terborgh, J., & Andresen, E. (1998). The composition of Amazonian forets: patterns at local and regional scales. Journal of Tropical Ecology, 14, 645-664.
Tews, J., Brose, U., Grimm, V., et al. (2004). Animal species diversity driven by habitat heterogeneity/ diversity: the importance of keystone structures. Journal of Biogeography, 31, 79-92.
Tilman, D., Balzer, C., Hill, J., et al. (2011). Global food demand and the sustainable intensification of agriculture. Proceedings of National Academy of Sciences of the United States of America, 108, 20260-20264.
Thompson, I.D., Mackey, B., McNulty, S., et al. (2009). Forest resilience, biodiversity, and climate change. A synthesis of the biodiversity/resilience/stability relationship in forest ecosystems. Secretariat of the Convention on Biological Diversity, Montreal.
Tuomisto, H., Ruokolainen, K., & Yli-Hala, M. (2003). Dispersal, environment and floristic variation of western Amazonian forests. Science, 299, 241-244.
Zuquim, G., Costa, F.R.C., Prado, J., et al. (2009). Distribution of pteridophyte communities along environmental gradients in central Amazonia, Brazil. Biodiversity and Conservation, 18, 151-166.
Yu, C.M. (2004). Seqüestro florestal de carbono no Brasil: Dimensões políticas, socioeconômicas e ecológicas. São Paulo, Annablume.

Appendix 1

List of the native tree species distributed between the land use types in the 30 sampled plots on the farms. Eucalytus plantation (n = 18), pasture (n = 4), cerrado *s.s.* (n = 2), cerradão (n = 2) and semi-deciduous seasonal forest (n = 4).

Family/Species			Native Vegetation		
	Eucalyptus Plantation	Pasture	Cerrado s.s.	Cerradão	Semi-deciduous Seasonal Forest
Agavaceae					
-*Cordilyne terminalis* Kunth.	0	0	0	0	7
Anacardiaceae					
-*Lithraea molleoides* (Vell.) Engl.	0	0	0	0	27
-*Tapirira guianensis* Aubl.	0	13	17	106	293
-*Schinus terebenthifolius* Raddi	0	0	0	1	20
Annonaceae					
-*Annona cacans* Warm.	0	0	2	0	0
-*Annona coriacea* Mart.	5	2	1	0	0
-*Annona crassiflora* Mart.	0	0	2	1	0
-*Guatteria nigrescens* Mart.	0	0	0	5	6
-*Rollinia silvatica* (St. Hil.) Mart.	0	0	0	0	1
-*Xylopia brasiliensis* Spreng.	0	0	0	77	
Apocynaceae					
-*Aspidosperma australe* M. Arg.	0	0	0	2	9
-*Aspidosperma subincanum* Mart.	0	0	0	0	1
-*Aspdosperma tomentosum* Mart.	2	0	2	0	0
-*Peschiera fuchsiaefolia* Miers.	0	0	0	0	12
Aquifoliaceae					
-*Ilex cerasifolia* Reiss.	0	0	0	1	1
-*Ilex dumosa* Reissek	0	0	1	0	0
Araliaceae					
-*Aralia warmingiana* (Marchal) J. Wen	0	0	0	0	7
-*Schefflera morototoni* (Aubl.) Maguire, Steyerm. & Frodin	0	0	0	3	0
-*Schefflera vinosa* (Cham. & Schltdl.) Frodin & Fiaschi	0	0	7	0	0
Arecaceae					
-*Geonoma brevispatha* Barb. Rodr.	0	0	0	1	0
-*Syagrus romanzoffiana* (Cham.) Glassm.	0	0	0	0	8

Family/Species			Native Vegetation		
	Eucalyptus Plantation	Pasture	Cerrado s.s.	Cerradão	Semi-deciduous Seasonal Forest
Asteraceae					
-*Dasyphyllum brasiliense* (Spreng.) Cabrera	0	0	0	0	8
-*Gochnatia polymorpha* (Less.) Cabr.	17	7	10	2	47
-*Vernonia discolor* (Spreng.) Less.	0	0	1	2	0
-*Vernonia polyanthes* Less.	0	0	0	1	8
Bignoniaceae					
-*Adenocalymma dusenii* Kraenzl.	2	0	0	3	44
-*Arrabidaea triplinervia* (Mart. ex DC.) Baill. ex Bureau	2	0	0	2	78
-*Jacaranda micrantha* Cham.	0	0	0	0	1
-*Tabebuia alba* (Cham.) Sandw.	2	1	1	12	0
-*Tabebuia heptaphylla* (Vell.) Toledo	0	0	0	0	2
Boraginaceae					
-*Cordia ecalyculata* Vell.	0	0	0	0	3
-*Cordia trichotoma* (Vell.) Arrab. ex Steud.	0	0	10	3	0
-*Cordia sellowiana* Cham.	1	0	0	0	81
-*Cordia superba* Cham.	0	0	0	0	1
-*Patagonula americana* L.	1	0	0	0	20
Burseraceae					
-*Protium heptaphyllum* (Aubl.) Marchand	0	0	0	0	5
Calophyllaceae					
-*Kielmeyera coriacea* Mart. & Zucc.	0	0	1	3	0
Caryocaraceae					
-*Caryocar brasiliensis* Camb.	6	3	6	0	0
Cecropiaceae					
-*Cecropia pachystachya* Trec.	0	0	0	1	49
Celastraceae					
-*Austroplenckia populnea* (Reiss.) Lund.	2	1	22	61	0
-*Maytenus robusta* Reiss.	0	0	0	11	13
Clusiaceae					
-*Clusia criuva* Cambess.	0	0	0	1	2

Family/Species	Native Vegetation				
	Eucalyptus Plantation	Pasture	Cerrado s.s.	Cerradão	Semi-deciduous Seasonal Forest
-*Garcinia gardneriana* (Planch. & Triana) Zappi	0	0	0	0	10
Combretaceae					
-*Terminalia brasiliensis* Mart.	1	0	0	2	16
Dilleniaceae					
-*Davilla elliptica* A. St.-Hil.	30	0	0	0	0
Euphorbiaceae					
-*Actinostemon concolor* (Spreng.) Müll. Arg.	0	0	0	1	0
-*Alchornea triplinervia* (Spreng.) M. Arg.	3	0	2	33	9
-*Croton floribundus* Spreng.	0	0	1	8	93
-*Sapium glandulatum* (Vell.) Pax	0	0	0	0	1
-*Sebastiania brasiliensis* Spreng.	0	0	0	1	0
-*Sebastiania commersoniana* (Baill.) Smith & Downs.	0	0	0	1	21
Erythroxylaceae					
-*Erythroxylum deciduum* A. St.-Hil.	0	1	5	1	0
Fabaceae					
-*Acacia* sp. Mill.	0	0	0	0	1
-*Acosmium subelegans* (Mohlenbr.) Yakovlev	18	45	128	0	1
-*Albizia hassleri* (Chodat) Burkart	0	0	0	0	63
-*Anadenanthera falcata* (Benth.) Speg.	36	3	137	0	0
-*Andira fraxinifolia* Benth.	0	0	0	5	2
-*Andira legalis* (Vell.) Toledo	2	3	0	0	0
-*Bauhinia bongardii* Steud.	3	0	0	0	23
-*Bauhinia forficata* Link	0	0	0	0	48
-*Bauhinia rufa* Graham	0	0	5	1	0
-*Caesalpinia ferrea* Mart. ex Tul.	0	3	0	0	0
-*Cassia ferruginea* (Schrad.) Schrad. ex DC.	0	0	0	0	1
-*Centrolobium tomentosum* Guillemin ex Benth.	0	0	0	0	1
-*Copaifera langsdorffii* Desf.	4	2	34	100	3
-*Dalbergia frutescens* (Vell.) Britton	0	0	0	3	23

Family/Species			Native Vegetation		
	Eucalyptus Plantation	Pasture	Cerrado s.s.	Cerradão	Semi-deciduous Seasonal Forest
-*Dalbergia miscolobium* Benth.	12	0	80	0	0
-*Dalbergia villosa* Benth.	0	0	0	0	14
-*Dimorphandra mollis* Benth.	2	4	63	0	1
-*Enterolobium contortisiliquum* (Vell.) Morong.	0	0	0	0	9
-*Enterolobium gummiferum* (Mart.) Macbride	0	0	2	0	0
-*Inga marginata* Kunth	0	0	0	0	3
-*Inga uruguensis* Hook & Arn.	0	0	0	0	1
-*Lonchocarpus muehlbergianus* Hassl.	0	0	0	0	76
-*Machaerium aculeatum* Raddi	0	0	0	1	1
-*Machaerium acutifolium* Vog.	0	0	47	2	0
-*Machaerium brasiliense* Vog.	0	0	1	8	7
-*Macherium hirtum* (Vell.) Stellfeld	0	0	0	0	5
-*Macherium nyctitans* (Vell.) Benth.	0	0	0	0	5
-*Macherium stipitatum* (DC.) Vog.	7	0	0	38	14
-*Macherium villosum* Vog.	2	16	2	0	14
-*Mimosa bimucronata* (DC.) Kuntze	0	0	0	0	6
-*Parapiptadenia rigida* (Benth.) Brenan	0	0	1	0	53
-*Peltophorum dubium* (Spreng.) Taub.	0	0	0	2	0
-*Piptadenia gonoacantha* (Mart.) J.F. Macbr.	1	0	0	0	35
-*Pithecolobium inculare* (Vell.) Benth.	0	0	0	0	1
-*Plathymenia reticulata* Benth.	0	0	0	0	1
-*Platypodium elegans* Vog.	1	1	4	1	16
-*Pterocarpus violaceus* Vogel	0	0	0	0	2
-*Stryphnodendron adstringens* (Mart.) Coville	6	6	28	0	0
Lamiaceae					
-*Aegiphila sellowiana* Cham.	7	5	0	0	8
-*Vitex polygama* Cham.	1	0	0	0	0
-*Vitex montevidensis* Cham.	0	0	0	0	15
Lauraceae					
-*Endlicheria paniculata* (Spreng.) J.F. Macbr.	0	0	0	1	2

Family/Species			Native Vegetation		
	Eucalyptus Plantation	Pasture	Cerrado s.s.	Cerradão	Semi-deciduous Seasonal Forest
-*Nectandra cissiflora* Ness	0	0	0	0	10
-*Nectandra grandiflora* Nees	1	0	0	11	3
-*Nectandra lanceolata* Nees	0	0	0	1	14
-*Nectandra megapotamica* (Spreng.) Mez	0	0	1	3	89
-*Ocotea corymbosa* (Meissn.) Mez	1	1	44	34	0
-*Ocotea pulchella* Mart.	6	1	73	7	0
-*Persea venosa* Ness & Mart. ex Ness	0	0	0	0	1
-*Persea pyrifolia* (D. Don) Spreng.	1	1	0	21	4
Lecythidaceae					
-*Cariniana estrellensis* (Raddi) Kuntze	0	0	0	0	6
Lythraceae					
-*Lafoensia pacari* A. St.-Hil	2	1	0	0	0
Loganiaceae					
-*Strychnos pseudoquina* St. Hill.	0	0	0	2	0
Malpighiaceae					
-*Byrsonima lancifolia* Kunth	0	0	0	1	0
Malvaceae					
-*Chorisia speciosa* St. Hill.	0	0	0	2	1
-*Eriotheca gracilipes* (K. Schum.) A. Robyns	0	0	50	0	0
-*Luehea divaricata* Mart.	0	0	0	0	44
-*Pseudobombax grandiflorum* (Cav.) A. Rob.	0	0	0	0	1
Melastomataceae					
-*Miconia chartacea* Triana	0	0	0	264	0
-*Miconia cinnamomifolia* (DC.) Naud.	0	5	2	46	1
-*Miconia lepidota* Schrank & Mart. ex DC.	0	0	0	26	0
-*Miconia rubiginosa* (Bonpl.) DC.	2	2	4	0	0
-*Miconia sellowiana* Naudin	0	0	1	8	0
-*Miconia theizans* (Bonpl.) Cogn.	0	5	93	5	8
-*Tibouchina mutabilis* Cong.	0	0	1	0	3
Meliaceae					
-*Cabralea canjerana* (Vell.) Mart.	0	0	0	1	10

Family/Species			Native Vegetation		
	Eucalyptus Plantation	Pasture	Cerrado s.s.	Cerradão	Semi-deciduous Seasonal Forest
-*Cedrela fissilis* Vell.	0	0	0	0	33
-*Guarea macrophylla* Vahl	0	0	0	0	3
-*Trichilia elegans* A. Juss.	0	0	0	0	11
-*Trichilia pallida* Swartz	0	0	0	0	1
-*Trichilia silvatica* C. DC.	0	0	0	0	1
Moraceae					
-*Brosimum glaziovii* Taub.	0	0	0	10	0
-*Ficus guaranitica* Schodat	0	1	0	2	0
-*Maclura tinctoria* (L.) D. Don ex Steud.	0	0	0	0	2
-*Sorocea bonplandii* (Baill.) W.C. Burger, Lanj. & Wess. Boer	0	0	0	0	1
Myrsinaceae					
-*Rapanea ferruginea* (Ruiz et Pav.) Mez	1	1	4	17	43
-*Rapanea guianensis* Aubl.	0	0	0	0	1
-*Rapanea umbellata* (Mart ex DC.) Mez	1	8	105	70	157
Myrtaceae					
-*Calyptranthes concinna* DC.	0	0	0	2	0
-*Calyptranthes clusiifolia* (Miq.) O. Berg	0	0	1	140	0
-*Campomanesia guaviroba* (DC.) Kiaersk	0	0	0	65	16
-*Campomanesia guazumifolia* (Cambess.) O. Berg	0	0	0	0	1
-*Campomanesia xanthocarpha* Berg	0	0	1	1	156
-*Eugenia* sp. L.	0	0	2	0	0
-*Eugenia brasiliensis* Lam.	0	0	0	66	2
-*Eugenia dodoneaefolia* Camb.	0	0	0	0	2
-*Eugenia involucrata* DC.	0	0	0	0	16
-*Eugenia pyriformis* Camb.	0	0	0	2	1
-*Eugenia ramboi* D. Legrand	1	0	0	0	2
-*Eugenia uniflora* L.	8	1	0	0	12
-*Hexachlamys edulis* (O. Berg) Kausel & D. Legrand	0	0	0	0	26
-*Myrcia* sp. DC.	0	0	0	0	1

Family/Species			Native Vegetation		
	Eucalyptus Plantation	Pasture	Cerrado s.s.	Cerradão	Semi-deciduous Seasonal Forest
-*Myrcia bella* Cambess	5	82	145	5	0
-*Myrcia crassifolia* (Miq.) Kiaersk	0	0	1	0	0
-*Myrcia fallax* (Rich.) DC.	0	0	43	66	3
-*Myrcia floribunda* Miq.	0	0	0	0	2
-*Myrcia langsdorffii* O. Berg	0	0	0	0	1
-*Myrcia lingua* (O. Berg.) Mattos & D. Legrand	0	15	7	0	0
-*Myrciaria tenella* (DC.) O. Berg	0	0	2	0	0
-*Plinia rivularis* (Cambess.) Rotman	0	0	0	1	1
-*Psidium guajava* L.	0	2	0	0	11
-*Psidium myrtodes* O. Berg	0	1	0	0	0
-*Psidium rufum* DC.	0	0	0	0	28
Nyctaginaceae					
-*Bougainvillea glabra* Choisy	0	0	0	0	3
Ochnaceae					
-*Ouratea spectabilis* (Mart.) Engl.	1	8	150	0	0
Opiliaceae					
-*Agonandra macedoi* Toledo	0	0	0	0	3
Peraceae					
-*Pera glabrata* (Schott) Baill.	1	50	188	102	17
Pinaceae					
-*Pinus elliotti* Engelm.	0	1	61	0	0
Phytolaccaceae					
-*Gallesia integrifolia* (Spreng.) Harms	0	0	0	1	2
Polygonaceae					
-*Coccoloba mollis* Casar.	0	0	0	0	16
-*Coccoloba rosea* Meisn.	0	0	0	0	22
-*Ruprechtia laxiflora* Meisn.	0	0	0	0	1
-*Triplaris brasiliana* Cham.	0	0	0	0	2
Proteaceae					
-*Roupala brasiliensis* Klotz.	0	7	29	1	1
Rhamnaceae					
-*Colubrina glandulosa* Perkins	0	0	0	0	1
-*Rhamnidium elaeocarpum* Reissek	0	1	0	0	0

Family/Species			Native Vegetation		
	Eucalyptus Plantation	Pasture	Cerrado s.s.	Cerradão	Semi-deciduous Seasonal Forest
Rosaceae					
-*Prunus sellowii* Koehne	0	0	0	0	18
Rubiaceae					
-*Amaioua guianensis* Aubl.	0	0	0	173	2
-*Chomelia obtusa* Cham. & Schltdl.	0	0	0	1	6
-*Faramea cyanea* Müll. Arg.	0	0	0	0	1
-*Genipa americana* L.	0	0	0	0	3
-*Randia armata* (Sw.) DC.	0	0	0	0	1
-*Rudgea jasminioides* (Cham.) Müll. Arg.	0	0	0	0	6
-*Tocoyena formosa* (Cham. & Schltdl.) K. Schum.	0	0	1	0	0
Rutaceae					
-*Citrus limonum* Risso	0	0	0	0	2
-*Esenbeckia febrifuga* (A. St. Hil.) A. Juss. ex Mart.	0	0	0	0	4
-*Metrodorea nigra* A. St.-Hil.	0	0	0	0	1
-*Zanthoxylum rhoifolium* Lam.	1	2	0	0	16
-*Zanthoxylum riedelianum* Engl.	2	0	0	0	8
Salicaceae					
-*Casearia sylvestris* Sw.	3	0	1	0	165
Sapindaceae					
-*Allophyllus edulis* (St. Hil.) Radlk.	0	0	0	7	65
-*Cupania vernalis* Camb.	0	0	0	10	75
-*Matayba elaeagnoides* Radlk.	1	0	0	0	20
Sapotaceae					
-*Chrysophyllum marginatum* (Hook. et Arn.) Radlk.	0	0	0	0	10
Siparunaceae					
-*Siparuna guianensis* Aubl.	0	0	0	41	0
Sterculiaceae					
-*Guazuma ulmifolia* Lam.	0	0	0	0	7
Styraceae					
-*Styrax camporum* Pohl.	0	1	33	4	0
-*Styrax pohlii* A. DC.	0	1	2	0	0

Family/Species			Native Vegetation		
	Eucalyptus Plantation	Pasture	Cerrado s.s.	Cerradão	Semi-deciduous Seasonal Forest
Symplocaceae					
-*Symplocos pubescens* Klotzsch ex Benth.	0	0	0	0	21
Thymeliaceae					
-*Daphnopsis racemosa* Griseb.	0	0	0	0	26
Ulmaceae					
-*Clethra scabra* Pers.	0	0	0	0	2
-*Trema micrantha* (L.) Blume	0	0	0	0	1
Verbenaceae					
-*Aloysia virgata* (Ruiz & Pav.) Juss.	1	0	0	0	5
Vochysiaceae					
-*Qualea dichotoma* (Warm.) Stafl.	3	1	14	3	1
-*Qualea grandiflora* Mart.	5	2	3	2	0
-*Qualea parviflora* Mart.	0	0	6	0	0
-*Vochysia tucanorum* Mart.	1	6	5	14	2

Giancarlo Conde Xavier Oliveira

7 Changes in Species and Genetic Diversity of Crops in Agricultural Systems, with an Emphasis on the State of São Paulo

Abstract: While agriculture is one of the major agents of negative impact on the diversity of natural ecosystems and the main cause of wild habitat shrinkage, the fact that crop biodiversity itself is dynamic should not be forgotten. The vicissitudinous behavior of economy, the whimsical oscillations of diet composition, human migrations, technological advances in plant breeding and even the biological peculiarities of the crops can have the effect of fostering the increase or decline of species diversity in agriculture at several geographic scales, as well as of changing the genetic diversity of crop species. Some of these factors, such as mechanization and selection methods, are intrinsic to the various agricultural models adopted, while others, including migration from the countryside to the cities and the fortuitous introduction of devastating pests, are historically contingent. In this chapter, we use a few indexes to quantify the changes in crop species diversity and crop intraspecific diversity that took place in Brazil in the last decades, at several spatial scales, with emphasis on the southeastern region, and investigate the main phenomena that have caused them.

7.1 Introduction

Since its inception, about 10,000 years ago, during the first historical instances of the Agricultural Revolution, farming activities have demonstrated their increasing power to affect the environment, notably by claiming land formerly occupied by natural vegetation. The destruction of wild habitats by agriculture and other forms of impact potentialized by its proximity, such as hunting and gathering, have depleted the species diversity of many natural ecosystems and provoked losses in the genetic diversity of many animal and plant species.

It is important, however, to investigate the processes of loss and gain of biological diversity within the agroecosystem itself. The domestication of plants and animals and the subsequent microevolutionary changes in the genetic composition of crops and domestic animals caused by germplasm transference and more or less scientific forms of breeding altered their resistance to thousands of pathogens and other parasites that invaded the farms. In the same token, the phytosociological and demographic revolution introduced by the farm structure created a new environment amenable to be colonized by feral animals, weedy plants, rhizosphere, phylosphere, endophytic and soil microbes which may have different densities in wild habitats.

The complexity of the specific biota associated with crops notwithstanding, it should not be forgotten that the species diversity and genetic variation of the crops themselves are dynamic variables. This type of change is the focus of this chapter, while the impact of agriculture on wild ecosystems and the farm-associated biota is dealt with elsewhere. The crop species composition of two or more agroecosystems can be as distinct as grasslands in a dairy district, a grain-producing area or a cocoa plantation, and this is ecologically relevant. The cocoa plantation in composed of only one crop, but needs the maintenance of the original tropical forest around and above it. The grassland and the grain zone may both be composed of a few crop species, but the management of the latter will probably leave the soil seasonally uncovered.

An ideal study of crop diversity evolution should encompass the following approaches: (1) the dynamics of crop species diversity along a considerable period of time; (2) the history of land occupation by all the landraces/cultivars/hybrids of all crops ever planted to a given region, alongside the genealogical relationships among those materials; (3) a description of the genetic composition of all the crops of the focused region, in terms both of molecular markers and of phenotypic traits, during a long historical period. Needless to say, such a huge endeavor is unworkable even for a small region, and for Southwestern Brazil, which is the geographic scope here, the incompletion of systematic data is an additional hindrance. Thus, two restricted approaches were selected to tackle some angles of the problem in this chapter. Firstly, the question about crop species dynamics over time was investigated with an analysis of governmental agricultural statistics, whose drawbacks are described in the next section. The aim of this analysis was to reveal trends (or the absence thereof) in the evolution of crop diversity in Brazilian Southeast (SE) agriculture in the last two decades. How is crop diversity changing in the Southeast? Are the change trends collinear among the states? Do they agree with the trends in other Brazilian regions and in the country as a unit? Particularly in São Paulo State, is there heterogeneity in the trends presented at the state, agricultural region and municipality levels? If so, is there any apparent explanation for that? Is there any strong relationship between total area planted and crop species diversity in the crop portion of the SE agrobiome or its subunits?

The approach related to item (2) above is beyond the scope of this chapter and will not be explored in any level of detail, but would be worth an independent (and probably laborious) review. In fact, the information on cultivar genealogy and the historical record of the total area and geographic distribution of varieties and genotypes in the region is scattered and available for just a few major crops, and even for these, the surveys are sometimes published by agribusiness associations in pamphlets not largely distributed, instead of peer-reviewed journals. Some very general predictions can be advanced, however, about the genetic base of the major crops. The domestication and diversity centers of most of the main crops cultivated in Brazil are located in other regions (Kloppenburg, 2004): sugarcane, rice and bananas come from SE Asia, coffee comes from NE Africa, soybeans were domesticated in

China, *Brachiaria* and sorghum are African, citrus species come from southern Asia and eucalyptus was brought from Australia. Most garden vegetables used in Brazil originated in the Mediterranean countries, Asia, Africa and the Andes. This disconnection between economic rank and domestication origin is not restricted to Brazil. Maize, wheat and oranges were not domesticated in the USA, sunflowers are not native to Russia, nor are grapes to France, rubber trees to Malaysia or tulips to Holland, and the list goes on. The genetic consequence is that generally these countries, including Brazil, are heavily dependent (more heavily in the case of First World countries, though) on genetic resources that are located in distant regions and are not easily available due to political barriers. Pepper, for instance, which is an important crop in Pará, has an extremely narrow genetic base because India, its diversity center, restricts germplasm exportation. Thus, the genetic base of most of major crops in Brazil and in the SE is rather narrow and the common varieties are usually closely related. Successful modern cultivars are generally adopted throughout, narrowing even more the genetic base of the crops. Some species, such as sugarcane, are managed as vegetatively reproducing crops. Others, such as rice and soybeans, are autogamous, and the preferred materials of many species, like maize, are homogeneous hybrids. All these factors, allied to the habit of using only one variety per producing unit (which is contrary to all evolutionary, ecological and epidemiological knowledge), contribute to genetic depletion and vulnerability. Systematic studies at a population genetics level only exist for germplasm collections (e.g. Priolli et al., 2002 for soybeans) and even so, they are not repeated over time to detect historical change.

As a case study related to item (3) above, the recent history of the swidden agriculture system in the coastal regions of São Paulo State is reviewed and some results from ongoing research in our lab are presented. These kinds of small farms, known locally as *"roças"* have been the subject of several ethnogenetic studies since the 1990's and are evolving quickly, as we were able to witness.

7.2 Historical Variation in Crop Species Diversity in Agricultural Systems at Several Geographic Scales

7.2.1 Material and Methods

7.2.1.1 The Diversity Index and Limitations of the Data Available

Ecological diversity indexes may be used to compare communities both in space and in time. Here, species richness and the Shannon-Weaver index (*H'*) are calculated to compare geographic regions among themselves and to detect differences in diversity among different times in the same region. Diversity indexes generally use numbers of individuals of each species present in a sample or in an entire population consisting of several species. Brillouin's H and Shannon-Weaver's (or

Shannon-Wiener's, or simply Shannon's) *H'* are two of the commonest indexes. Neither is strictly appropriate for the kind of data available for answering our questions about diversity in Southeast agriculture. There are no official data on numbers of individuals of most crop species, and the conversion of other types of data into numbers is impractical, hindering the use of Brillouin's H index, which uses absolute numbers. The governmental statistics most amenable to being subjected to a heterodox use of *H'* are the areas planted to each species, because this index uses fractions instead of absolute numbers and we can use fractions of the total area planted as a surrogate. The Gini index would be also improper, because it needs two numerical variables, and our data relate a categorical variable, such as municipality, region or state, to numerical variables.

H' is defined as

$$H' = -\sum_{i=1}^{S} p_i \ln p_i$$

where:

n_i is the value of the area planted to species i
N is the sum total value of the area planted to all crop species
p_i is the relative value of the area planted to crop species i, calculated by n_i/N
S is the number of crop species, or species richness

Since p_i is a positive fraction, $\ln p_i$ is negative and so is $p_i \ln p_i$. The minus signal before the sum symbol makes the index conveniently positive again. The Shannon-Wiener index can be used for any number of species, and its maximum value is obtained when all crop species are planted in equal areas. The maximum value is not fixed, though, and varies with S, according to the equation $H'_{max} = \ln S$ (Poole, 1974). Two or more regions may have evenly distributed areas and consequently locally maximal, but different, *H'* values, if their *S* values are different, so that *relative H'* values may be more suitable for comparisons among regions. The relative Shannon-Wiener index is calculated by $H'rel = H'/ H'_{max}$. The lower the value of *H'rel*, the more distant the region is from an even distribution of crop productions. Absolute *H'* thus can be decomposed into the two basic diversity indices S and *H'rel*.

The same area may contain very different numbers of individuals for different species (think of orange trees and maize, for instance), so that the use of area may seem arguable. However, ecological consequences of diversity such as soil protection, formation of inter-deme matrix for wild animal mobility and the capability to provide shelter for animals, to mention a few, are best estimated by areas than by numbers of individuals, which depend on density. Using the total areas planted to each species in a region is equivalent to collecting the whole population as a sample. In this case, there is no sampling error or standard deviation, and all differences are significant.

7.2.1.2 Planted Area Data at the National, Regional and State Levels

The area data come from statistics collected and published by IBGE (Brazilian Institute of Statistics and Geography, a federal agency), and by the LUPA project of the Agriculture State Secretary of São Paulo. IBGE data were obtained from the Historical and Statistical Series web page (IBGE, 2008), and the tables examined refer individually to (1) Brazil as a whole; the official Brazilian geographic regions, viz., (2) North; (3) Northeast; (4) Central-West; (5) Southeast; (6) South; the four states of the Southeast region, viz., (7) Minas Gerais; (8) Rio de Janeiro; (9) Espírito Santo; and (10) São Paulo. The tables consolidate data collected from questionnaires applied at a household level and are located at the Produção Agrícola Municipal page (PAM: Municipal Agricultural Production) < Agropecuária (Agriculture and Animal Husbandry) < Temas e Subtemas (Themes and Subthemes). The annual/biennial and perennial crop planted area tables were merged in individual tables for each of the geographical units listed above. Each table shows the planted area of each of the 65 major species in the country. Some species are absent in some regions and in others there are surely plenty of planted species besides the 65 registered, but their area would be negligible at the geographical levels considered. The annual data available covers the 1990-2010 period.

7.2.1.3 Planted Area Data at the State, Agricultural Region and Municipality Level in São Paulo State

The LUPA (Censitary Survey on Agricultural Production Unities of the State of São Paulo) (São Paulo, 2008) shows crop planted area (among others) data collected by CATI (the rural extension state agency) at both a municipal and a regional level during the 1995/1996 and the 2007/2008 crop years, allowing a historical trend study along 12 years. The State of São Paulo is divided into 40 agricultural regions, or Regional Development Offices (EDR's; Tab. 7.1; Fig. 7.6) and this division provides a convenient intermediate geographical level between the municipality and the state. The LUPA region-based tables were used for the *H'* calculations, but not for obtaining the species richness values (S); rather, the municipal lists of crop species were consolidated into regional lists. The reason for this strategy is that the merged municipal data include more species, totaling a number much higher than the maximum S set in the regional tables (23 and 30 for the crop years examined) and yields much more exact regional S values. The areas per crop species were used as weights to calculate the weighted means of S, *H'* and *H'rel* per region depicted in the graphics our discussion was based upon. Area by species richness relationships were assessed by the Pearson correlation coefficient. Depending on the particular data used, the coefficients have different interpretations. The coefficients based on the IBGE data measure the correlation of species richness with the total cultivated areas in the focused geographic unit (country, region or state) over different years (1990-2010). The coefficients based on the LUPA data correlate species richness with the areas of the 40 agricultural regions of São Paulo in the same year.

Tab. 7.1: Names and code numbers of the 40 agricultural regions of the State of São Paulo.

Region number	Region name	Region number	Region name	Region number	Region name	Region number	Region name
1	São Paulo	11	Avaré	21	Ribeirão Preto	31	Tupã
2	Registro	12	Botucatu	22	Franca	32	Araçatuba
3	Mogi das Cruzes	13	Piracicaba	23	Assis	33	São José do Rio Preto
4	Pindamonhangaba	14	Limeira	24	Marília	34	Presidente Venceslau
5	Guaratinguetá	15	Mogi Mirim	25	Lins	35	Dracena
6	Itapeva	16	São João da Boa Vista	26	Catanduva	36	Andradina
7	Itapetininga	17	Ourinhos	27	Jaboticabal	37	General Salgado
8	Sorocaba	18	Bauru	28	Barretos	38	Jales
9	Campinas	19	Jaú	29	Orlândia	39	Fernandópolis
10	Bragança Paulista	20	Araraquara	30	Presidente Prudente	40	Votuporanga

Paradoxically, while there is a plethora of governmental data available which should not be left unused, their collecting efforts were not ecologically guided and are not ideal for this study. Some minor species, for instance, appear grouped under collective categories such as "other vegetables", "other forestry species", "other ornamentals", "*Brachiaria*" (a genus composed of several species) and so on, generally occupying small areas, except for *Brachiaria*. There are two solutions to this problem and neither is optimal: either omitting such data or treating them as a single species. The latter was selected here, because, although this still introduces a slight error in the calculation of both S and *H'*, no gap is left in the land cover, which is our target after all. "It is better to have an approximate answer to the right question than an exact answer to the wrong question", as John Tukey, one of the genial founding fathers of statistics, once wrote (Salsburg, 2001).

7.2.2 Results and Discussion

7.2.2.1 Species Diversity and Richness at the National, Regional and State Levels

At the national level, the two diversity components, species richness and the relative Shannon index (*H'rel*) varied in opposite directions over the two decades from 1990 to 2010 (Fig. 7.1A and 7.2A). *H'rel* dropped about 10% while the crop list had an increase of two species, after some oscillation. This means that the evenness in planted areas diminished because some crops had their area greatly increased while others suffered a deep reduction in area. The crop species mostly responsible for this variation were soybeans, sugarcane, maize, sorghum, persimmon, tobacco, cashew nuts, rubber, mate and sunflower, on the increasing side; and common beans, rice, coffee, cotton, wheat, cassava, castor beans, cocoa and orange, on the decreasing side. The overall national indices, however, are a complex composition of the regional tendencies, and some regions have a greater impact than others. Results for the five regions are presented here in order to put the Southeast in context. All the regions except the Northeast experienced a species enrichment of three or four crops. The distribution of areas became more uneven in the three regions with more advanced agriculture, *viz.*, Southeast, South and Central-West, where some major commodity crops such as sugarcane, soybeans and forage crops (not a commodity in themselves, but associated with the meat agroindustry) suffered an expansion accompanied by a relative shrinkage of minor crops (Fig. 7.1 D, E, F). The Northeast suffered a strong oscillation, especially before 2002, and the North presented a V-shaped graphic, with a depression in *H'rel* by the middle of the period (Fig. 7.1 B, C). The greatest decline in *H'rel* among the regions was found in the Southeast (about 0.10, roughly twice as much as that in the Central-West and the South).

Going down a level further, the declining *H'rel* values of the Southeast were shown to result from the sum of a declining trend in all of its four states, but São Paulo had the steepest reduction (Fig. 7.3) and the only one not presenting any relevant oscillation in the period.

The rise in the species richness in São Paulo, Rio de Janeiro and Minas Gerais was slight, and so was the decline in Espírito Santo. In relative terms, changes in S were negligible (Fig. 7.4), with some oscillation occurring due to reduction or increase in the area of minor crops from year to year. The explanation for the total S increase of 4 species in the Southeast is that different crops were added to each state.

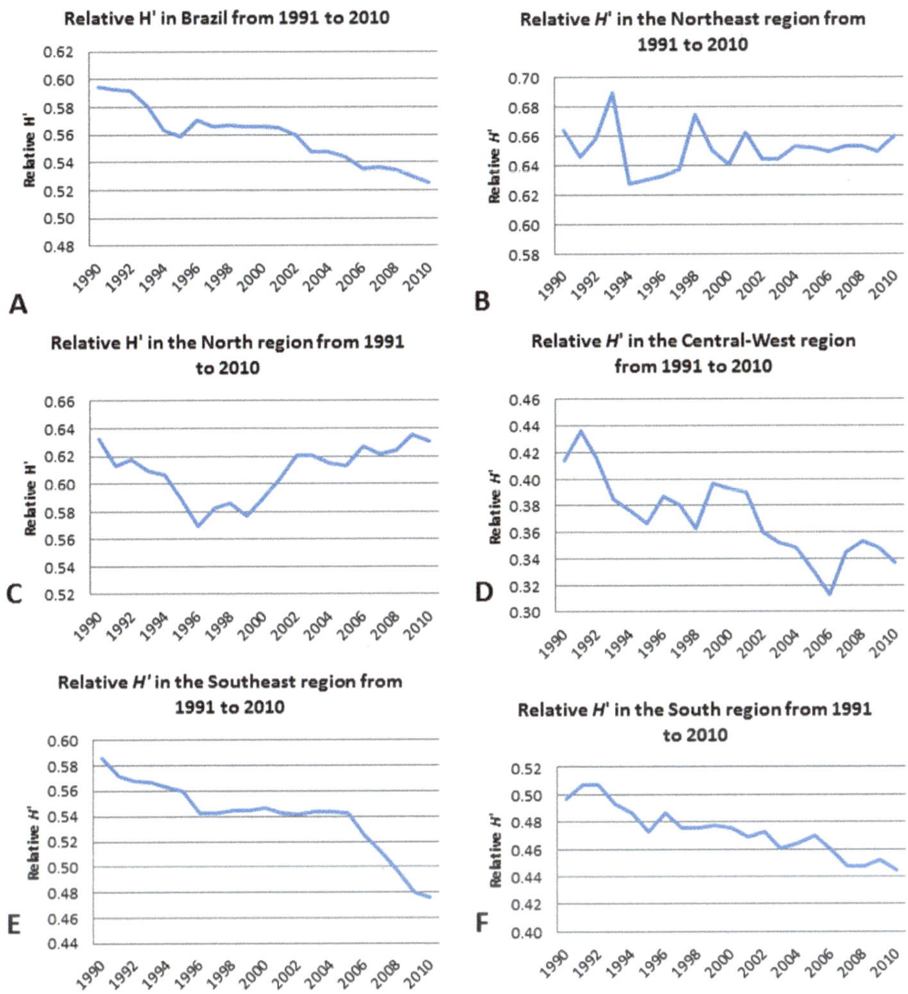

Fig. 7.1: Relative *H'* values for the IBGE annual data on crop area distribution (1991-2010) in (A) Brazil; (B) Northeast; (C) North; (D) Central-West; (E) Southeast; (F) South. Relative *H'* was calculated by dividing absolute *H'* by logS.

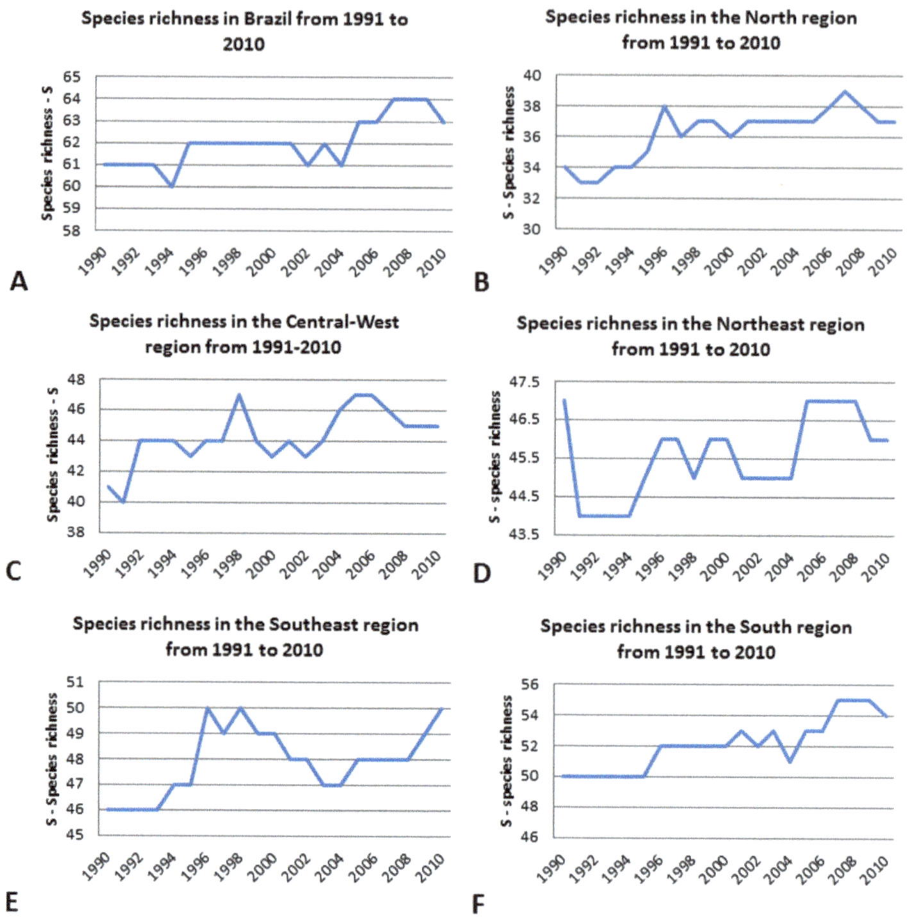

Fig. 7.2: Species richness (S) values for the IBGE annual data on crop area distribution (1991-2010) in (A) Brazil; (B) Northeast; (C) North; (D) Central-West; (E) Southeast; (F) South.

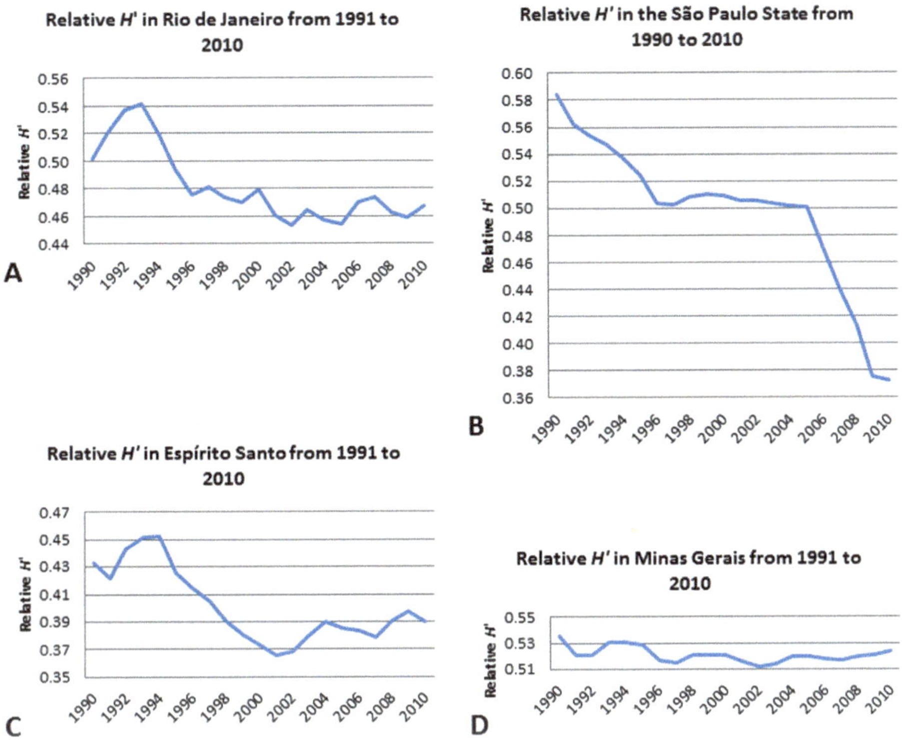

Fig. 7.3: Relative *H'* values for the IBGE annual data on crop area distribution (1991-2010) in (A) Rio de Janeiro; (B) São Paulo; (C) Espírito Santo; (D) Minas Gerais. Relative *H'* was calculated dividing absolute *H'* by log S.

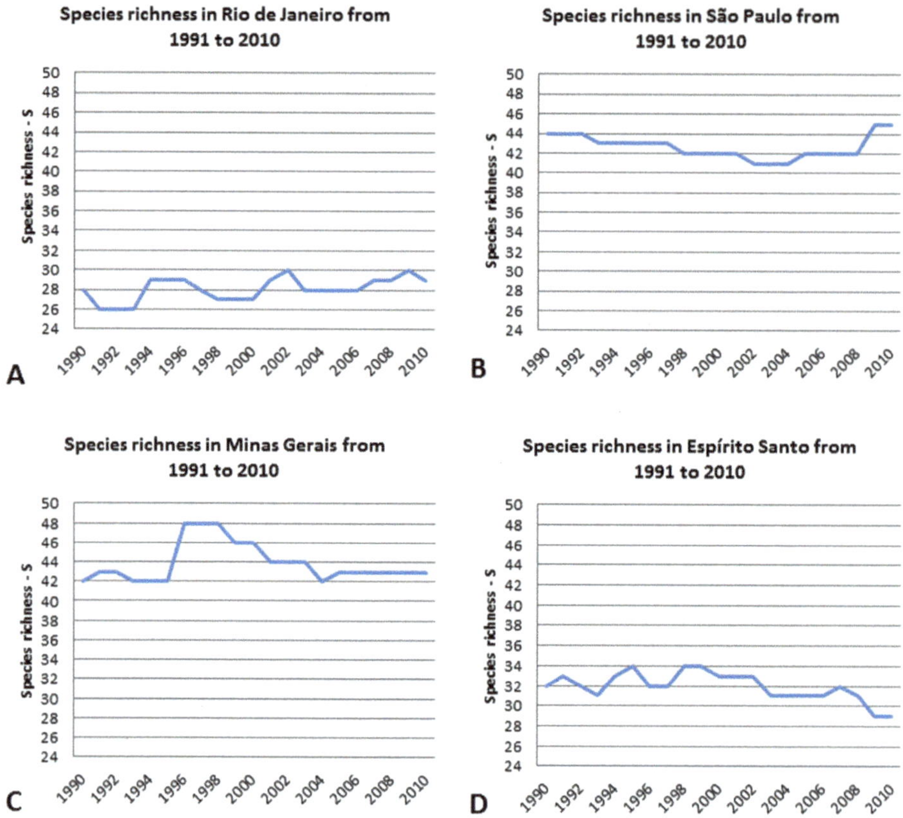

Fig. 7.4: Species richness (S) values for the IBGE annual data on crop area distribution (1990-2010) in (A) Rio de Janeiro; (B) São Paulo; (C) Minas Gerais; (D) Espírito Santo.

7.2.2.2 Species Diversity and Richness at the State, Agricultural Region and Municipality Level in São Paulo State

The agricultural parameters examined showed considerable micro- (at the level of municipality) and mesogeographic (at the level of agricultural region) variation in the state of São Paulo. The total and mean area planted increased with variable magnitude in most regions of São Paulo (ca. 1.4 million ha in the whole state) and increased more in its southeastern moiety (Fig. 7.5 A, regions 1 to 16). Since this moiety includes the regions containing most of the continuous remnants of original vegetation of the state, this implies that the area growth may have been based on Atlantic Forest land claimed by agriculture. Formerly abandoned land may also have been brought back to cultivation. In the northwestern moiety almost all of the original vegetation has been removed long ago and little was left for further clearing.

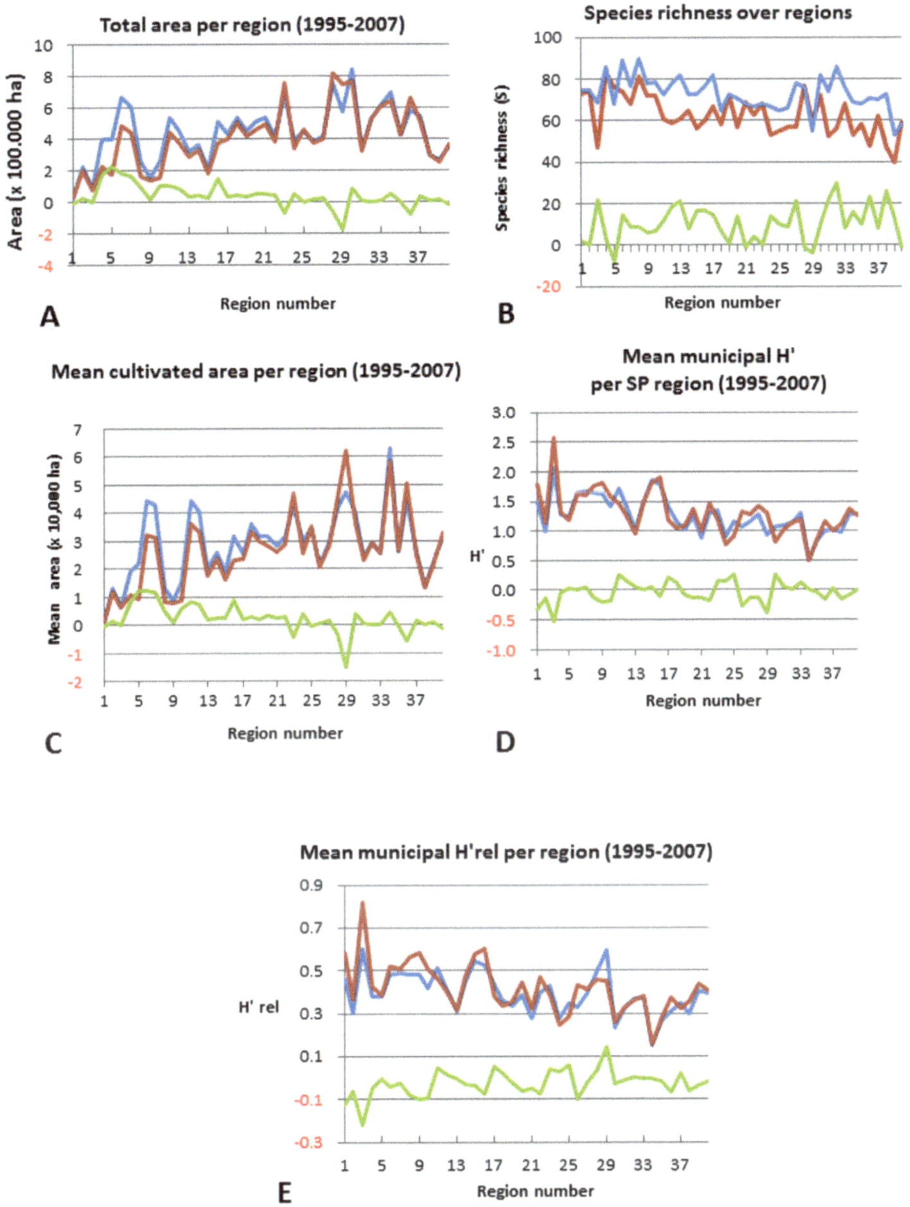

Fig. 7.5: Distribution of descriptive ecological parameters over the 40 official agricultural regions of the State of São Paulo (EDR's). (A) Total cultivated area per region. (B) Species richness (S). (C) Mean municipal cultivated area in hectares. (D) Mean municipal H' values, weighted by municipal total cultivated area. (E) Mean municipal relative H' values, weighted by municipal total cultivated area. For the meaning of the region numbers, see table 7.1. All the graphics show the data from the 1995/1996 (1995 for short, in red) and from the 2007/2008 (2007 for short, in blue) surveys, as well as the variation from 1995 to 2007 (in green).

Both temporal and geographic variation can be observed in species richness in São Paulo (Fig. 7.5 B). S values increased in most regions, up to *ca.* 50% in some. It is noteworthy that more species are cultivated nowadays in the regions near the coast (ca. 75-90) and this number steadily declines northwestwards until it reaches minimum figures of 53, 55 and 58 in Fernandópolis, Orlândia and Votuporanga, respectively. The coastal regions and the regions on the southeastern side of the Peripherical Depression have a higher proportion of small farms and vegetable gardens, which include the Green Belt around the Greater São Paulo and normally are richer in species, and are associated with the high demand of vegetables by the megalopolis.

The high heterogeneity in *H'* and *H'rel* across the 40 regions of São Paulo demonstrates the need to analyze a grain finer than the state level in order to obtain a more detailed picture of the complexity beneath more comprehensive indices. There was a high level of congruence between *H'* and *H'rel* values, so that treating only *H'rel* in the discussion will suffice. There is an intriguing spatial segregation of diversity trends across the state. Four combinations of *H'rel* and S are possible but two of them were found only once each. Most regions displayed either increase in both *H'rel* and S or simultaneous decrease in *H'rel* and increase in S (Fig. 7.6). The regions with an increasing trend in *H'rel* and S are located mainly in what could be called the Central-West of São Paulo (merging parts of the Central and West agricultural macroregions, according to the classification of Camargo et al. (2008), from Botucatu to Araçatuba). This is caused by a more even distribution of land among the crops, rather than by augmenting the number of species. The area with decreasing evenness, measured by *H'rel*, spreads around the Central-West, in the remainder of the state, except for Orlândia and Barretos. Some regions suffered a particularly strong decrease in *H'rel*, and they are concentrated in the Greater São Paulo and the regions around it (Fig. 7.6). The explanation for this decrease is the increase in the number of crops with smaller planted areas in contrast to a small number of crops present in larger areas (e.g. eucalyptus, banana, *Brachiaria*, "other forage grasses" and lettuce in Region 1).

Although they are not fully accountable for the changes in crop cover diversity in SP, the recent expansion of sugarcane plantations and the retraction of the area planted to citrus can be considered key factors (Olivette et al., 2010). These authors studied changes in area planted to sugarcane caused by recent economic stimuli and divided the municipalities into four groups, according to the percent increase in that area. Group 1 includes mostly municipalities in the West and Central regions, many of which have had a decline in *H'rel*, according to our study. Sugarcane has been replacing grains and pastures in this group – in the Jaú EDR, for instance, it replaced cultivated pastures and maize (Camargo et al., 2010) and, because of a government's technical recommendation, an expansion in the rubber tree plantations has been stimulated by sugarcane. Sugarcane had a less expressive expansion in Group 2, accompanied by an increase in eucalyptus cover and a decrease in pastures, which may explain the reduction in *H'rel* in the municipalities of the West, East and Southwest belonging to this group. In the municipalities of Group 3 (mainly in the North and Central regions), where sugarcane cultivation is traditional and already dominant, there was an expansion in the area covered by sugarcane, with a correspondent retraction of citrus (for instance, orange and tangerine in the Limeira EDR (Camargo et al., 2010)), other fruits and coffee, which are associated with the variation of *H'rel* in EDR's such as Araraquara, Catanduva, Limeira, São José do Rio Preto and Araçatuba. Changes in sugarcane cover did not influence Group 4 municipalities, mainly in the East, Southwest, Central and South.

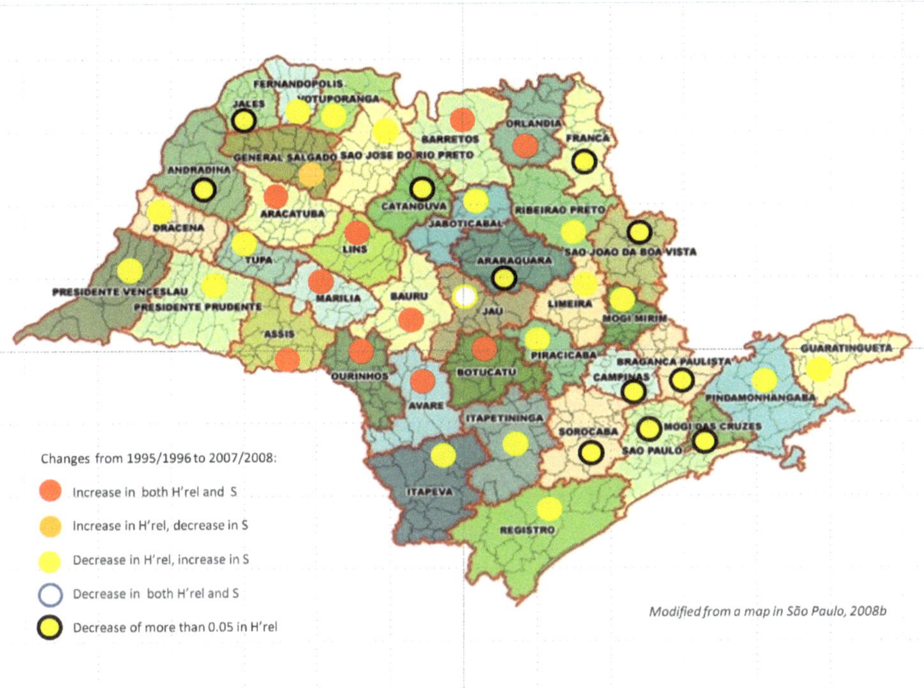

Fig. 7.6: Geographical distribution of crop diversity change in terms of combinations of *H'rel* and S across the 40 agricultural regions (EDR's) of São Paulo from the 1995/1996 LUPA to the 2007/2009 LUPA. S increased in all regions with a yellow circle, irrespective of the thickness of the contour.

Correlation coefficients varied considerably among geographic units. For most of the larger units, correlation was moderate and positive (0.662 for Brazil; 0.694 for the North; 0.634 for the Central-West; 0.563 for the South), meaning that the number of crop species *increased* with the increase in area cultivated. The Northeast had a small and positive correlation (0.121). Area and species richness changes were practically disconnected (r = -0.077) in the Southeast as a whole, but this figure disguises a strong heterogeneity for this parameter among the states of the region (+0.336 for SP, -0.604

for RJ, +0.166 for ES and -0.820 for MG). This means that in Rio de Janeiro and Minas Gerais, as cultivated area changed in one direction, the number of species changed in the opposite direction. The explanation lies in the expansion of the major crops on new tracts and on land previously planted to minor crops. Quite the opposite happened in Espírito Santo, where a reduction from 0.88 million ha to 0.67 million ha was accompanied by a reduction of 3 crops from 1990 to 2010, and São Paulo, where an increase in cultivated land from 6.3 million ha to 7.7 million ha was associated with an increase of one crop species from 1990 to 2010. For the LUPA-based analyses in São Paulo, a different interpretation must be given to the correlation coefficients (-0.236 in 1995 and -0.187 in 2007). In this case, *in the same year*, the larger the cultivated land area in an agricultural region (EDR), the fewer crop species were planted. The values were small, though, and would not permit very accurate predictions of S based on area.

The relationship between area (A) and species richness (S) is one of the major concerns of community ecology. Most studies agree that there is a positive correlation between A and S in natural ecosystems (e.g. Begon et al., 2006; Poole, 1974). A debate has been held in the realm of Island Biogeography about the relative importance of the heterogeneity of habitat and the inherent effect of island dimension by itself in determining the impact of the area factor. The number of habitat types tends to increase with area because of the greater probability of occurrence of different altitudes, soil types, rivers, microclimates, providing conditions for the colonization of the island by a variety of species adapted to each of them. Large islands, irrespective of the number of habitats, are a larger target for immigration of individuals of new species from the continent and from nearby islands, and their populations are less prone to random extinction. Distances among islands and between an island and the continent diminish the migration rates and interact with island size. Habitat heterogeneity and the inherent island size effect can be told apart through statistical procedures and the latter has often been proved to be important (Begon et al., 2006). "Island" is a broad concept and can be applied to any kind of patch, including agricultural fields, but Island Biogeography must suffer adaptations to be applied to agroecosystems because the migration factor, at least within national limits, is negligible – seeds and other propagules can be easily transported by humans among regions irrespective of distance. A second peculiarity of the agroecosystems is that humans can modify extensively the environmental variables of crop fields through fertilization, irrigation, pesticide application, controlling crop cycles and rotating crops, among other procedures that partially neutralize habitat heterogeneity. Crops can also be modified genetically and new cultivars can become adapted to new environments in ecological time. Humans decide which crops are planted and how their artificial demography is set up. All these factors weaken considerably the applicability of Island Biogeography to agroecosystems and can explain the low S × A correlation coefficients, especially in micro- and mesogeographical scales such as those of the LUPA data. At the nation and macroregional level, such as those of the Brazil, North, Southeast, Central-West and South data, habitat heterogeneity can still impose problems that agronomic techniques are not yet able to tackle, so that wheat is not planted in the Amazon, and São Paulo is the southernmost limit for the fructification of cashew. The negative, if very low, correlations shown in the LUPA data of São Paulo can be explained by the fact that S declines along the southeast-northwest direction and cultivated land area increases in the same direction (graphic not shown). Municipalities located in the southeastern portion of the state are either highly urbanized or rich in protected areas located mainly in mountainous terrain, while those in the flat plateaus of the north/northwest have more land available to agriculture, which explains the difference in area.

7.3 Historical Diversity Variation in Swidden Agriculture of São Paulo

As we noticed in section 7.1, agricultural practices affect and deteriorate the original ecosystems in several degrees. On one end is highly technological agriculture, generally producing commodities such as soybeans and sugarcane, with high disturbing power. On the other end is swidden agriculture, also known as shifting or slash-and-burn agriculture, which provokes a paradoxically much lower impact on the forests. Under swidden agriculture, plots of forest are slashed and burned, which opens room for cultivation and adds mineral nutrients to the soil. After a few years, when soil fertility has been exhausted, the plots (called "*roças*" in Brazil) are abandoned for some years (the fallow) and cultivation shifts to other plots. During the fallow (*alqueive*, in Portuguese), the soil recovers its fertility and the forest regenerates, and the plot is again ready for slashing and burning, closing the cycle.

Swidden agriculture, as it is observed in tropical humid areas around the world, has some special differences in relation to commercial, technified agriculture. Firstly, the plots cultivated are very small, and many of them are below the size treshold surveyed by LUPA, for instance, so that the yield of some crops is underestimated and some species are sometimes absent from the crop list of a region (Tab. 7.2). Secondly, most of the swidden production is intended for subsistence of familiar groups of agriculturists, being only partially marketed. Thirdly, in striking opposition to large-scale agriculture, swidden farmers have a cultural penchant to maintaining both intraspecific genetic/phenotypic variation and interspecific diversity within their plots. Different landraces (often labeled "ethnovarieties", even when the farmers do not form any ethnic group distinguishable from the surrounding population) are used for different purposes – some cassavas are better for flour, others for cooking, and some yams are preferred for soups rather than for flour, for instance.

Tab. 7.2: Area (ha) used for the cultivation of cassava, yams and sweet potato in five municipalities of the Registro agricultural region and in four municipalities of the Pindamonhangaba agricultural region (both in São Paulo State) according to the LUPA 1995/1996 and 2007/2008 surveys. Data from São Paulo (2008a).

Region	Crop								
	Cassava			Yams			Sweet potato		
	1995	2007	Δ	1995	2007	Δ	1995	2007	Δ
Registro Region									
Iguape	194.5	148.1	-46.4	-	-	-	-	-	-
Cananeia	12.9	46.7	33.8	-	-	-	-	-	-
Iporanga	33.6	21.6	-12.0	-	-	-	1.2	1.1	-0.1
Ilha Comprida	-	-	-	-	-	-	-	-	-
Eldorado	36.1	169.3	133.2	-	-	-	-	-	-
Pindamonhangaba Region									
Ubatuba	69.1	207.7	138.6	18.2	24.3	6.1	-	-	-
Caraguatatuba	18.7	15	-3.7	-	2.4	2.4	-	-	-
São Luís do Paraitinga	39.7	3.1	-36.6	-	-	-	-	-	-
Ilha Bela	7.6	26.7	19.1	-	-	-	0.7	-	-0.7

In Brazil, swidden agriculture is practiced typically by indigenous-european more or less mixed populations, called "*caboclos*" in Amazonia and "*caiçaras*" on the coastal communities, mainly in the Southeast. In the Amazon, the plots are located normally near the rivers and lakes and in the Southeast they are near the coast and rivers. "*Quilombolas*", descendants of escaped African slaves living in understandably remote communities, also practice swidden agriculture. Swidden agriculture has been the object of an increasing number of anthropological, ethnobotanic, agronomic and genetic studies since the 1990's and many features common to both Amazonian and Southeastern "*roças*" have been found (Martins & Oliveira, 2009).

In the "*roças*" species richness is beneficial and consciously fostered. Plant species of different architectures are put together so that roots of different depths explore the soil water and nutrients more evenly and shoots do not compete strongly for light. Diversity also reduces the spread of disease and insect infestation. Nevertheless, the crop assemblage is far from even. *H'*, had it ever been estimated, would be low, for cassava is predominant (Peroni & Hanazaki, 2002). About 40 crop species are common in *caiçara* "*roças*", and subterranean organs are the three top elements: cassava (*Manihot esculenta*), yams (*Dioscorea* spp.) and sweet potato (*Ipomoea batatas*) (Siqueira and Veasey, 2009). Next are rice, banana, sugarcane, common beans, squash, *taiá* (*Xanthosoma sagittifolium*; Araceae), maize, mangarito (*Xanthosoma mafaffa*.), cucumber, tomato, pineapple, watermelon and arrowroot, and others with lesser yields.

Genetic studies have concentrated on cassava, yams and sweet potato (Bressan et al., 2011, 2014; Siqueira et al., 2012, 2014; Nascimento et al., 2013; Peroni et al.,

2007). The genetic composition and structure of the populations of these species present in *roças* have been assessed with different markers, including morphology (Mezzette et al., 2013), microsatellites (Siqueira et al., 2009; Siqueira et al., 2011; Elias et al., 2004), AFLP's (Mühlen et al., 2000) and isozymes (Faraldo et al., 2000). Genetic partition analyses, such as AMOVA (Analysis of Molecular Variance), for instance, usually find most of the variation within *roças* and very little variation among *roças*. The reason suggested by the interviews with the agriculturists is not only the already cited cultural interest in variation *per se* but also the habitual interchanges of germplasm among relatives and friends (e.g. Elias et al., 2004), sometimes living in other municipalities. No - or very little (Veasey et al., 2008, in sweet potato) - correlation between geographic and genetic distances was detected, however, perhaps because of long-distance interchanges and erroneous assumptions of linear geographic distances among the communities in the analysis. A normally huge amount of genetic variation is found in *roças*, and more so in those owned by the agriculturist. Variation is measured in terms both of population genetics parameters and number of phenotypically distinct varieties.

All the microevolutionary processes, such as mutation (crops with sexual/asexual mixed reproductive system, such as cassava, sweet potato and yams, accumulate many mutations), gene flow, introgression from wild relatives (when present), selection, hybridization between varieties, drift and their interactions with many different reproductive systems occur in the *roças* (Cury, 1993; Sambatti et al., 2001). Farmers do not use seeds for propagating sweet potato, cassava or yams, and most are not even aware of the existence of seeds, but natural crosses occur and do produce seeds (da Silva et al., 2003). Intervarietal hybrids are sometimes found in *roças* and are recognized as new varieties by the farmers. The swidden system helps to create and keep allelic and genotypic variation and applies divergent selection in several directions, so that genetic uniformization is avoided. A striking contrast between the levels of intraspecific variation of traditional swidden varieties and commercial cultivars can be seen in the sweet potatoes collected by our current project in municipalities of the Registro and Pindamonhangaba Agricultural Regions of the State of São Paulo (Fig. 7.7) and in commercial sweet potatoes for sale in a supermarket (Fig. 7.8).

Inspite of all the data produced and analyzed in the last two decades, only gross estimates of changes in species diversity and genetic variation in the swidden system of the coastal plains of São Paulo can be produced. One of the reasons is the methodology used to collect data about the so-called neglected crops in projects like LUPA. At the scale of *roça*, the estimates of area, yield and species richness are not reliable. For both the 1995/1996 and the 2007/2008 LUPA surveys, for instance, there is no recorded area used for yam production in Iguape, Cananeia, Iporanga, Ilha Comprida and Eldorado, places where yams are known to be produced in *roças* (Tab. 7.2). Likewise, sweet potato is also known to be produced in *roças* of all these municipalities plus Ubatuba but LUPA records exist only for Iporanga.

Fig. 7.7: Whole tuberous root and respective transversal section phenotype of 28 out of the ca. 80 sweet potato (*Ipomoea batatas*) landraces collected in the Ribeira Valley and the North Coast, both in the State of São Paulo, Brazil. There is a huge variation in shape, skin color, pulp color and skin texture. A large amount of leaf variation is also present (not shown). Photos courtesy of Hendrie F. Nunes and Gláucia B.R. Moreira.

Fig. 7.8: The only two sweet potato varieties available in a supermarket in Piracicaba, SP, Brazil. The genetic diversity in large-scale systems is just a minute fraction of that conserved by swidden farmers. Photos courtesy of Hendrie F. Nunes and Gláucia B.R. Moreira.

The genetic and ethnobotanic studies performed so far were innovational and disclosed a large amount of basic information about the *roças*, but they described diversity at a particular time in history. Studies following the genetic and floristic composition of the same *roças* through a period long enough to permit the detection of significant change are lacking. Indirect information on landrace loss over time was retrieved, however, from interviews in which the agriculturists listed the landraces they had stopped planting for some reason (Peroni & Hanazaki, 2002). A Loss Index value of 30.6 over 50 species (Peroni & Hanazaki, 2002) indicate that many local variety extinctions have occurred in the last 30 years. Among the main reasons alleged to explain the current species and genetic erosion are the restrictions imposed by environmental laws, which prohibit the slashing of forests; the attraction of the new generations of potential agriculturists by urban activities, tourism and fishing; the migration of rural community dwellers to cities; and an increasing market dependence, which causes genetic homogenization (Peroni & Hanazaki, 2002). Our ongoing research in the Ribeira Valley and the northern coast of São Paulo reveals that a large amount of genetic variation with phenotypic expression is still present in sweet potatoes produced in traditional small-holding agriculture (Fig. 7.7). However, the gradual collapse of this system is undeniable. Many of the agriculturists visited in the early 2000's (Veasey et al., 2008) and visited again in 2012 still cultivate many of the traditional crops, but now the area planted is restricted to small home gardens instead of more distant, larger *roças*, which were abandoned (Hendrie F. Nunes & Gláucia B.R. Moreira, personal communication). The reduction in size of the planted field has an effect not only on the number of species but also on the number of individuals of each species that can be cultivated. The lower the population size, the fewer varieties can be conserved on-farm (unfortunately, literature has adopted the use of "on-farm", instead of "*in fundo*", which would be preferable for uniformity's sake, since the equally Latin locutions "*in situ*" and "*ex situ*" have been largely established in the conservation nomenclature). Indeed, many of the agriculturists keep only one or two varieties of sweet potato, introducing a local bottleneck and preventing much of the former potential for recombination. Moreover, the interruption of the slash-and-burn cycles means that the natural fertilization of the soil is no longer available and the varieties selected for the home gardens are only those adapted to low-fertility. The expectation of many of the researchers working on swidden agriculture in São Paulo is that within a few decades this system will be practically extinct.

7.4 Final Considerations

All components of agriculture are in constant change. The internal dynamic nature of the farming activities as well as its incessant impact on surrounding natural ecosystems can be assessed in many ways and by many indicators, and one of the most telling is the disconnection between the current main crops produced in certain municipalities and their eponym crops or wild plant or animal species, which were once abundant locally: jaboticabas, potatoes, *indaiá* palms, *buriti* palms, *araçás*, bamboos, *pitangas*, cedar and coffee are not as easy to find as they used to be in Jaboticabal, Batatais, Indaiatuba, Buritizal, Araçatuba, Taquaritinga, Pitangueiras, Cedral and both Cafelândia and Rubiácea, and macaws and monkeys are rare in Araras and Quatá. In the last few decades, crop species diversity, as measured by the relative Shannon-Wiener index, or evenness, declined in Brazil, and particularly in São Paulo, as a result of an increasing disequilibrium among the most and the least extensively cultivated plants. The changes in diversity, though, are structured spatially, and some municipalities, EDR's, states and regions do not follow the trends of the geographic units hierarchically placed above them. Species richness has not changed conspicuously in the country, including São Paulo, although the consumption of some regional species, such as Amazonian fruits, have become more common outside its traditional states. In São Paulo, a few trends were identified. Commercial agriculture has become more diverse in the Central-West Region of the state, and less diverse in the remaining regions, more acutely around the Greater São Paulo. Some commodities, led by sugarcane, have received stimuli for expansion, claiming land from other crops. In the coastal regions, where some neglected crops, such as cassava, yams and sweet potato, are cultivated, social pressures such as urbanization, other more attractive economic activities and environmental laws are imperiling the maintenance of the swidden system and the conservation of the germplasm associated to it. In the near future, this germplasm will probably be kept only in *ex situ* collections.

Diversity indexes, as the Shannon-Wiener index, were showed to be useful to depict a synthetic panorama of the changes in the unevenness of land occupation by the several crop components of the agrobiome, complementing the more common monospecific historic studies on the distribution of planted area. Unevenness, though, should not be interpreted in a hasty, simplistic way. Some degree of unevenness is inevitable and even desirable in a spatially optimized agricultural region, because different species are needed in different quantities by humans. It is not to be expected that parsley or garlic, which are used primarily as seasoning, be planted in areas equivalent to those planted to sugarcane or eucalyptus, which are associated to the energy and paper industries. But extreme imbalance is generally a symptom of short-sighted planning or lack thereof. The recent importation of massive amounts of products widely utilized by Brazilians, such as tomatoes and rice, for instance, indicates a lack of equilibrium in the distribution of relative areas planted to the major crops of the country, a situation that should be corrected.

Acknowledgements: I thank Hendrie F. Nunes and Gláucia B. R. Moreira for kindly providing the sweet potato pictures and information from their ongoing projects on swidden crop diversity; Eduardo de A. Bressan for revising the text and expunging it from many errors, as well as for suggesting several improvements; Carlos R. Macedônio for having helped format the figures with his computer wizardry; and CAPES (the Ministry of Education of Brazil) and CNPq (the Ministry of Science, Technology and Innovation of Brazil) for financial support.

Bibliography

Begon, M., Townsend, C.R., & Harper, J.L. (2006). Ecology. 4th ed. Oxford: Blackwell Publishing Ltd.

Bressan, E.A., Briner Neto, T., Zucchi, M.I., et al. (2014). Genetic structure and diversity in the *Dioscorea cayenensis/D. rotundata* complex revealed by morphological and isozyme markers. Genetics and Molecular Research, 13 (1), 425-437.

Bressan, E.A., Briner Neto, T., Zucchi, M.I., et al. (2011). Morphological variation and isozyme diversity in *Dioscorea alata* L. landraces from Vale do Ribeira, Brazil. Scientia Agricola, 68(4), 494-502.

de Camargo, A.M.M.P., Caser, D.V., de Camargo, F.P., et al. (2008). Dinâmica e tendência da expansão da cana-de-açúcar sobre as demais atividades agropecuárias, Estado de São Paulo, 2001-2006. Informações Econômicas, 38 (3), 47-66.

Cury, R. (1993). Evolutionary dynamics and characterization of cassava germplasm *(Manihot esculenta* Crantz) in the autochthonous agriculture of southern São Paulo State [Dinâmica

evolutiva e caracterização de germoplasma de mandioca *(Manihot esculenta* Crantz), na agricultura autóctone do Sul do Estado de São Paulo]. Piracicaba: Escola Superior de Agricultura "Luiz de Queiroz", University of São Paulo (In Portuguese).
Elias, M., Muhlen, G.S., Mckey, D., et al. (2004). Genetic diversity of traditional South American landraces of cassava (*Manihot esculenta* Crantz): an analysis using microsatellites. Economic Botany, 58 (2), 242-256.
Faraldo, M.I.F., da Silva, R.M., Ando, A., et al. (2000). Variabilidade genética de etnovariedades de mandioca em regiões geográficas do Brasil. Scientia Agricola, 57 (3), 499-505.
IBGE (2008). Séries históricas e estatísticas. http://seriesestatisticas.ibge.gov.br/lista _tema.aspx?op=0&de=83&no=1, Accessed on 30/03/2014.
Kloppenburg, J.R. (2004). First the seeds. The political economy of plant biotechnology. Madison, Wisconsin: The University of Wisconsin Press.
Martins, P.S., & Oliveira, G.C.X. (2009). Dinâmica evolutiva em *roças* de caboclos amazônicos. In I.C.G. Vieira, J.M.C. Silva, D. Oren, et al. (Eds.), Diversidade biológica e cultural da Amazônia (p.373-391). Belém: Museu Paraense Emílio Goeldi.
Mühlen, G.S., Martins, P.S., & Ando, A. (2000). Variabilidade genética de etnovariedades de mandioca, avaliada por marcadores de DNA. Scientia Agricola, 57 (2), 319-328.
Mezette, T.F., Blumer, C.G., & Veasey, E.A. (2013). Morphological and molecular diversity among cassava genotypes. Pesquisa Agropecuária Brasileira, 48 (5), 510-518.
Nascimento, W.F., Rodrigues, J.F., Koehler, S., et al. (2013). Spatially structured genetic diversity of the Amerindian yam (*Dioscorea trifida* L.) assessed by SSR and ISSR markers in Southern Brazil. Genetic Resources and Crop Evolution, 60, 2405–2420.
Olivette, M.P.A, Nachiluk, K., & Francisco, V.L.F.S. (2010). Análise comparativa da área plantada com cana-de-açúcar frente aos principais grupos de culturas nos municípios paulistas, 1996-2008. Informações Econômicas, 40 (2), 42-59.
Peroni, N., & Hanazaki, N. (2002). Current and lost diversity of cultivated varieties, especially cassava, under swidden cultivation systems in the Brazilian Atlantic Forest. Agriculture, Ecosystems and Environment, 92, 171–183.
Peroni, N., Kageyama, P.Y., & Begossi, A. (2007). Molecular differentiation, diversity, and folk classification of "sweet" and "bitter" cassava (*Manihot esculenta*) in Caiçara and Caboclo management systems (Brazil). Genetic Resources and Crop Evolution, 54, 1333–1349.
Poole, R.W. (1974). An introduction to quantitative ecology. New York, NY: Mcgraw-Hill, Inc.
Priolli, R.H.G., Mendes-Junior, C.T., Arantes, N.E., et al. (2002). Characterization of Brazilian soybean cultivars using microsatellite markers. Genetics and Molecular Biology, 25 (2), 185-193.
Salsburg, D. (2001). The lady tasting tea: how statistics revolutionized science in the twentieth century. New York, NY: Owl Books.
Sambatti, J.B.M., Martins, P. S., & Ando, A. (2001). Folk taxonomy and evolutionary dynamics of cassava: a case study in Ubatuba, Brazil. Economic Botany, 55 (1), 93-105. New York, NY: Springer & New York Bot. Garden Press.
São Paulo (2008a). Levantamento censitário de unidades de produção agrícola do Estado de São Paulo - LUPA 2007/ 2008. São Paulo, SP: Secretaria de Agricultura e Abastecimento. Coordenadoria de Assistência Técnica Integral. Instituto de Economia Agrícola (SAA/CATI/IEA). Available at: http://www. cati.sp.gov.br/projetolupa, Access on 30/03/2014.
São Paulo (2008b). Levantamento censitário de unidades de produção agrícola do Estado de São Paulo - LUPA 2007/ 2008. São Paulo, SP: Secretaria de Agricultura e Abastecimento. Coordenadoria de Assistência Técnica Integral. Instituto de Economia Agrícola (SAA/CATI/IEA). Site of the map showing the 40 regions of the State of São Paulo, Projeto LUPA. Available at http://www.cati.sp.gov.br/projetolupa/dadosregionais.php, Accessed on 10/04/2014.
da Silva, R.M., Bandel, G., & Martins, P.S. (2003). Mating system in an experimental garden composed of cassava (*Manihot esculenta* Crantz) ethnovarieties. Euphytica, 134, 127–135.

Siqueira, M.V.B.M., Bonatelli, M.L., Günther, T., et al. (2014). Water yam (*Dioscorea alata* L.) diversity pattern in Brazil: an analysis with SSR and morphological markers. Genetic Resources and Crop Evolution, 61, 611–624.

Siqueira, M.V.B.M., Borges, A., Valle, T.L., et al. (2011). A comparative genetic diversity assessment of industrial and household Brazilian cassava varieties using SSR markers. Bragantia, 70 (4), 745-752.

Siqueira, M.V.B.M., Dequigiovanni, G., Corazon-Guivin, M.A., et al. (2012). DNA fingerprinting of water yam (*Dioscorea alata*) cultivars in Brazil based on microsatellite markers. Horticultura Brasileira, 30, 653-659.

Siqueira, M.V.B.M., Queiroz-Silva, J.R., Bressan, E.A., et al. (2009). Genetic characterization of cassava (*Manihot esculenta*) landraces in Brazil assessed with simple sequence repeats. Genetics and Molecular Biology, 32 (1), 104-110.

Siqueira, M.V.B.M., & Veasey, E.A. (2009). Raíces y tubérculos tropicales olvidados o subutilizados en Brasil. Revista Colombiana de Ciencias Hortícolas, 3 (1), 110-125.

Veasey, E.A., Borges, A., Rosa, M.S., et al. (2008). Genetic diversity in Brazilian sweet potato (*Ipomoea batatas* (L.) Lam., Solanales, Convolvulaceae) landraces assessed with microsatellite markers. Genetics and Molecular Biology, 31 (3), 725-733.

Cristina Yuri Vidal, Julia Raquel Mangueira, Fabiano Turini Farah, Débora Cristina Rother, Ricardo Ribeiro Rodrigues

8 Biodiversity Conservation of Forests and their Ecological Restoration in Highly-modified Landscapes

8.1 Introduction

Tropical forests contain between one-half and two-thirds of the Earth's terrestrial biodiversity (Gardner et al., 2010) but remain subject to increasingly severe deforestation and forest degradation. In regions where deforestation occurred long ago, conversion into agricultural use or other human activity has typically resulted in highly fragmented landscapes (Gardner et al., 2009) where the reduction in forest habitat has had a drastic effect on biodiversity loss. The consequences of this loss towards global environmental changes remain unclear (Sutherland et al., 2013).

In the Brazilian Atlantic Forest most of the forest cover (> 80%) exists as small fragments (< 50 ha) (Ribeiro et al., 2009). A few large forest remnants are concentrated in some parts of the original Atlantic Forest zone, but most of what is left occurs within small fragmented remnants.

Some tropical ecosystems have lost so much of their forest cover that they have already attained the theoretical fragmentation threshold, a concept describing the point at which species extinction rates dramatically increase due to the collapse of both core habitats and the structural connectivity of the landscape (Fahrig, 2003; Ribeiro et al., 2009; Lôbo et al., 2011; Estavillo et al., 2013).

Although it is hard to extrapolate a more general biodiversity threshold for tropical forests (Estavillo et al., 2013), some studies based on fauna communities have concluded that landscapes with forest cover below 20-30% present little hope for conservation efforts and a low effectiveness for ecological restoration projects (Estavillo et al., 2013; Tambosi et al., 2013). However, evaluation of plant communities within these highly-deforested landscapes demonstrates that tropical forest remnants still hold high floristic diversity (Rodrigues & Nave, 2004; Kotcheftkoff-Henriques et al., 2005; Sabino, 2011; Mangueira, 2012). Further research into these fragmentation thresholds with the focus on plant communities remains highly necessary.

Despite the dominant idea that only primary forests, large fragments or landscapes with forest cover above 30% are able to consolidate biodiversity conservation (Gibson et al., 2011; Laurance et al., 2012; Estavillo et al., 2013), we believe that even smaller and isolated forests can and do play an important role in the maintenance of regional endemic biodiversity, with unreplaceable value for conservation. Along with ecological restoration initiatives, we propose that a conservation model within highly-modified landscapes should be developed.

The main purpose of this chapter is to present and discuss possible strategies to promote biodiversity conservation via a combination of protection and adaptive management for forest remnants and through complementary ecological restoration efforts.

8.2 Conservation Management of Forest Remnants for Biodiversity

8.2.1 Forest Remnants and their Value for Conservation

It is known that the current worldwide net of protected environmental reserves will be insufficient to effectively compensate for the imminent global climate changes (Laurance et al., 2012; Lemes et al., 2014) and that they are mostly inadequate to ensure the conservation of tropical forest biodiversity (Putz et al., 2001). In the Brazilian Atlantic Forest only 7-9% of the remaining forest cover is currently under some protection regime (Ribeiro et al., 2009; Lemes et al., 2014), reinforcing that we cannot afford to undervalue and neglect the biodiversity that exists outside protected areas (e.g. forest fragments within private lands).

For the purposes of this chapter we consider forest remnants to be all forest habitats that are still present within highly-modified landscapes. These habitats include i) old-growth forest fragments that have never experienced clear-cutting or other severe natural impacts and ii) secondary forests that are regenerating (FAO, 2003) after extraction, fire or the abandonment of croplands and pastures, among other previous land-uses. Highly-modified landscapes are therefore typically an agro-mosaic containing forest fragments of various origin (i.e. old-growth forest and regenerating forests) embedded within agricultural and urban land (Gardner et al., 2009).

Despite this environmental context, some studies have shown that many tropical forest species can tolerate these degraded conditions and persist within secondary tropical fragments (Dent & Wright, 2009; Norden et al., 2009). Several studies exist documenting that small forest fragments or even biodiversity-friendly crops (Ranganathan et al., 2008) are able to retain high levels of biodiversity within several taxa (i.e. birds, mammals, vascular plants) (Hawes et al., 2008; Norden et al., 2009; Sabino, 2011; Thornton et al., 2011; Mangueira, 2012).

In São Paulo countryside Kotchetkoff-Henriques (2005) evaluated the floristic composition of 95 fragments and found a total of 509 tree species belonging to 71 botanic families. In the same state, Mangueira (2012) sampled 225 shrub and tree species within the forest regeneration layer of twelve forest fragments embedded in sugar-cane and pasture matrices. In the same fragments Sabino (2011) sampled 163 canopy shrub and tree species, 11 of them included on the global red-list of endangered species (IUCN, 2011). These studies suggest that even when faced with recurring impacts and degradation, forest fragments within consolidated and intensive agricultural matrices can sustain a high number of species. The processes that maintain such biodiversity on highly degraded forest remnants are complex and remain poorly understood (Gardner et al., 2009).

It is known that some areas around the world are experiencing a change in their land use, whereby abandoned areas are regenerating and enhancing total forest cover (FAO, 2011). This "forest transition" phenomena (Mather, 1992) is also observed in the Atlantic Forest (Farinaci, 2012). In the countryside of São Paulo state Ferraz et al. (2014) found regenerating areas within sugar-cane and pasture landscapes, even after decades of agricultural exploration. Despite the importance of these results and the hopes that they trigger, the authors highlight the fact that whilst these areas are regenerating old-growth forests in other regions are being replaced by secondary forests, representing a concerning overall decrease in habitat quality and ecosystem services.

Although we must recognize that the total biodiversity of pristine forest will never be supported by regenerating forests (Dent & Wright, 2009; Gibson et al., 2011), small forests remnants might play an important conservation role concerning the enhancement of landscape connectivity and the support of certain species, especially those endemic (Turner & Corlett, 1996; Chazdon et al., 2009; Dent & Wright, 2009; Gibson et al., 2013). For this reason conservation planning at a regional level must consider the conservation management of forest fragments in combination with ecological restoration efforts, the aim being an increase in forest cover and an enhancement of the habitat quality within existing fragments.

8.2.2 Main Problems and Challenges Related to Conservation of Fragmented Forest Remnants

The consequences of the ongoing level of habitat loss and of the human disturbance of natural ecosystems have generated discussions within the scientific community. Several studies have addressed the effects of the fragmentation process at a wider ecosystem scale (Metzger, 2001; Ribeiro et al., 2009) and on different taxonomic groups (Metzger, 2000; Tabarelli et al., 2008; Banks-Leite et al., 2010; Pardini et al., 2010; Lôbo et al., 2011; Arroyo-Rodríguez et al., 2013; Estavillo et al., 2013).

Beyond habitat loss, a series of changes in plant communities is related to this fragmentation process and the recurrent degradation. In the Atlantic Forest most remnants are experiencing arrested succession after a range of disturbances (Tabarelli & Gascon, 2005), including severe edge effects, selective logging, recurrent fires, cattle grazing, hunting, biological invasions and pollution (Metzger, 2000; Tabanez & Viana, 2000).

In the short-term these disturbances affect environmental conditions, favouring colonization of the canopy by hyperabundant lianas (Laurance et al., 2006) and by pioneer tree species in habitat edges (Laurance et al., 2006; Lôbo et al., 2011). They also cause an increase in the levels of floristic homogenization (reduced β-diversity) (Arroyo-Rodríguez et al., 2014), invasion by exotic species (Tabarelli et al., 1999; Benítez-Malvido & Martínez-Ramos, 2003) and increase on the vulnerability to extinction in species with low dispersal ability, small population sizes, and specialization to restricted habitats (Henle et al., 2004). Over the long-term, key ecological processes (e.g. pollination, seed dispersal, plant recruitment and animal dispersal throughout the landscape) can be strongly affected by fragmentation (Rozza et al., 2007; Guimarães et al., 2008; Lenz et al., 2011).

The biodiversity conservation potential of fragmented forest remnants requires large-scale studies (Chazdon et al., 2009) since it still unclear how much and for how long tropical forest biodiversity can persist in human-modified landscapes under present land uses (Gardner et al., 2009). As previously emphasized, in several regions these small and isolated remnants constitute the only refuge for numerous species, creating a need to protect and integrate these forest fragments through ecosystem restoration initiatives. It is important to point out that forest fragments undergoing disturbance effects can persist in an alternative steady state if no adaptive management is applied to improve biodiversity conservation (Brancalion et al., 2013).

8.2.3 The Adaptive Management of Fragmented Forest Remnants

Despite their relevance, little is known about the dynamics of forest fragments located within agricultural landscapes (Tabarelli et al., 2008; Farah et al., 2014) (Fig. 8.1). Most knowledge on this topic was developed in regions where the forest still represents the most significant component of the landscape (i.e. Amazonian regions). Thus, depending on the total landscape forest cover, different approaches should be considered.

For the Brazilian Amazonian forests, where remaining forest cover is predominant, the adoption of a more biodiversity-sensitive management towards economic production (Putz et al., 2001) should be considered when promoting land-use practices compatible with biodiversity maintenance. Traditionally, the 'natural forest management' concept proposes a low or reduced impact logging of tropical forests where tree harvesting and biodiversity conservation are compatible due to the high inherent regeneration capacity of the still-forested landscape (Bawa & Seidler, 1997).

On the other hand, in biomes like the Atlantic Forest where fragments are notably small and isolated, our recommendation is to adopt a restoration-centered adaptive management (Gunderson, 2000), excluding economic exploitation and focusing efforts towards biodiversity conservation. This approach must be careful and based on periodic evaluations that allow for the recommendation of actions based on observed data, which characterizes 'adaptive management'. Thus, the recommendations are flexible and adaptable to the uncertainties inherent to the process of managing complex systems.

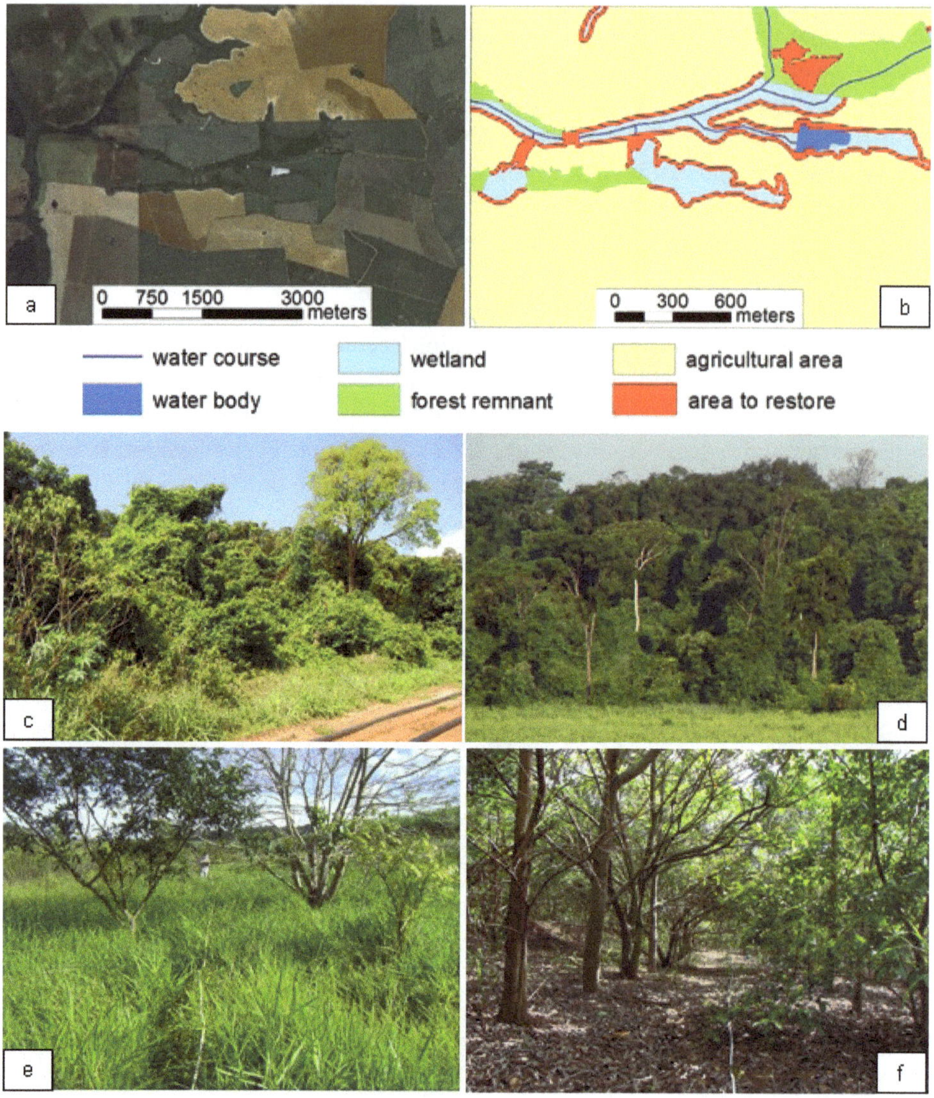

Fig. 8.1: (A) Agro-mosaic landscape with sparse natural remnants; (B) Environmental Planning Program map showing areas with need for conservation and restoration; (C) disturbed forest fragment; (D) surviving fragment of an old-growth forest; (E) 4-year-old restoration project showing arrested sucession; (F) relatively more successful 3-year old project with continuing development of the target plant community.

The conservation of ecosystem biodiversity depends upon the maintenance of its structure and processes, particularly regarding rare species that support long-term ecosystem functioning (Bracken & Low, 2012). Sometimes the negative trajectory of an ecosystem with weakening structure and function must be actively reversed with management actions. Potential actions include the elimination of disturbance factors, improvement of the underlying or surrounding abiotic environment, control of invasive exotic organisms, control of any local organisms with hyper-abundant populations in the degraded habitat, reintroduction of absent local species belonging to important functional groups or to those more vulnerable to extinction (Rodrigues et al. 2011; Leao et al., 2014). Basic information on species' biology (i.e. spatial density and distribution, microhabitats of occurrence, etc.) is essential to guide adaptive management recommendations.

Beside the exclusion of disturbance factors (e.g. invasive plants, hyper-abundant local plants, fire, etc.), the management success of degraded fragments depends upon a set of elements that define ecosystem resilience, such as soil integrity, existing seed bank and proximity to source areas that could provide propagules (Norden et al., 2009; Rother et al., 2013).

One example of adaptive management is Atlantic Forest fragments that are frequently affected by fire, where heliophytic lianas form an autochthonous group that can cause a severe imbalance within the plant community. They restrict the regeneration of shrubs and trees and contribute to bringing down adult trees (Farah et al., 2014). The management of these non-equilibrium plants within the degraded habitat can prevent further ecosystem collapse and lead to the return of forest dynamics (Rozza et al., 2007).

A complementary effort to enhance the viability of forest remnants is the reconnection of the fragments through ecological corridors, decreasing the degree of isolation among them. These corridors may be located along riparian zones and within sub-utilized agricultural areas (i.e. areas of low agricultural productivity), and can be established by forest restoration initiatives that use regional species and specific functional groups. Considering highly-modified landscapes it is essential to reintroduce a high number of species both through adaptive management and by restoration efforts. This strategy promotes an increase on the permeability of the agro-mosaic landscape to faunal seed dispersers, enabling biological fluxes to pass through forest patches. It also enhances the landscape towards a minimum 30% forest cover, indicated as a possible biodiversity threshold (Estavillo et al., 2013; Tambosi et al., 2013).

8.3 Ecological Restoration within Highly-modified Landscapes

8.3.1 Development of Ecological Restoration Techniques

When the level of fragmentation and degradation prevents an ecosystem´s natural recovery from disturbances, ecological restoration represents the main pathway to conserving biodiversity and related ecosystem services (Lamb et al., 2005). We recognize ecological restoration as the process of assisting the recovery of ecosystems, as defined by the Society for Ecological Restoration (SER, 2004).

Considering that only recently restoration ecology has matured as a discipline with international scientific support, ecological restoration (i.e. practice of restoring ecosystems) has tended to evolve on a trial-and-error basis, most of the time lacking a robust conceptual framework. This context, combined with the poor dissemination of results has created a barrier to the effective meta-analysis of combined project data, and rendered any generalized conclusions hard to extrapolate (Rodrigues et al., 2009).

Briefly summarizing the major conceptual and practical advances within Atlantic Forest Restoration, we note that early restoration projects were typically driven by concerns over water quality and soil protection, with little regard towards the maintenance of forest biodiversity (until 1980) (Rodrigues et al., 2009).

As knowledge of forest succession increased and evolved, different functional groups started to be considered within project planning, despite the majority of native species being poorly known and usually unavailable in tree nurseries. Efforts to directly "copy" the community composition and structure of remaining forests continued until 2000. Only recently restoration initiatives have sought to reestablish basic ecological processes that lead to self-sustaining ecosystems, via the introduction of native species with different functional traits (such as light requirements, dispersal syndromes and pollination systems). In addition to traditional tree-planting, other restoration techniques were considered, e.g. direct seeding, seed bank and saplings translocation etc (Lamb et al., 2005; Rodrigues et al., 2009). Currently the main challenges are to incorporate and guarantee both the intraspecific genetic diversity and a much wider floristic diversity, especially of non-tree species (Rodrigues et al., 2009).

It is important to emphasize that advances on restoration ecology and ecological restoration are never completely aligned. We can state that all sorts of practical issues hinder the implementation of ecological restoration. Common examples include the limited availability of local species in plant nurseries, the lack of non-tree species incorporated into restoration plantations and the high costs associated with these techniques. Other site specific circumstances, for instance the presence of invasive and highly competitive exotic C4 grasses, create an additional challenge for ecological restoration efforts.

An important aspect of the knowledge established thus far is that a few particular concepts and implementation techniques have been found to be crucial in the successful delivery of ecological restoration projects within high-use agricultural landscapes.

8.3.2 Planning of Ecological Restoration within High-use Agricultural Landscapes

To overcome the current scale of forest degradation it is essential to convince large numbers of landowners and land-managers to become involved with reforestation efforts (Lamb et al., 2005, Rodrigues et al., 2011). Our experiences within the sugarcane-dominated matrix of the interior of São Paulo state indicates two potential drivers for restoration: i) the enforcement of environmental laws and ii) the development of a market for environmentally-friendly products.

Evidence exists worldwide concerning the importance of environmental legislation in stimulating ecological restoration programs. For instance, approximately 60% of the projects evaluated by Ruiz-Jaen & Aide (2005) to assess restoration success were undertaken in order to comply with the law. In Brazil, the main environmental legislative instrument - known as the Forest Act – constitutes a key policy tool towards reconciling the production of goods on private lands with biodiversity conservation goals (in this case controlling deforestation and conserving forest remnants) and with provision of ecosystem goods and services (Calmon et al., 2011).

In seeking access to a high-demand international market via environmental certifications and financial credits, some landowners have adopted an Environmental Planning Program (EPP), developed to identify environmental irregularities and to recommend ecological restoration (Rodrigues et al., 2011). This program was first developed in 1999 and considers restoration efforts at two spatial scales: i) the local level and ii) the landscape level.

At a local level site specific methodologies are based upon local ecological resilience, the objectives being to reestablish the ecological functions and processes typical of high-diversity tropical forests. At a landscape level, the intention is to reconnect isolated habitats and reestablish biological fluxes via the maintenance and restoration of riparian zones, which constitute natural ecological corridors (Fig. 8.1).

To guarantee its implementation, the EPP also considers the need landowners have for clear guidance regarding how to ecologically restore ecosystems. Regional floristic composition is thus assessed and the location of trees used to provide seed for restoration nurseries is recorded in order to ensure the genetic diversity of seedlings used in the project. Additionally, comprehensive information on how to set up a tree nursery and all other technical support for restoration initiatives is made available. Recommendations on improvements to agricultural production are also provided whenever possible, as we contend that successful environmental planning is directly dependent on and related to appropriate agricultural planning.

Drawing upon this growing knowledge of landscape ecology, some connectivity metrics have been used to guide the planning of restoration through time and space (Tambosi et al., 2012), presuming that restoration efforts should be instigated preferentially within regions where overall connectivity would be significantly increased through these initiatives (e.g. restoring habitats that enhance connection among isolated forest remnants).

Whilst recognizing that realizing site-based restoration efforts at a landscape scale remains a huge challenge (Lamb et al., 2005), the EPP nevertheless sets a good example in how to achieve this scale shift. However, we concede that this addresses a very specific context involving the extensive rural properties of large landowners. Restoration efforts taking place amongst different rural communities that exist in other regions will present issues that must clearly be addressed in other ways.

8.4 How to Enable the Conservation Management and Ecological Restoration of Fragmented Forest Remnants?

8.4.1 Social and Economic Aspects

Forest remnants and areas undergoing restoration exist within human-modified landscapes, where many elements – social, economic, political, cultural and institutional - influence the ecological outcomes of these areas. When proposing conservation management or restoration plans it is important to recognize that these areas are part of a complex socio-ecological system, defined as integrated systems of ecosystems and human society with powerful reciprocal feedbacks and interdependence (Folke et al., 2002; Westley et al., 2002). We have already mentioned the importance of legal instruments and the role of their enforcement in making forest protection or restoration a reality. However, in order to be effective both conservation and restoration strategies must also address diverging interests among stakeholders – landowners, practitioners, scientists, government institutions and local community (Joly et al., 2010; Melo et al., 2013).

One of the most challenging social-economic aspects is the consolidation of a proper supply chain that would enable the scaling-up of restoration projects and support the management of fragments for conservation-value. Rural job creation would be an important consequence of a well-sustained supply chain providing there be a strong emphasis on training and institutional capacity-building. Both of which are in turn essential in promoting project quality and best practices amongst restoration efforts (Melo et al., 2013).

Unfortunately, conservation initiatives that focus on adaptive management of forest fragments are not yet widely spread in practice and little research exists that studies this issue on a long term basis, making it difficult to develop evidence-based recommendations. An aggravating factor is the lack of specific detailed legislation and the poor enforcement of laws regarding the conservation of non-riparian fragments; most of the focus is being directed towards riparian zones, where restoration initiatives are similarly concentrated.

Nevertheless, forest restoration policy is still substantially advanced in Brazil in spite of the gaps and challenges still to be addressed. It is known that the forgone opportunity cost of the land in combination with the cost of restoration presents a major obstacle towards the achievement of large scale efforts (Lamb et al., 2005; Melo et al., 2013). Beyond the establishment of a forest restoration supply chain (seed collectors, plant nurseries, implementation via skilled labour, etc.) a continuous improvement of restoration techniques must be pursued through close cooperation between scientists and practitioners.

Despite the difficulties and challenges facing the implementation of successful fragment conservation management and restoration projects, some good examples exist that show how to consider and incorporate socio-economic factors into planning and subsequently achieve more successful outcomes.

8.4.2 Instruments and Initiatives for the Conservation Management and Restoration of Fragmented Forests

The Brazilian Forest Act is the central federal environmental law that aims to ensure the continued provision of ecosystem services within private lands (Calmon et al., 2011), and each state can propose additional legislation to detail and legislate on other specific issues. There are currently 33 legislative initiatives from various Brazilian states (Santos et al., 2012) relating to Payment for Ecological Services (PES), a broadly known and recognized financial tool for delivering conservation management and restoration efforts. Federal Legislation concerning PES is currently being considered.

A further example is that in 2009 the São Paulo state government announced the "State Policy on Climate Changes" (Law Nº 13.798/2009). One year later a related law (Decree Nº 55947/10) established state government programs such as the "Green Economy Loan" and the "Forest Remnants Program". Through these programs, any rural landowner willing to conserve forest fragments or undertake forest restoration efforts could be eligible for PES or other financial support.

Another example of financial and economic support is the "Mata Atlantica" Initiative - funded by the Brazilian Bank of Development (BNDES) – that promotes the development of forest restoration and conservation projects within this specific biome. This initiative emerged from a partnership with the "Atlantic Forest Restoration Pact", to be described below.

The "Atlantic Forest Restoration Pact" (hereafter AFRP), created in 2009, is a social agreement that involves a great variety of stakeholders such as practitioners, scientists, landowners, managers, NGO´s and private companies, among others. These stakeholders work together to promote and increase the successful delivery of ecological restoration within the Atlantic Forest biome (Calmon et al., 2011; Melo et al., 2013). As a nationwide and collective effort the AFRP is intended to be inclusive, touching upon social, political and economic themes along with ecological ones. It has an audacious goal to restore 15 million hectares of tropical forest in 40 years and so far it has generated important lessons on how to move forward and achieve a large-scale restoration agenda (Melo et al., 2013).

At a smaller scale, it is important to remember that the Environmental Planning Program also stands as a successful example of an integrative approach, identifying land use irregularities and conflicts according to the Brazilian Forest Act and proposing conservation management and restoration actions within rural properties. By 2011, in São Paulo state, this program had been applied to more than 527,000 hectares of agro-mosaic landscapes and had been demonstrated as an effective tool for local conservation and restoration planning (Rodrigues et al., 2011).

A holistic conservation model that addresses ecosystems within highly-modified landscapes has to be adopted in order to effectively realize the potential conservation value of these fragmented and degraded habitats.

8.5 Final Considerations

Conservation of large areas of forests must be at the center of any biodiversity conservation plan, but both protection of smaller forest fragments and ecological restoration projects are crucial for the future of the Atlantic Forest within highly-modified landscapes.

Improving the conservation value of forest remnants is possible through adaptive management that focuses on biodiversity conservation. It is essential to reintroduce compromised functional groups while more studies are needed to allow evidence-based recommendations.

In complement to the conservation of forest fragments, ecological restoration is the main pathway towards the reestablishment of the local ecosystem in areas where forests have nearly vanished. In this situation, restoration of natural ecological corridors must reintroduce high diversity in order to enhance overall biodiversity.

To be effective, recommendations concerning the adaptive management of forest remnants and forest restoration need to be both planned through a multi-scale approach and delivered whilst constantly pursuing the refinement of technique. Because human modified landscapes are part of a complex socio-ecological system, conservation strategies need to take into account diverging interests among stakeholders – landowners, practitioners, scientists, government, NGO´s, etc.

We would encourage researchers to focus on this emerging topic - biodiversity conservation in human modified landscapes - and policy makers and practitioners to consider and address these issues with the objective of achieving tangible results.

Acknowledgements: We wish to thank all LERF/ESALQ/USP participants (past and current) engaged in conservation, restoration and environmental planning projects. We would also like to thank Simon Dunster for his advice regarding our use of English and Fernando Lamonato for the environmental planning figures. Ricardo Ribeiro Rodrigues wishes to thank CNPq and FAPESP for their financial support.

Bibliography

Arroyo-Rodríguez, V., Rös, M., Escobar, F., et al. (2013). Plant β-diversity in fragmented rain forests: testing floristic homogenization and differentiation hypotheses. Journal of Ecology, 101 (6), 1449–1458, DOI:10.1111/1365-2745.12153

Banks-Leite, C., Ewers, R. M., & Metzger, J.P. (2010). Edge effects as the principal cause of area effects on birds in fragmented secondary forest. *Oikos*, 119, 918-926, DOI:10.1111/j.1600-0706.2009.18061.x

Bawa, K.S., & Seidler, R. (1997). Natural Forest Management and Conservation of Biodiversity in Tropical Forests. Conservation Biology, 12 (1), 46-55, DOI:10.1111/j.1523-1739.1998.96480.x

Benayas, J.M.R., & Bullock, J.M. (2012). Restoration of Biodiversity and Ecosystem Services on Agricultural Land. Ecosystems, 15, 883-899, DOI:10.1007/s10021-012-9552-0

Benítez-Malvido, J., & Martínez-Ramos, M. (2003). Impact of forest fragmentation on understory plant species richness in Amazonia. Conservation Biology, 17, 389-400. DOI:10.1046/j.1523-1739.2003.01120.x

Brancalion, P.H.S., Melo, F.P.L., Tabarelli, M., et al. (2013). Biodiversity persistence in highly human-modified tropical landscapes depends on ecological restoration. Tropical Conservation Science, 6 (6), 705–710.

Bracken, M.E.S. & Low, N.H.N. (2012). Realistic losses of rare species disproportionately impact higher trophic levels. Ecology Letters, 15, 461-467, DOI: 10.1111/j.1461-0248.2012.01758.x

Calmon, M., Brancalion, P.H.S., Paese, A. et al. (2011). Emerging threats and opportunities for large-scale ecological restoration in the Atlantic Forest of Brazil. Restoration Ecology, 19 (20), 154-158, DOI: 10.1111/j.1526-100X.2011.00772.x

Chazdon, R.L., Peres, C.A., Dent, D.H. et al. (2009). The potential for species conservation in tropical secondary forests. Conservation Biology, 23 (6), 1406-1417, DOI: 10.1111/j.1523-1739.2009.01338.x

Dent, D.H. & Wright, S.J. (2009). The future of tropical species in secondary forests: A quantitative review. Biological Conservation, 142, 2833-2843, DOI: 10.1016/j.biocon.2009.05.035

Estavillo, C., Pardini, R. & Rocha, P.L.B. (2013). Forest loss and the biodiversity threshold: an evaluation considering species habitat requirements and the use of matrix habitats. PlosOne, 8(12), e82369, DOI: 10.1371/journal.pone.0082369

Fahrig, L. (2003). Effects of habitat fragmentation on biodiversity. Annual Review of Ecology Evolution and Systematics, 34, 487–515, DOI: 10.1146/annurev.ecolsys.34.011802.132419

FAO (2003). Food and Agriculture Organization of the United Nations. Workshop on Tropical Secondary Forest Management in Africa: Reality and Perspectives. Retrieved from http://www.fao.org/docrep/006/j0628e/J0628E00.htm#TopOfPage

FAO (2006). Food and Agriculture Organization of the United Nations. Global forest resources assessment 2005: progress toward sustainable forest management. *FAO Forestry Paper*, *147*, 1–348. Retrieved from http://www.fao.org/dOCrEP/008/a0400e/a0400e00.htm

FAO (2011). Food and Agriculture Organization of the United Nations. State of the world's forests 2011. Retrieved from http://www.fao.org/docrep/013/i2000e/i2000e00.htm

Farah, F.T., Rodrigues, R.R., Santos, F.A.M., et al.(2014). Forest destructuring as revealed by the temporal dynamics of fundamental species – Case study of Santa Genebra Forest in Brazil. *Ecological Indicators*, *37*, 40-44. DOI: 10.1016/j.ecolind.2013.09.011

Farinaci, J. S. (2012). As novas matas do estado de São Paulo : um estudo multiescalar sob a perspectiva da teoria da transição florestal. Tese (doutorado) - Instituto de Filosofia e Ciências Humanas. Campinas, SP: Universidade Estadual de Campinas.

Ferraz, S.F.B., Ferraz, K.M.P.M.B., Cassiano, C.C., et al. (2014). How good are tropical forest patches for ecosystem services provisioning? Landscape Ecology, 29 (2), 187–200, DOI: 10.1007/s10980-014-9988-z

Folke, C., Carpenter, S. R., Walker, B., et al. (2010). Resilience Thinking: Integrating Resilience, Adaptability and Transformability. Ecology and Society, 15 (4), 20. Retrieved from: http://www.ecologyandsociety.org/vol15/iss4/art20/

Gardner, T.A., Barlow, J., Chazdon, R., et al. (2009) Prospects for tropical forest biodiversity in a human-modified world. Ecology Letters, 12, 561–582, DOI: 10.1111/j.1461-0248.2009.01294.x

Gardner, T.A., Barlow, J., Sodhi, N.S., et al. (2010). A multi-region assessment of tropical forest biodiversity in a human-modified world. Biological Conservation, 143, 2293–2300, DOI:10.1016/j.biocon.2010.05.017

Gibson, L., Lee, T. M., Koh, L. P., et al. (2011). Primary forests are irreplaceable for sustaining tropical biodiversity. Nature, 478, DOI: 10.1038/nature10425

Gibson, L., Lynam, A. J., Bradshaw, C.J.A. et al. (2013). Near-complete extinction of native small mammals fauna 25 years after forest fragmentation. Science, 341 (6153), 1508-1510, DOI: 10.1126/science.1240495

Guimarães, P.R., Galetti, M., Jordano, P. (2008). Seed dispersal anachronisms: rethinking the fruits extinct megafauna ate. PLoS One, 3, e1745, DOI: 10.1371/journal.pone.0001745

Gunderson, L.H. (2000). Ecological resilience – in theory and application. Annual Review of Ecology and Systematics, 31, 425-439, DOI: 10.1146/annurev.ecolsys.31.1.425

Hawes, J., Barlow, J., Gardner, T.A., et al. (2008). The value of forest strips for understorey birds in an Amazonian plantation landscape. Biological Conservation, 141, 2262–2278, DOI: 10.1016/j.biocon.2008.06.017

Henle, K., Davies, K.F., Kleyer, M., *et al.* (2004). Predictors of species sensitivity to fragmentation. Biodiversity and Conservation, 13, 207–251, DOI: 10.1023/B:BIOC.0000004319.91643.9e

IUCN (2011). The IUCN Red List of Endangered Species. Retrieved from http://www.iucnredlist.org/

Joly, C.A., Rodrigues, R.R., Metzger, J.P., et al. (2010). Biodiversity Conservation Research, Training, and Policy in São Paulo. Science, 328 (5984), 1358-1359, DOI: 10.1126/science.1188639

Kotchetkoff-Henriques, O., Joly, C.A. & Bernaci, L.C. (2005). Relação entre o solo e a composição florística de remanescentes de vegetação natural no município de Ribeirão Preto, SP. Revista Brasileira de Botânica, 28 (3), 541-562.

Lamb, D., Erskine, P.D. & Parrotta, J.A. (2005). Restoration of degraded tropical forest landscapes. Science, 310, 1628–1632, DOI: 10.1126/science.1111773

Laurance, W.F., Nascimento, H.E. M., Laurance, S.G., et al. (2006). Rain forest fragmentation and the proliferation of successional trees. Ecology, 87, 469-482, DOI: 10.1890/05-0064

Laurance, W.F., Useche, D.C., Rendeiro, J., et al. (2012). Averting biodiversity collapse in tropical forest protected areas. Nature, 489, 290-294, DOI: 10.1038/nature11318

Leão, T.C.C., Fonseca, C.R., Peres, C.A., et al. (2014). Predicting Extinction Risk of Brazilian Atlantic Forest Angiosperms. Conservation Biology, 28 (5), 1349-1359, DOI: 10.1111/cobi.12286

Lemes, P., Melo, A.S., & Loyola, R.D. (2014). Climate change threatens protected áreas of the Atlantic Forest. Biodiversity Conservation, 23, 357-368, DOI: 10.1007/s10531-013-0605-2

Lenz, J., Fiedler, W., Caprano, T., et al. (2011). Seed-dispersal distributions by trumpeter hornbills in fragmented landscapes. Proceedings of the Royal Society of London B, 278, 2257–2264, DOI: 10.1098/rspb.2010.2383

Lôbo, D., Leão,T., Melo, F.P.L., et al. (2011). Forest fragmentation drives Atlantic forest of northeastern Brazil to biotic homogenization. Diversity and Distributions, 17, 287-296, DOI: 10.1111/j.1472-4642.2010.00739.x

Mangueira, J.R.S.A. (2012). A regeneração natural como indicadora de conservação, de sustentabilidade e como base do manejo adaptativo de fragmentos florestais remanescentes inseridos em diferentes matrizes agrícolas. Piracicaba, SP: Universidade de São Paulo.

Mather, A.S. (1992). The forest transition. Area, 24 (4), 367-379.

Melo, F.P.L., Pinto, S.R.R., Brancalion, P.H.S., et al. (2013). Priority setting for scalling-up tropical forest restoration projects: Early lessons from the Atlantic Forest Restoration Pact. Environmental Science and Policy, 33, 395-404, DOI: 10.1016/j.envsci.2013.07.013

Metzger, J.P. (2000). Tree functional group richness and landscape structure in a Brazilian tropical fragmented landscape. Ecological Applications, 10, 1147-1161, DOI: 10.1890/1051-0761(2000)010[1147:TFGRAL]2.0.CO;2

Metzger, J.P. (2001). O que é ecologia de paisagens? Biota Neotropica, 1 (1/2).

Norden, N., Chazdon, R.L., Chao, A., et al. (2009). Resilience of tropical rain forests: tree community reassembly in secondary forests. Ecology Letters, 12 (5), 385–94, DOI: 10.1111/j.1461-0248.2009.01292.x

Pardini, R., Bueno, A.A., Gardner, T.A., et al. (2010). Beyond the fragmentation threshold hypothesis: regime shifts in biodiversity across fragmented landscapes. PLoS One, 5, e13666, DOI: 10.1371/journal.pone.0013666

Putz, F.E., Blate, G.M., Redford, K.H., et al. (2001). Tropical Forest Management and Conservation of Biodiversity: An Overview. Conservation Biology, 15 (1), 7-20.

Ranganathan, J., Daniels, R.J.R., Chandran, M.D.S., et al. (2008). How biodiversity can live with agriculture. Proceedings of the National Academy of Sciences of the United States of America, 105, 17852–17854.

Ribeiro, M.C., Metzger, J.P., Martensen, A.C., et al. (2009). The Brazilian Atlantic Forest: How much is left, and how is the remaining forest distributed? Implications for conservation. Biological Conservation, 142 (6), 1141-1153, DOI: 10.1016/j.biocon.2009.02.021

Rodrigues, R.R. & Nave, A.G. (2004). Heterogeneidade florística das matas ciliares. In R.R. Rodrigues & H.F. Leitão Filho (Eds.), Matas ciliares: conservação e recuperação (p.45-71). São Paulo: EDUSP.

Rodrigues, R.R., Lima, R.,Gandolfi, S., et al. (2009). On the restoration of high diversity forests: 30 years of experience in the Brazilian Atlantic Forest. Biological Conservation, 142, 1242-1251, DOI: 10.1016/j.biocon.2008.12.008

Rodrigues, R.R., Gandolfi, S., Nave, A.G, et al. (2011). Large-scale ecological restoration of high-diversity tropical forests in SE Brazil. Forest Ecology and Management, 261 (10), 1605–1613, DOI: 10.1016/j.foreco.2010.07.005

Rother, D.C., Jordano, P., Rodrigues, R.R., et al. (2013). Demographic bottlenecks in tropical plant regeneration: A comparative analysis of causal influences. Perspectives in Plant Ecology Evolution and Systematics, 15, 86-96, DOI: 10.1016/j.ppees.2012.12.004

Rozza, A.F., Farah, F.T., Rodrigues, R.R. (2007). Ecological Management of degraded forest fragments. In R.R. Rodrigues, S.V. Martins, S. Gandolfi (Eds.), High Diversity Forest restoration in Degraded Areas. New York: Nova Science Publishers.

Ruiz-Jaen, M.C., & Aide, T.M. (2005). Restoration success: how is it been measured? Restoration Ecology, 13, 569–577, DOI: 10.1111/j.1526-100X.2005.00072.x

Sabino, A.P.S. (2011). Composição florística e estrutural de bordas de fragmentos florestais inseridos em matrizes agrícolas. Araras, SP: UFSCAR.

Santos, P., Brito, B., Maschietto, F., et al. (2012). Marco regulatório sobre pagamento por serviços ambientais no Brasil. Belém, PA: IMAZON, FGV.

Sutherland, W.J., Bardsley, S., Clout, M., et al. (2013). A horizon scan of global conservation issues for 2013. Trends in Ecology and Evolution, 28 (1), 16-22, DOI: 10.1016/j.tree.2012.10.022

Tabanez, A.A.J., & Viana, M. (2000). Patch structure within Brazilian Atlantic forest fragments and implications for conservation. Biotropica, 32, 925-933, DOI: 10.1111/j.1744-7429.2000.tb00630.x

Tabarelli, M., Mantovani, W. & Peres, C.A. (1999). Effects of habitat fragmentation and plant guild structure in the montane Atlantic forest of southeastern Brazil. Biological Conservation, 91, 119-127, DOI: 10.1016/S0006-3207(99)00085-3

Tabarelli, M., & Gascon, C. (2005). Lessons from fragmentation research: improving management and policy guidelines for biodiversity conservation. Conservation Biology, 19, 734-739, DOI: 10.1111/j.1523-1739.2005.00698.x

Tabarelli, M., Lopes, A., & Peres, C. (2008). Edge-effects drive forest fragments towards an early-successional system. Biotropica, 40, 657–661, DOI: 10.1111/j.1744-7429.2008.00454.x

Tambosi, L.R., Silva, M.M. & Rodrigues, R.R. (2012). Adequação ambiental de propriedades rurais e priorização da restauração florestal visando otimizar o ganho de conectividade da paisagem. In Conservação da Biodiversidade com SIG (p.24-38). São Paulo: Oficina de textos.

Tambosi, L.R., Martensen, A.C., Ribeiro, M.C., et al. (2013). A framework to optimize biodiversity restoration efforts based on habitat amount and landscape connectivity. Restoration Ecology, 22 (2), 169-177, DOI: 10.1111/rec.12049.

Thornton, D.H., Branch, L.C., & Sunquist, M.E. (2011). The relative influence of habitat loss and fragmentation: do tropical mammals meet the temperate paradigm? Ecological Applications, 21, 2324–2333.

Turner, I.M., & Corlett, R.T. (1996). The conservation value of small, isolated fragments of lowland tropical rain forest. Trends in Ecology and Evolution, 11 (8), 330-333, DOI: 10.1016/0169-5347(96)10046-X

Westley, F., Capenter, S.R., Brock, W.A., et al. (2002). Why systems of people and nature are not just social and ecological systems. In L.H. Gunderson & C.S. Holling C. S. (Eds.), Panarchy: understanding transformations in human and natural systems. Washington: Island Press.

Flavio de Barros Molina, Bruno de Oliveira Ferronato, Franco Leandro de Souza

9 Freshwater Turtles in Natural, Rural and Urban Ecosystems in São Paulo State, Southeastern Brazil

9.1 São Paulo State Landscapes and its Native Turtle Species in Natural, Rural and Urban Environments

São Paulo is a southeastern Brazilian State inhabited by more than 44 million people and has a demographic density of 166.23 inhabitants/km² (Instituto Brasileiro de Geografia e Estatística, 2015). In pre-colonial times the State was completely covered by the Atlantic Forest, a Neotropical rain forest (86%), and by the Cerrado, a tropical savanna-like vegetation (14%) (Durigan et al., 2007; Leite, 2007). Both ecosystems are classified as hotspots of biodiversity (Myers et al., 2000). Due to anthropogenic activities related to the extensive urbanization/industrialization and the spread of agricultural frontiers, only 13.94% of the original area is still preserved within a landscape matrix dominated by croplands, pastures and urban areas (Kronka et al., 2005; Tabarelli et al., 2010; Verdade et al., 2012).

The Atlantic Forest occupied an area of more than 1.5 million km² (20% of the Brazilian territory) extending in latitude from 4° to 32° S and in altitude from sea level to 2,900 m. It presents heterogeneous environmental conditions due to the action of different climatic belts and soil types (Tabarelli et al., 2005). Diversity of species is high (e.g. 20,000 species of plants) including a substantial degree of endemism (e.g. 8,000 species of plants) (Myers et al., 2000). Today the forest is restricted to approximately 11.7% of its original distribution, but 32 to 40% of this area is represented by secondary forests and small forest fragments, most covering less than 50 ha (Ribeiro et al., 2009).

Almost 60% of the remaining vegetation in São Paulo State is composed by ombrophilous dense forests located in the Coastal Province and parts of the Atlantic Plateau (Kronka et al., 2005). This area is part of the Serra do Mar Corridor, a recognized center of species endemism. The presence of steep slopes, unsuitable for agriculture, guaranteed a lower rate of deforestation. The most serious problems impacting the forests in the Serra do Mar Corridor are intensive land use, hunting, logging, poaching, water pollution, and the establishment of invasive species (Aguiar et al., 2003). The Serra do Mar drains the Southeastern Atlantic Basin rivers that run to the Atlantic Ocean and is the habitat for five species of side-necked turtles (Chelidae): *Acanthochelys radiolata*, *A. spixii*, *Hydromedusa maximiliani*, *H. tectifera*, and *Phrynops geoffroanus* (Iverson, 1992; Souza, 2005).

The Cerrado (*sensu lato*) occupied an area of 2 million km² (21% of the Brazilian territory) distributed mainly in central Brazil and comprise a set of physiognomically distinct habitats ranging from dense grasslands to an almost closed woodland (Klink

& Machado, 2005; Durigan et al., 2007). Species diversity is high (e.g. 10,000 species of plants) including a substantial degree of endemism (e.g. 4,400 species of plants) (Myers et al., 2000; Klink & Machado, 2005). Today the Cerrado is restricted to about 45% of its original area (Klink & Machado, 2005).

The remaining Cerrado in the State of São Paulo is composed of thousands of fragments, some covering less than 20 ha. This represents less than 6% of the States' remaining vegetation (Durigan, 2006; Durigan et al., 2007; Leite, 2007). These fragments are situated mainly to the west of the Atlantic Plateau and have the coldest and wettest climate in the Cerrado domain (Durigan, 2006). The main threats to these fragments are uncontrolled fires and invasive species (Durigan, 2006; Durigan et al., 2007) that can affect native populations in several ways, including genetic isolation. Rivers that run to the west of the Atlantic Plateau belong to the Parana River Basin where five chelidae species can be found: *Acanthochelys spixii*, *Hydromedusa maximiliani*, *H. tectifera*, *Mesoclemmys vanderhaegei*, and *Phrynops geoffroanus* (Iverson, 1992; Souza, 2005). Two invasive freshwater turtle species are also found in both biomes: *Trachemys dorbigni* and *T. scripta elegans* (Emydidae) (Ferronato et al., 2009; Santos et al., 2009).

Different studies indicate that populations of *H. tectifera* and *P. geoffroanus* can survive in polluted waters (Souza & Abe, 1999, 2000, 2001; Ribas & Monteiro-Filho, 2002; Ferronato et al., 2009) and they appear to be tolerant to anthropogenic altered landscapes that include agricultural, silvicultural and pasture mosaics.

In Argentina, Lescano et al. (2008) estimated a *H. tectifera* population size of 153 individuals in a 2 km stretch of stream, resulting in an estimated population density of 219 turtles/ha river and a mean biomass of 96.1 kg/ha of river. No such information is available for Brazilian populations. Though *H. tectifera* was already found in polluted urban rivers (Ferronato et al., 2009; Molina, personal observation; Ribas & Monteiro-Filho, 2002), no data about population dynamics was collected in such localities. It is not known if these anthropically impacted populations are genetically and demographically well adapted. Clutch size ranges from 5 to 15 ellipsoid eggs and incubation time ranges from 70 to 128 days (Cabrera, 1998; Chinen et al., 2004; Fagundes & Bager, 2007; Lescano et al., 2007; Souza et al., 2006). These are general reproductive data obtained both in nature and in captivity and these data indicate that *H. tectifera* reproductive potential is not as high as the one observed in *Trachemys* invasive species.

Phrynops geoffroanus is abundant in urban areas, although few records are available. In a study conducted in a polluted urban river in Ribeirão Preto city, São Paulo State, 170 - 230 turtles/ha of river with a biomass of 255 - 345 kg/ha of river was found for this species, representing one of the highest recorded densities for a Neotropical freshwater turtle (Souza & Abe, 1999, 2000). These results were hypothesized to be related to an association of different urban factors favorable to the turtles, such as an abundance of sewage and organic wastes produced by humans that are used as food, an absence of predators, and an increased availability of nesting sites (Souza & Abe, 2000). Domestic waste discharge into urban rivers can act as food supplementation for some organisms, including freshwater turtles such as *P. geoffroanus*. Supplementation in turn, minimizes dispersal behavior and reduces home range size (Souza et al., 2008). In areas with little anthropic interference sporadic observations suggest that species abundance is low. Without further research about this species from across distinct habitats, comparisons with other habitats are not possible. Clutch size and incubation time do not seem to be good predictors of the competitive potential of *P. geoffroanus* when compared to *Trachemys* species. Clutch size ranges from 7 to 26 ellipsoid eggs that hatched in 149 to 331 days (Guix et al., 1989; Molina, 1989, 1998; Souza & Abe, 2001).

Although geographical distribution records confirm that *M. vanderhaegei* can be common in natural and anthropic habitats, including urban areas (Brandão et al., 2002; Brito et al., 2012; Iverson, 1992; Souza et al., 2000), there are few records on demographic parameters for the species in the State of São Paulo. Using capture-recapture methods at a silvicultural system in Angatuba city, São Paulo State, Marques et al. (2013) estimated the population of *M. vanderhaegei* at a single 39.6 ha lagoon as 20 individuals (range: 12 – 60), representing 0.65 individual/ha (range: 0.3 – 1.5 individual/ha) and a biomass of 492 g/ha (range: 227 – 1,136 g/ha). The sex ratio was approximately 1:1. In the streams found in Cerrado habitat in Central Brazil, Brito et al. (2009) also reported an unbiased sex ratio. Population estimates in a 1 km section of river ranged from 11.6 to 26.7 individuals (Brito et al., 2009). Clutch size, in captivity, ranges from 1 to 14 eggs, and at least some females can lay more than one clutch per reproductive period. Hatching was observed after an incubation period of 300 days (Corazza & Molina, 2004). As observed for the other chelid species, it is possible that *M. vanderhaegei* will not tolerate competition with *Trachemys* species.

To the best of our knowledge, the other three turtle species have not been reported in urban or agricultural ecosystems in São Paulo State. *Hydromedusa maximiliani* is clearly an inhabitant of pristine aquatic environments and probably is jeopardized by the intense degradation of the Atlantic Forest (Souza & Martins, 2009). The same situation may be true of *A. spixii* (Brasil et al., 2011). We do not have any information about the tolerance of *A. radiolata* to habitat degradation.

Acanthochelys radiolata lives in lentic waters, such as marshes (Mocelin et al., 2008), and seems to be restricted to the remaining fragments of the Brazilian Atlantic rainforest (Souza, 2005; TTWG, 2014). In the last few decades, Atlantic rainforest devastation was severe and persistent threats, such as the socioeconomic needs for more agricultural lands, residential areas, and land settlements, are real (Tabarelli et al., 2005). A point of concern for the future of *A. radiolata* is the fact that it is not known if it can survive in the polluted waters of urbanized ecosystems. Another important question is related to the possible impact that invasive turtle species, such as *T. dorbigni* and *T. s. elegans* already detected in São Paulo State (Ferronato et al., 2009; Martins et al., 2014; Santos et al., 2009), can pose to the survival of *A. radiolata*. Data obtained in captivity indicate that *A. radiolata* has a low reproductive rate (1 to

6 eggs per clutch, according to Molina, 1998) and a long period of egg incubation (135 days, according to Mocelin et al., 2008). Such reproductive traits apparently do not make it a good competitor.

Acanthochelys spixii is also an inhabitant of lentic waters (Bujes, 2010; Bujes & Verrastro, 2008), being found in fragments of the Atlantic rainforest and Cerrado ecosystems (Souza, 2005; TTWG, 2014). Although many authors have commented on its geographical distribution (Brandão et al., 2002; D'Amato & Morato, 1991; Ribas & Monteiro Filho, 2002) only one study analyzed demographic parameters for a Brazilian population. Fraxe Neto et al. (2011) used a capture-recapture method to estimate the population of *A. spixii* at a single 4 ha lentic pond in the Parque Nacional de Brasília (Distrito Federal), a Cerrado area in central Brazil. Despite being a protected area, the park is totally surrounded by rural and urban areas, including a large waste dump. The average monthly population size was estimated at 30 adult turtles (range: 10 – 35) and the sex ratio was approximately 1:1. The annual population growth rate varied depending upon the year, in 2005 - 2006 it was 1.37, while in 2006-2007 it was 0.59 (Fraxe Neto et al. 2011). *A. spixii* is probably not well adapted for competition with *Trachemys* species. Data obtained in captivity indicate that it has a low reproductive rate (1 to 4 eggs per clutch, according to Molina, 1998).

Hydromedusa maximiliani is typical of the mountainous Atlantic Forest, being found especially in the streams of primary forests. When in sympatry with *Hydromedusa tectifera*, it is found in water courses above 600 m of elevation (Souza, 2005; Souza & Martins, 2009). It prefers shallow streams with rocky or sandy bottoms and cold, clear, waters. These animals are thermoconformers, probably a result of the dense canopy of the Atlantic Forest allowing little sunlight to reach the surface. (Souza & Abe, 1998; Souza & Martins, 2009). Population size at Parque Estadual de Carlos Botelho and Parque Estadual da Serra do Mar, both in São Paulo State, are similar: respectively 21 and 25 specimens/km river (Famelli et al., 2011; Souza & Abe, 1997a). In some areas, population density and biomass can be high, respectively 190 specimens/ha of river and 40 kg/ha of river (Souza & Abe, 1997a; Souza & Martins, 2009). Sex ratios are female biased from 0.8/1 (Famelli et al., 2011) to 1/2 (Martins & Souza, 2009). It is a sedentary species that is active mainly when the weather is hot and humid (Souza & Abe, 1997a; Souza & Martins, 2009; Souza et al., 2002). Reproductive data suggests that this species is not well adapted to compete with *Trachemys* species. It has a low reproductive rate (1 to 3 eggs per clutch, according to Famelli et al., 2014) and a long period of egg incubation (250-300 days, according to Souza & Martins, 2009).

After 20 years of population study in its natural habitat, *H. maximiliani* is one of the best known Brazilian freshwater turtles. A great deal is known about it at the local scale although much is yet to be learned. Ecological and genetic approaches to be conducted at the regional scale are priorities. From a conservation perspective, this freshwater turtle species seems to be restricted to pristine, mountainous, Atlantic rainforest habitats. Famelli et al. (2012) found that a population inhabiting a Conservation Unit in São Paulo State (Parque Estadual Carlos Botelho), exhibited a stable population according to population viability analysis models. However, if scenarios of habitat fragmentation are included in the models, the probability of extinction can approach 99%. This suggests that populations from forest fragments (a common landscape throughout the species geographical range) may actually be threatened. Thus, for species that do not tolerate pronounced habitat modification, such as *H. maximiliani* and probably *A. radiolata* and *A. spixii*, protected areas and increased connectivity among fragmented habitats must be viewed as a practical management strategy. It is interesting to note that more than a century ago *H. maximiliani* and *A. spixii* were still living in São Paulo city, in neighborhoods such as Ipiranga and Vila Prudente (Luederwaldt, 1926).

9.2 Turtle Invasive Species in Natural, Rural and Urban Environments in São Paulo State

Santos et al. (2009) recorded, for the first time, the presence of *T. dorbigni* in a natural habitat in the State of São Paulo and this finding could represent a recent colonization or a human introduction into the region. The illegal removal of the eggs of *T. dorbigni* from within its native range for sale into the pet trade has been reported elsewhere (Bager, 2003; Fagundes, 2007). The commercialization of *T. dorbigni* in the pet trade is common in Brazil (Molina & Rocha, 1987). The subsequent release of this turtle, by the pet owners, in urban areas has been documented (Molina, 2006; Santos et al., 2009). The species is found in urban areas in six states outside its native range in Brazil (Santos et al., 2009). The occurrence of *T. dorbigni* in polluted waterways within its natural range (Bujes & Verrastro, 2008; Fagundes et al., 2010a) demonstrates its environmental tolerance and makes the establishment of individuals in invaded areas in urban ponds more likely.

The density and biomass of the species in its native range can reach 7.61 turtles/ha and 8.71 kg/ha, respectively (Bujes et al., 2011). It is still not clear how well populations of *T. dorbigni* are persisting in invaded regions and how abundant they are. In natural areas clutch size ranges from 4 to 19 eggs (Bager et al., 2007; Fagundes et al., 2010b; Krause et al., 1982) and in captivity incubation time ranges from 54 to 120 days (Molina & Gomes, 1998). In Rio Grande do Sul State, females can lay more than one clutch per season (Bager et al., 2007; Bujes, 2010). These reproductive parameters, though very simple, suggest that the species is more prepared to win an eventual competition with chelid species. Not only is clutch size bigger than the numbers observed for *A. radiolata*, *A. spixii*, *H. maximiliani*, and *M. vanderhaegei*, but incubation time is the shortest observed. Another important point is the capacity *T. dorbigni* has for producing multiple clutches per breeding season. Besides that, *T. dorbigni* has a generalist diet that is better prepared to survive in a highly competitive environment. It is an opportunistic omnivore (Bujes et al., 2007) while at least *H.*

maximiliani, *H. tectifera* and *M. vanderhaegei* are mainly or exclusively carnivores (Souza & Abe, 1997b, 1998; Bonino et al., 2009; Brito et al., 2009; Souza & Martins, 2009; Alcalde et al., 2010; Novelli et al., 2013).

It has been demonstrated that *T. dorbigni* can live and breed in polluted rivers (Fagundes et al., 2010b), in addition to intensively farmed areas (Bager et al., 2007). This illustrates *T. dorbigni*'s resilience and attributes that can help it to establish breeding populations outside its native range. Successful hatching was already detected in São Paulo city (Molina, personal observation). Future research should evaluate the full potential of *T. dorbigni* to invade and become established in alien habitats, as the species possesses life history traits similar to the well-recognized invasive *T. s. elegans* (Chen & Lue, 1998; Cadi & Joly, 2004; Thomson et al., 2010).

It is still not clear how large the populations of the invasive *T. s. elegans* are in ecosystems in São Paulo State. In fact, there is no population estimate of *T. s. elegans* in Brazil, although the species has recently been recorded as feral in the country (Ferronato et al., 2009; Bujes, 2011). Investigations have shown that *T. s. elegans* is an opportunist omnivore in habitats where they have been released (Chen & Lue, 1998; Pérez-Santigosa et al., 2011). In Piracicaba, Brazil, *T. s. elegans* was observed laying eggs in the surroundings of an artificial pond (Ferronato et al., 2009). Breeding populations of *T. s. elegans* have been confirmed in several countries outside its native range, such as Australia, France, Spain, and Taiwan (Chen & Lue, 1998; Cadi et al., 2004; Burgin, 2006; Pérez-Santigosa et al., 2008). In addition, it has been demonstrated that *T. s. elegans* can have multiple clutches in the invaded habitats, perhaps aiding populations of *T. s. elegans* outside of its native range to persist (Pérez-Santigosa et al., 2008).

In Taiwan (Chen & Lue, 1998) and California (Thomson et al., 2010), *T. s. elegans* seems to be more abundant in urban areas rather than in natural areas, suggesting that the distribution of this taxon may be driven primarily by introductions rather than by expansion of established populations (Thomson et al., 2010). In some urban areas, *T. s. elegans* shows a plastic behavior related to feeding and reproduction that could facilitate the initial establishment of populations of this exotic species (Pérez-Santigosa et al., 2008). In Brazil, at least two native freshwater turtle species, *P. geoffroanus* and *H. tectifera* are associated with habitats that have been heavily impacted by humans. Apparently they can survive in these areas, in some cases, sympatric with the exotic *T. s. elegans* (Ferronato et al., 2009). Nonetheless, no information is available for *T. s. elegans* diet in natural and rural sites in São Paulo State, and no study has evaluated the role of dietary overlap and competition between invasive and native Brazilian turtle species.

9.3 Final Considerations

World biodiversity is threatened due to alterations in natural ecosystems by human actions, including agriculture expansion (Hamer & McDonnell, 2010; Dobrovolski et al., 2011; Kowarik, 2011; Verdade et al., 2012). In the last 300 years, several human activities have resulted in a conversion of approximately 50% of natural habitats into anthropic ones (Ellis & Ramankutty, 2008; Ellis et al., 2010). Such a phenomenon implies that today, a considerable part of the research on biodiversity will focus on areas with human impact (Kowarik, 2011; Martin et al., 2012). Urbanization impoverishes and splits available area for animals and plants, simplifying habitats and reducing resources (Ellis & Ramankutty, 2008; Hamer & McDonnell 2008; McKinney 2008).

Future studies on dietary overlap and competition are recommended to understand the population dynamics of these invasive *Trachemys* species in Brazil, and fully evaluate its potential to expand from the sites where they are released by pet owners into pristine ecosystems. The fact that these species are successful generalists may represent an extra impact to the survival of chelid native species in highly altered environments, such as São Paulo State's urban and agricultural ecosystems.

Acknowledgements: We thank the editors for the invitation to participate in this book. We also are grateful to Craig Berg, Richard Sajdak and two anonymous reviewers for critically commenting on the manuscript. FLS is a researcher at CNPq (process number 303006/2014-5).

Bibliography

Aguiar, A.P., Chiarello, A.G., Mendes, S.L., & Matos, E.N. (2003). The Central and Serra do Mar Corridors in the Brazilian Atlantic Forest. In C. Galindo-Leal & I.G. Câmara (Eds.), The Atlantic Forest of South America: Biodiversity status, threats, and outlook (p.118-132). Washington, D.C.: Island Press.

Alcalde, L., Derocco, N.N., & Rosset, S.D. (2010). Feeding in syntopy: Diet of *Hydromedusa tectifera* and *Phrynops hilarii* (Chelidae). Chelonian Conservation and Biology, 9, 33-44.

Bager, A. (2003). Aspectos da biologia e ecologia da tartaruga tigre d'agua, *Trachemys dorbigni*, (Testudines, Emydidae) no extremo sul do estado do Rio Grande do Sul – Brasil. Unpublished PhD thesis, Universidade Federal do Rio Grande do Sul, Porto Alegre, RS, Brazil.

Bager, A., Freitas, T.R.O., & Krause, L. (2007). Nesting ecology of a population of *Trachemys dorbigni* (Emydidae) in southern Brazil. Herpetologica, 63, 56-65.

Bonino, M.F., Lescano, J.N., Haro, J.G., & Leynaud, G.C. (2009). Diet of *Hydromedusa tectifera* (Testudines-Chelidae) in a mountain stream of Córdoba province, Argentina. Amphibia-Reptilia, 30, 545-554.

Brandão, R.A., Zerbini, G.J., Sebben, A., & Molina, F.B. (2002). Notes on the distribution and habitats of *Acanthochelys spixii* and *Phrynops vanderhaegei* (Testudines, Chelidae) in Central Brazil. Boletín de la Asociación Herpetológica Española, 13, 11-15.

Brasil, M.A., Horta, G.F., Fraxe Neto, H.J., Barros, T.O., & Colli, G.R. (2011). Feeding ecology of *Acanthochelys spixii* (Testudines, Chelidae) in the Cerrado of Central Brazil. Chelonian Conservation and Biology, 10, 91-101.

Brito, E.S., Strüssmann, C., Kawashita-Ribeiro, R.A., Morais, D.H., Ávila, R.W., & Campos, V.A. (2012). New records and distribution extensions of three species of *Mesoclemmys* Gray, 1863 (Testudines: Chelidae) in Mato Grosso state, Brazil, with observations on terrestrial movements. Check List, 8, 294-297.

Brito, E.S., Strüssmann, C., & Penha, J.M.F. (2009). Population structure of *Mesoclemmys vanderhaegei* (Bour, 1973) (Testudines:Chelidae) in the Cerrado of Chapada dos Guimarães, Mato Grosso, Brazil. Biota Neotropica, 9, 245-248.

Bujes, C.S. (2010). Os Testudines continentais do Rio Grande do Sul, Brasil: taxonomia, história natural e conservação. Iheringia, Série Zoologia, 100, 413-424.

Bujes, C.S. (2011). Chelonia-Project – Study group for freshwater turtle conservation and biology in southern Brazil: introduction of *Trachemys scripta elegans* in the Jacuí Delta. Turtle and Tortoise Newsletter, 15, 14-18.

Bujes, C.S., Ely, I., & Verrastro, L. (2007). *Trachemys dorbigni* (Brazilian slider). Diet. Herpetological Review, 38, 335.

Bujes, C.S., Molina, F.B., & Verrastro, L. (2011). Population characteristics of *Trachemys dorbigni*, (Testudines, Emydidae) from Delta do Jacuí State Park, Rio Grande do Sul, southern Brazil. South American Journal of Herpetology, 6, 27-34.

Bujes, C.S., & Verrastro, L. (2008). Quelônios do Delta do Jacuí, RS, Brasil: uso de hábitats e conservação. Natureza e Conservação, 6, 47-60.

Burgin, S. (2006). Confirmation of an established population of exotic turtles in urban Sydney. Australian Zoologist, 33, 379-384.

Cabrera, M.R. (1998). Las Tortugas Continentales de Sudamérica Austral. Cordoba, Argentina: Author.

Cadi, A., & Joly, P. (2004). Impact of the introduction of the red-eared slider (*Trachemys scripta elegans*) on survival rates of the European pond turtle (*Emys orbicularis*). Biodiversity and Conservation, 13, 2511-2518.

Cadi, A., Delmas, V., Prévot-Julliard, A.-C., Joly, P., Pieau, C., & Girondot, M. (2004). Successful reproduction of the introduced slider turtle (*Trachemys scripta elegans*) in the South of France. Aquatic Conservation: Marine and Freshwater Ecosystems, 14, 237-246.

Chen, T.H., & Lue, K Y. (1998). Ecological notes on feral populations of *Trachemys scripta elegans* in northern Taiwan. Chelonian Conservation and Biology, 3, 87-90.

Chinen, S., Lisboa, C.S., & Molina, F.B. (2004). Biologia reprodutiva de *Hydromedusa tectifera* em cativeiro (Testudines, Chelidae). Arquivos do Instituto Biológico, 71, 401-403.

Corazza, S.S., & Molina, F.B. (2004). Biologia reprodutiva e conservação ex-situ de *Bufocephala vanderhaegei* (Testudines, Chelidae). Arquivos do Instituto Biológico, 71, 407-410.

D'Amato, A.F., & Morato, S.A.A. (1991). Notas biológicas e localidades de registro de *Platemys spixii* (Duméril & Bibron, 1835) (Testudines: Chelidae) para o estado do Paraná, Brasil. Acta Biológica Leopoldensia, 13, 119-130.

Dobrovolski, R., Loyola, R.D., De Marco Jr, P., & Diniz-Filho, J.A.F. (2011). Agricultural expansion can menace Brazilian protected areas during the 21st century. Natureza & Conservação, 9, 208-213.

Durigan, G. (2006). Observations on the Southern Cerrados and their relationship with the core area. In R.T. Pennington, G.P. Lewis & J.A. Ratter (Eds.), Neotropical savannas and dry forests: Diversity, biogeography, and conservation (p.67-77). Boca Raton, FL: Taylor and Francis.

Durigan, G., Siqueira, M.F., & Franco, G.A.D.C. (2007). Threats to the Cerrado remnants of the state of São Paulo, Brazil. Scientia Agricola, 64, 355-363.

Ellis, E.C., & Ramankutty, N. (2008). Putting people in the map: anthropogenic biomes of the world. Frontiers in Ecology and the Environment, 6, 439-447.

Ellis, E.C., Klein Goldewijk, K., Siebert, S., Lightman, D., & Ramankutty, N. (2010). Anthropogenic transformation of the biomes, 1700 to 2000. Global Ecology and Biogeography, 19, 589-606.

Fagundes, C.K. (2007). Dinâmica populacional de *Trachemys dorbigni*, (Testudines: Emydidae) em ambiente antrópico em Pelotas. MSc dissertation. Santa Maria, RS, Brazil: Universidade Federal de Santa Maria.

Fagundes, C.K., & Bager, A. (2007). Ecologia reprodutiva de *Hydromedusa tecifera* (Testudines: Chelidae) no sul do Brasil. Biota Neotropica, 7, 179-184.

Fagundes, C.K., Bager, A., & Cechin, S.T.Z. (2010a). *Trachemys dorbigni* in an anthropogenic environment in southern Brazil: I) Sexual dimorphism and population estimates. Herpetological Journal, 20, 185-193.

Fagundes, C.K., Bager, A., & Cechin, S.T.Z. (2010b). *Trachemys dorbigni* in an anthropogenic environment in southern Brazil: II) Reproductive ecology. Herpetological Journal, 20, 195-199.

Famelli, S., Bertoluci, J., Molina, F.B., & Matarazzo-Neuberger, W.M. (2011). Structure of a population of *Hydromedusa maximiliani* (Testudines, Chelidae) from Parque Estadual da Serra do Mar, an Atlantic Rainforest Preserve in Southeastern Brazil. Chelonian Conservation and Biology, 10, 132-137.

Famelli, S., Pinheiro, S.C.P., Souza, F.L., Chiaravalloti, R.M., & Bertoluci, J. (2012). Population viability analysis of a long-lived organism, the freshwater turtle *Hydromedusa maximiliani* (Testudines, Chelidae). Chelonian Conservation and Biology, 11,162-169.

Famelli, S., Adriano, L.R., Pinheiro, S.C.P., Souza, F.L., & Bertoluci, J.A. (2014). Reproductive biology of the freshwater turtle *Hydromedusa maximiliani* (Chelidae) from southeastern Brazil. Chelonian Conservation and Biology, 13, 81-88.

Ferronato, B.O., Marques, T.S., Guardia, I., Longo, A.L.B., Piña, C.I., Bertoluci, J., & Verdade, L.M. (2009). The turtle *Trachemys scripta elegans* (Testudines, Emydidae) as an invasive species in a polluted stream of southeastern Brazil. Herpetological Bulletin, 109, 29-34.

Fraxe Neto, H.J., Brasil, M.A., Horta, G.F., Barros, T.O., Falcon, G.B., & Colli, G.R. (2011). Demography of *Acanthochelys spixii* (Testudines, Chelidae) in the Brazilian Cerrado. Chelonian Conservation and Biology, 10, 82-90.

Guix, J.C., Salvatti, M., Peroni, M.A., & Lima-Verde, J.S. (1989). Aspectos da reprodução de *Phrynops geoffroanus* (Schweigger, 1812) em cativeiro (Testudines, Chelidae). Grupo de Estudos Ecológicos, Série Documentos, 1, 1-19.

Hamer, A.J., & McDonnell, M.J. (2008). Amphibian ecology and conservation in the urbanizing world: A review. Biological Conservation, 141, 2432-2449.

Hamer, A.J., & McDonnell, M.J. (2010). The response of herpetofauna to urbanization: Inferring patterns of persistence from wildlife databases. Austral Ecology, 3, 568-580.

Instituto Brasileiro de Geografia e Estatística. (n.d.). São Paulo. Retrieved February 27, 2015, from http://www.ibge.gov.br/estadosat/perfil.php?sigla=sp

Iverson, J.B. (1992). A revised checklist with distribution maps of the turtles of the world. Richmond, IN: J.B. Iverson.

Klink, C.A., & Machado, R.B. (2005). A conservação do Cerrado brasileiro. Megadiversidade, 1, 147-155.

Kowarik, I. (2011). Novel urban ecosystems, biodiversity, and conservation. Environmental Pollution, 159, 1974-1983.

Krause, L., Gomes, N., & Leyser, K.L. (1982). Observações sobre a nidificação e desenvolvimento de *Chrysemys dorbigni* (Duméril & Bibron, 1835) (Testudines, Emydidae) na Estação Ecológica do Taim, Rio Grande do Sul. Revista Brasileira de Zoologia, 1, 79-90.

Kronka, F.J.N., Nalon, M.A., Matsukuma, C.K., et al. (2005). Monitoramento da vegetação natural e do reflorestamento no estado de São Paulo. In J.C.N. Epiphanio & G.J.F. Banon (Eds.), Anais do XII Simpósio Brasileiro de Sensoriamento Remoto (p.1569-1576). São José dos Campos, SP: Instituto Nacional de Pesquisas Espaciais.

Leite, M. (2007). Nos caminhos da biodiversidade paulista. São Paulo: Secretaria do Meio Ambiente.

Lescano, J.N., Bonino, M.F., & Leynaud, G.C. (2007). *Hydromedusa tectifera*. Copulation. Herpetological Review, 38, 332.

Lescano, J.N., Bonino, M.F., & Leynaud, G.C. (2008). Density, population structure and activity pattern of *Hydromedusa tectifera* (Testudines-Chelidae) in a mountain stream of Córdoba province, Argentina. Amphibia-Reptilia, 29, 505-512.

Luederwaldt, H. (1926). Os chelonios brasileiros, com a lista das espécies do Museu Paulista. Revista do Museu Paulista, 14, 405-479.

Marques, T.S., Lara, N.R.F., Bassetti, L.A.B., Ferronato, B.O., Malvásio, A., & Verdade, L.M. (2013). Population structure of *Mesoclemmys vanderhaegei* (Testudines, Chelidae) in a silvicultural system in southeastern Brazil. Herpetology Notes, 6, 179-182.

Martin, L.J., Blossey, B., & Ellis, E. (2012). Mapping where ecologists work: Biases in the global distribution of terrestrial ecological observations. Frontiers in Ecology and the Environment, 10, 195-201.

Martins, F.I., & Souza, F.L. (2009). Demographic parameters of the neotropical freshwater turtle *Hydromedusa maximiliani* (Chelidae). Herpetologica, 65, 82-91.

Martins, R.A.; Assalim, A.M. & Molina, F.B. (2014). The presence of the red-eared slider, *Trachemys scripta elegans* (Wied, 1838) (Testudines, Emydidae), an invasive species, in the Paraibuna river basin, Southeastern Brazil. Herpetology Notes, 7, 437-441.

McKinney, M.L. (2008). Effects of urbanization on species richness: A review of plants and animals. Urban Ecosystems, 11, 161-176.

Mocelin, M.A., Fernandes, R., Porto, M., & Fernandes, D.S. (2008). Reproductive biology and notes on natural history of the side-necked turtle *Acanthochelys radiolata* (Mikan, 1820) in captivity (Testudines: Chelidae). South American Journal of Herpetology, 3, 223-228.

Molina, F.B. (1989). Observações sobre a biologia e o comportamento de *Phrynops geoffroanus* (Schweigger, 1812) em cativeiro (Reptilia, Testudines, Chelidae). MSc dissertation. São Paulo, SP, Brazil: Universidade de São Paulo.

Molina, F.B. (1998). Comportamento e biologia reprodutiva dos cágados *Phrynops geoffroanus*, *Acanthochelys radiolata* e *Acanthochelys spixii* (Testudines, Chelidae) em cativeiro. Revista de Etologia, special number, 25-40.

Molina, F.B., (2006). Ambientes urbanos e a fauna de répteis no Estado de São Paulo: entre desconhecidos, desaparecidos e indesejáveis. In R. Glezer & M.S.M. Mantovani (Eds.), Parques urbanos e meio ambiente: Desafios de uso (p.203-222). São Paulo, SP: Parque Cientec/USP, PEFI.

Molina, F.B., & Gomes, N.M. (1998). Observações sobre a incubação artificial dos ovos e o processo de eclosão em *Trachemys dorbignyi* (Reptilia, Testudines, Emydidae). Revista Brasileira de Zoologia,15, 135-143.

Molina, F.B., & Rocha, M.B. (1987). Tartaruga japonesa: biologia e manejo. Aquacultura, 3, 27-28.

Myers, N., Mittermeier, R.A., Mittermeier, C.G., Fonseca, G.A.B., & Kent, J. (2000). Biodiversity hotspots for conservation priorities. Nature, 403, 853-858.

Novelli, I.A., Gomides, S.C., Brugiolo, S.S.S., & Sousa, B.M. (2013). Alimentary habits of *Hydromedusa maximiliani* (Mikan, 1820) and its relation to prey availability in the environment. Herpetology Notes, 6, 503-511.

Pérez-Santigosa, N., Díaz-Paniagua, C., & Hidalgo-Vila, J. (2008). The reproductive ecology of exotic *Trachemys scripta elegans* in an invaded area of southern Europe. Aquatic Conservation: Marine and Freshwater Ecosystems, 18, 1302-1310.

Pérez-Santigosa, N., Florencio, M., Hidalgo-Vila, J., & Díaz-Paniagua, C. (2011). Does the exotic invader turtle, *Trachemys scripta elegans*, compete for food with coexisting native turtles? Amphibia-Reptilia, 32, 167-175.

Ribas, E.R., & Monteiro Filho, E.L.A. (2002). Distribuição e habitat das tartarugas de água-doce (Testudines, Chelidae) do estado do Paraná, Brasil. Biociências, 10, 15-32.

Ribeiro, M.C., Metzger, J.P., Martensen, A.C., Ponzoni, F.J., & Hirota, M.M. (2009). The Brazilian Atlantic Forest: How much is left, and how is the remaining forest distributed? Implications for conservation. Biological Conservation, 142, 1141-1153.

Santos, T.G., Vasconcelos, T.S., Molina, F.B., & Zaher, H. (2009). First record of *Trachemys dorbigni* (Duméril & Bibron, 1835) (Testudines, Emydidae) in a remnant of Mesophytic Semideciduous Forest of Sao Paulo State, southeastern Brazil. Herpetological Bulletin, 108, 27-30.

Souza, F.L. (2005). Geographical distribution patterns of South American side-necked turtles (Chelidae), with emphasis on Brazilian species. Revista Española de Herpetología, 19, 33-46.

Souza, F.L., & Abe, A.S. (1997a). Population structure, activity and conservation of the Neotropical freshwater turtle, *Hydromedusa maximiliani*, in Brazil. Chelonian Conservation and Biology, 2, 521-525.

Souza, F.L., & Abe, A.S. (1997b). Seasonal variation in the feeding habits of *Hydromedusa maximiliani* (Testudines, Chelidae). Boletín de la Asociación Herpetológica Española, 8, 17-20.

Souza, F.L., & Abe, A.S. (1998). Resource partitioning by the Neotropical freshwater turtle, *Hydromedusa maximiliani*. Journal of Herpetology, 32, 106-112.

Souza, F.L., & Abe, A.S. (1999). Um sobrevivente em rios poluídos. Ciência Hoje, 25, 59-61.

Souza, F.L., & Abe, A.S. (2000). Feeding ecology, density and biomass of the freshwater turtle, *Phrynops geoffroanus*, inhabiting a polluted urban river in south-eastern Brazil. Journal of Zoology, 252, 437-446.

Souza, F.L., & Abe, A.S. (2001). Population structure and reproductive aspects of the freshwater turtle, *Phrynops geoffroanus*, inhabiting an urban river in southeastern Brazil. Studies on Neotropical Fauna and Environment, 36, 57-62.

Souza, F.L., & Martins, F.I. (2009). *Hydromedusa maximiliani* (Mikan, 1825) – Maximilian's snake-necked turtle, Brazilian snake-necked turtle. Chelonian Research Monographs, 5, 285-327.

Souza, F.L., Martins, M., & Sawaya, R.J. (2000). A new record and observations of Vanderhaege's Toad-Headed Turtle, *Phrynops vanderhaegei* (Testudines, Chelidae) in SE Brazil. Boletín de la Asociación Herpetológica Española, 11, 85-88.

Souza, F.L., da Cunha, A.F., de Oliveira, M.A., Pereira, G.A.G., & dos Reis, S.F. (2002). Estimating dispersal and gene flow in the neotropical freshwater turtle *Hydromedusa maximiliani* (Chelidae) by combining ecological and genetic methods. Genetics and Molecular Biology, 25, 151-155.

Souza, F.L., Giraldelli, G.R., & Martins, T.A. (2006). Reproductive aspects of Brazilian side-necked-turtles (Chelidae). Boletin de la Asociación Herpetológica Española, 17, 28-34.

Souza, F.L., Raizer, J., Costa, H.T.M., & Martins, F.I. (2008). Dispersal of *Phrynops geoffroanus* (Chelidae) in an urban river in Central Brazil. Chelonian Conservation and Biology, 7, 257-261.

Tabarelli, M., Pinto, L.P., Silva, J.M.C., Hirota, M., & Bedê, L. (2005). Challenges and opportunities for biodiversity conservation in the Brazilian Atlantic Forest. Conservation Biology, 19, 695-700.

Tabarelli, M., Aguiar, A.V., Ribeiro, M.C., Metzger, J.P., & Peres, C.A. (2010). Prospects for biodiversity conservation in the Atlantic Forest: Lessons from aging human-modified landscapes. Biological Conservation, 143, 2328-2340.

Thomson, R.C., Spinks, P.Q., & Shaffer, H.B. (2010). Distribution and abundance of invasive red-eared sliders (*Trachemys scripta elegans*) in California's Sacramento River Basin and possible impacts on native western pond turtles (*Emys marmorata*). Chelonian Conservation and Biology, 9, 297-302.

Turtle Taxonomy Working Group [van Dijk, P.P., Iverson, J.B., Rhodin, A.G. J., Shaffer, H.B., & Bour, R.]. (2014). Turtles of the world, 7th edition: Annotated checklist of taxonomy, synonymy, distribution with maps, and conservation status. In A.G.J. Rhodin, P.C.H. Pritchard, P.P. van Dijk et al. (Eds.), Conservation biology of freshwater turtles and tortoises: A compilation project of the IUCN/SSC Tortoise and Freshwater Turtle Specialist Group. Chelonian Research Monographs 5(7):000.329-479, DOI:10.3854/crm.5.000.checklist.v7.2014.

Verdade, L.M., Gheler-Costa, C., Penteado, M., & Dotta, G. (2012). The impacts of sugarcane expansion on wildlife in the state of São Paulo, Brazil. Journal of Sustainable Bioenergy Systems, 2, 138-144.

Rodney Kozlowiski de Azevedo, Vanessa Doro Abdallah

10 Fish Parasites and their Use as Environmental Research Indicators

Abstract: Brazil is the fifth largest country in the world and presents the greatest diversity of species among all of the megadiverse countries, contributing approximately 14% of the world's biota. Currently, the biodiversity of the saltwater and freshwater ecosystems in Latin America have been threatened, mainly by environmental problems resulting from the degradation of ecosystems. Disruptive activities (monoculture, reforesting and cattle raising) in the drainage areas of rivers have reduced the native and riparian forests, thus triggering changes in the water quality and exposing the watershed to increasing erosion processes which lead to the siltation of the river bed, besides polluting its waters. From another perspective, the fish parasites have been indicated as excellent models for studies of parasite ecology and environmental indicators, considering that the features of the aquatic ecosystem, especially the physical-chemical characteristics of water, influence the composition and structure of parasites communities. Parasites are ubiquitous, though often invisible, components of ecosystems. They influence the survival and reproduction of individual hosts, the dynamics of host populations, and the structure of entire communities. Parasites are significant regulators of host populations and are potent agents that maintain the integrity and stability of ecosystems. Parasites, especially those that have complex life cycles involving more than one obligate host, are indicators of stable trophic structure in ecosystems. This is because all the biotic components necessary for completion of the life cycle must co-occur regularly in order to maintain any given parasite species. Knowing the complement of parasite species inhabiting any given host thus provides a means of rapid assessment of the breadth and form of trophic interactions of host species. Parasites are key to understanding the context of global change. Recently there has also been an increasing interest in the relationship between parasitism and pollution in the aquatic environment. Parasites may offer advantages over currently-used bioindicators including a more widespread distribution and a higher accumulation potential. Introduced parasites can have unpredictable and deleterious impacts on native species of hosts. It is therefore important to be able to quickly distinguish native from introduced parasite species.

10.1 Introduction

Parasites by definition have a dependence on their host for their survival and growth. Behind this simple conception lies a complexity that reflects the degree of dependence, of virulence, or both, and how these vary in time and space. The notion of dependence is relative, not absolute. A particular parasite may lie somewhere along the scale of 0% metabolic dependence (or being absolutely free-living) to 100% metabolically dependent (or 'totally' parasitic). An ecologic rule specifies that an obligatory parasite should not kill (cause the ultimate harm to) its host to benefit from the adapted long-term symbiosis. Highly virulent parasites may lose their virulence, eventually resulting in a mutualistic relationship with the host. Host–parasite relations may switch from being apparently benign to injurious, that is, disease causing. This may occur through context-dependent virulence in response to host condition, or through a 'strategic' expression of virulence aimed at enhancing transmission. These shifts occur at different temporal scales, either within the generation time of the host, or over perhaps extraordinarily long periods in an evolutionary sense. Thus any Host–parasite relationship is a dynamic one, shifting along a continuum precisely definable only at an instant in time and space. Much like a waltz, the 'partners' are constantly adapting to each other's 'moves' in responses to the presence, or potential presence, of each other. This adaptive response can be behavioural, genotypic or phenotypic, implicating each of the partners, and the interaction between them, environmental noise and spatial population structures. An added dimension to this waltz is at what stage the injurious action occurs, and to whom (Hortwiz & Wilcox, 2005).

The parasitism, under all its forms and guises, is a profoundly important ecological process. In the context of wildlife and human infections, parasitism is not simply a pathogenic relationship requiring treatment, but rather a process that through multiple agencies contributes to within and between species diversity, community structure and diversity, and therefore to the ability of organisms to respond to change (Hortwiz & Wilcox, 2005).

Host–parasite relationships are the 'unseen' part of an ecological community affecting, as do predator–prey relationships and inter-specific competition, species distribution and abundance and thus community composition. Parasites and their hosts are part of ecological communities, and just as they cannot be considered in isolation from one another they cannot be separated from the communities of which they form a part. Nor can Host–parasite relationships or their ecological communities be fully understood outside the ecosystem context within which both exist. We can conceive of Host–parasite relationships as being embedded. Hosts have other parasites, competitors and predators and each parasite has other hosts or vectors, and so on. Hosts are linked to other species and non-living properties of their surroundings, and so are parasites, depending on their life cycle. All of them are influenced by their environment (Hortwiz & Wilcox, 2005).

Brazil is the fifth largest country in the world and has the highest species diversity of all of the megadiversity countries, accounting for roughly 14 % of the world's biota (Muniz-Pereira et al., 2009). In recent years, parasites have been recognized as an important component of global biodiversity (Poulin & Morand, 2004). Given the integral roles played by parasites in natural ecosystems, identifying hotspots of high parasite diversity, as well as areas of relatively low parasite diversity, is crucial for a complete understanding of the functioning of the biosphere (Luque & Poulin, 2007). Currently, the biodiversity of the freshwater ecosystems of Latin America is threatened, mainly by environmental problems resulting from the degradation of the ecosystems. In this context, parasite biodiversity can be very important because parasitism plays key roles in ecosystems, regulating the abundance or density of host populations, stabilizing food webs and structuring animal communities. Thus, a good knowledge of parasite diversity and whether or not it is declining is crucial for environmental management and conservation (Luque & Poulin, 2007). Therefore, biotic diversity studies are presently conducted with a sense of urgency and they underscore the need for baseline biological inventories and surveys. The faunal survey is critical, because even the most elementary information is lacking for many groups in many parts of the Neotropics (Moravec, 1998).

Habitat destruction is the single most important cause of the loss of rainforest biodiversity and is directly related to human population growth. Industrial, agricultural and waste-based pollutants can have catastrophic effects on many species. Those species which are more tolerant of pollution will survive; those requiring pristine environments (water, air, food) will not. Thus, pollution can act as a selective agent. Pollution of water in lakes and rivers has degraded waters so that many freshwater ecosystems are dying. Since almost 12% of animal species live in these ecosystems, and most others depend on them to some degree, this is a very serious matter. In developing countries approximately 90% of wastewater is discharged, untreated, directly into waterways (Sthapak, 2013).

As has also occurred worldwide, siltation in Brazilian watercourses has followed agriculturalization with neither environmental planning nor preservation. Traditional water protection for human uses does not necessarily protect biological diversity and integrity of aquatic ecosystems (Casatti, 2004). Among several consequences of vegetation removal, siltation is one that most directly affects warm water fish assemblages. A variety of mechanisms by which siltation affects fishes have been described, including physiological stress from clogged gills, egg and larvae smothering, changes in normal feeding, and other activities that depend on vision. At broader temporal and spatial scales, siltation can cause local extinction of species that depend on coarse substrates for reproduction and feeding, and also nektonic species that require a minimum water volume for foraging (Casatti, 2004).

The relevance of parasites in ecosystem organization, and in maintaining baseline ecological dynamics can themselves be considered a service. Parasites may additionally aid conservation efforts by providing information about the status of ecosystems. Due to their conspicuous roles in trophic networks, parasites can be indicators of food web structure. Studies of parasites can also shed light on the host's evolutionary and demographic history, migratory patterns, and help identify host origins. Finally, some intestinal helminths can bioaccumulate heavy metals, potentially removing significant amounts from the host's tissues (Gomez & Nichols, 2013).

10.2 A Brief History of the Parasitology of Fish in Brazil

Fish parasitology and pathology are fields of growing importance in view of a worldwide expansion of pisciculture efforts. It would seem to be inevitable that world populations will come to depend more and more on artificially cultured fish as naturally occurring fish populations become depleted. Fish parasitology and pathology are not really separate areas of study since most fish parasites cause mild to severe pathological alterations in their hosts (Thatcher, 2006).

The following major groups of animal parasites invade Neotropical freshwater fishes: Protozoa, Myxozoa, Ciliophora, Monogenea, Digenea, Cestoda, Nematoda, Acanthocephala, Crustacea, Hirudinea and Pentastomida.

Brazil shows a highly diversified freshwater fish fauna with approximately 2600 species of freshwater fish and 1300 species of marine fish (Buckup et al., 2007). The number of parasites described is low when matched with the fish species registered in Brazil. There are many studies of fish parasitology in the Neotropical region and more precisely in Brazil, focusing on different areas such as ecology, taxonomy and biodiversity. Some states in Brazil have been more greatly explored such as the Amazon, Paraná, São Paulo, Rio de Janeiro and Mato Grosso do Sul because of the importance of the great rivers found in these states. Below is a brief summary of groups of fish parasites with higher species richness.

In relation to class Trematoda, the first compilation work of species was performed by Viana (1924). Subsequently Travassos et al. (1969) published a list with approximately 500 species of trematodes parasites of animals from Brazil. Other compilations of records were published by Thatcher (1991, 1993, 2006). Finally, the most current study was performed by Kohn et al. (2007) which includes all published reports of trematodes parasites of fish from South America with more than 500 species from Brazil.

Monogenea species compilations for South America have been done by Thatcher (1991), Kohn & Cohen (1998), Kohn & Paiva (2000) and Thatcher (2006). Most recently Cohen et al. (2013) published a catalog of monogeneans parasites from South America, referring to 437 species from Brazil.

In relation to Crustacea the following studies should be mentioned: Rego (1983-1984), Boxshall & Montú (1997), Young (1998). Other important contributions to the knowledge of the diversity of parasitic Crustacea in Brazil were the studies of Thatcher (2006), Luque &Tavares (2007) and Eiras et al. (2010). Most recently Luque et al. (2013) published a list of Crustacea parasitizing fish from Brazil. This study increased the number of parasitic crustaceans currently known in Brazil to a total of 251 species.

Travassos et al. (1928) performed the original work which recorded numerous nematodes from freshwater fish. Subsequently, a large number of papers by many authors, especially in recent years, have added to our knowledge of this fauna. Compilations of records have been published by Thatcher (1991, 2006), restricted to the Amazon River basin, and by Vicente et al. (1985), supplemented by Vicente & Pinto

(1999). Moravec (1998), in his monograph on nematode parasites of Neotropical freshwater fish, also included numerous records from Brazil. More recently, Muniz-Pereira et al. (2009) recorded nematode species in a list of helminth parasites of threatened vertebrate species from Brazil; and Eiras et al. (2010) published a list of the nematode species associated with Brazilian freshwater fish. Finally, the most current study was performed by Luque et al. (2011) which produced a list of Nematoda associated with the fishes of Brazil, comprising 74 genera distributed in 142 named and 75 undetermined species of adult and larval nematodes.

10.3 Parasites as Bioindicators and Models for Studies of Ecology

An increase in urban and agricultural landscapes near forests threatens the preservation of biodiversity by reducing stream acidity and increasing dissolved solid concentrations, allowing for the invasion of exotic species (Zampella & Bunnell, 1998).

In recent years aquatic ecosystems suffered a permanent increase in pollution caused by industrialization and urbanization. Simultaneously, humans attempt to study every component of the ecosystem in order to understand the consequences of such external stress. In general, ecosystems are complex systems consisting of a number of mutual interacting components. Observed independently, each part (component) of a given ecosystem represents a piece of a puzzle. Combining each of the puzzle pieces should deliver an entire picture of the ecosystem condition. In the field of ecological monitoring, researchers are trying to study as many parts of a given ecosystem as possible in order to detect external stress factors, which mostly occur in the form of contamination. The chemical (all external substances, which naturally do not belong to the system) or physical (thermal, noise, radioactive, etc.) contamination itself can induce changes in the ecosystem's functionality and structure, which affects its overall performance. Therefore, ecological monitoring is mostly aimed at studying the changes that can be assessed after exploring in detail the balance between the system components. Worth noting is that all these groups (components) have a basic common characteristic – they are an inseparable part of aquatic ecosystems. But there are still some components less investigated, one of them being the group of fish parasites (Nachev, 2010).

In general, accumulation indicators are organisms, which are able to accumulate substances (in most cases toxic) from the surrounding environment within their bodies and thus deliver information about the bioavailability of the given substance and its environment contents. Good indicators must be sensitive to environmental alteration so that changes in their numbers can be used as a warning of deteriorating conditions before the majority of less sensitive organisms are seriously affected. Parasites may rank among the most sensitive of bioindicators because parasitic infections of fish reflect the health of the entire aquatic community (Marcogliese & Price, 1997). Various experimental and field studies demonstrated the sentinel features of parasites (Nachev, 2010). Parasites are attracting increasing interest from ecologists as potential indicators of environmental quality due to the variety of ways in which they respond to anthropogenic pollution. In environmental impact studies, certain organisms provide valuable information about the chemical state of their environment not through their presence or absence but instead through their ability to concentrate environmental toxins within their tissues. Free living invertebrates, notably bivalve mollusks, are commonly employed in this role as "sentinel organisms" to monitor the concentrations of bioavailable metals in aquatic ecosystems. Also certain parasites, particularly intestinal acanthocephalans of fish, can accumulate heavy metals to concentrations to orders of magnitude higher than those in the host tissues or the environment. A number of experimental studies demonstrate a clear time dependent accumulation of lead for acanthocephalans in their final hosts. These investigations provide evidence that the extremely high metal concentrations in the intestinal acanthocephalans of fish are not the result of a slow process of accumulation but instead a relatively rapid uptake to a steady-state level. Thus, metal concentrations in adult acanthocephalans respond rapidly to changes in environmental exposure of their hosts (Sures, 2001).

The presumption that aquatic parasites have no relation to environmental conditions prevailed for quite a while, which stands in disagreement with the parasite's specific biology. Fish parasites were always underrated by field ecologists in aquatic monitoring, because they lacked in most cases "direct" connection with the ambient water medium. They were observed mostly from the perspective of water born diseases or breakout infection events in fish populations, and not on the disturbed environment conditions. In the last couple of decades, after gathering more detailed information concerning these aspects, many studies showed that fish parasite communities also react to alterations in conditions. Furthermore these alterations resemble those of free living organisms. The first evidence was delivered by impact surveys on some ectoparasitic species of fish, particularly on monogenean trematodes. They are common fish parasites occurring on gills and skin, therefore they are in permanent contact with the surrounding environment. By observing the presence or absence of monogeneans and the diversity characteristics of their communities, it is possible to obtain valuable information about the alternation in environment factors. Thus, their close relationship to eutrophication processes was demonstrated, as well as to other pollution sources like effluents from the industry (e.g. pulp and paper mills). This relationship was mostly expressed by reduced species richness and unequal distribution of abundances. However, this parasite group exhibits some features similar to free living organisms, which are also in permanent contact with the surrounding environment (Nachev, 2010).

Endoparasitic assemblages, although "embodied" in the host, may also have a relationship to pollution. Thus, the first step to achieve an understanding of the interaction between parasites and environmental factors is to get an overview of parasite transmission. Despite the high variety and complexity in transmission, the larger part of the endoparasites exhibit stages affected by the environment conditions. The direct effect is normally expressed by lethal reactions of the free living larval stages (e.g. Metacercaria) or adults, whereas the indirect impact is addressed on the intermediate or final host – the pollution could drive the suitable intermediate and final hosts to extinction. It can also affect the host physiology and thus the infected host, as well as the parasites, may suffer more from environmental exposure. In both cases the pollution leads to changes in the diversity and richness of parasite communities and thus parasites can be used as effect indicators. For that reason the parasite communities are more frequently analyzed in respect to pollution in the last decades (Nachev, 2010).

Aquatic pollution is still a problem in many freshwater and marine environments. It causes negative effects for the health of the respective organisms. According to their effects pollutants can either be lethal or sublethal. The effects may manifest immediately (acute toxicity) or after prolonged exposure to the pollutant (chronic toxicity) (Sures, 2008).

There is mounting evidence that human activity facilitates the emergence of infectious diseases. A variety of anthropogenic pollutants are known to acutely or chronically influence fish parasites. These pollutants include heavy metals, petroleum hydrocarbons, pesticides, pulp mill and thermal effluents, domestic sewage and waste, sedimentation, and agricultural and indrustrial toxins (Landsberg, 1998). Agriculture is increasingly being demonstrated as a threat to aquatic organism survival, with many studies showing links between agricultural pollution and parasitism. For example, pesticides can immunosuppress aquatic organisms, ultimately increasing transmission and pathology due to infection. Although most research has centered on water pollution, there is evidence that the surrounding landscape is significant to transmission (King et al., 2010).

In addition to effects on free-living organisms, pollutants also affect the health of parasites and consequently their occurrence and distribution. In recent years, there has been increasing awareness that parasitism should be investigated in light of the respective environmental conditions. These studies show that pollution can increase or decrease parasitism depending on an uncountable number of interacting variables (Sures, 2008).

For heteroxenous metazoan parasites, environmental conditions must be favourable for all host levels (intermediate and final hosts) and for eventually free living stages of the parasites. For monoxenous metazoans, only the ecological demands of the host and the parasite must be realized. Parasites with direct life cycles are normally ectoparasites. They are in permanent contact with the water and are therefore likely to have developed a resistance to changes in water quality. Consequently, populations of such monoxenous parasites are expected to be less affected by changing environmental conditions compared to heteroxenous parasites. Apart from the fact that adverse environmental conditions, such as water pollution, might have deleterious effects on intermediate hosts of heteroxenous parasites and thus reduce parasite abundance, these substances might also have immunosuppressive effects leading to higher infection rates (Sures, 2008).

Effects of pollution on parasites may be positive or negative: pollution may increase parasitism, or it may be fatal for certain parasite species, leading to a decrease in parasitism. Generally, infections with ectoparasites tend to increase, whereas infections with endoparasitic helminthes tend to decrease with increasing levels of pollution. Hence, parasites may be used as bioindicators to monitor pollution (Sures, 2005).

Differences in the susceptibility of heteroxenous and monoxenous parasites can be used to indicate environmental pollution. In a study by Diamant et al. (1999) the ratio between heteroxenous and monoxenous (H/M) parasites was analyzed, as well as parasite species richness (SH/SM) and species diversity in rabbitfish (*Siganus rivulatus*) collected from several sites in the Red Sea. Higher H/M ratios and SH/SM ratios were found in the rabbitfish collected at the ecologically stable habitat of coral reef compared to rabbitfish from sandy habitat or mariculture-impacted sandy habitat.

The advantage of using population fluctuations in parasites to indicate environmental changes, such as pollution, is mainly that conclusions are not based only on a single species but on a combination of species: host(s) and parasite(s). If a certain parasite is present, the ecological demands of all organisms involved in the life cycle must be fulfilled (Sures, 2008).

King et al. (2010) tested the hypothesis that agricultural activity can affect parasite abundance, species richness and diversity in native bullfrogs, *Lithobates* (= *Rana*) *catesbeianus*, and demonstrated that agricultural activity was related to the abundance of many amphibian parasites, as well as to species richness and diversity, and that both pesticide contamination and landscape development can affect parasitism via different mechanisms.

10.4 Introduction of Species and their Deleterious Impacts

The introduction of species in new environments by anthropogenic action, on purpose or accidentally, has received great attention. One of the reasons for this attention is the capacity of introduced species to become invasive, causing extensive loss of biodiversity through biotic homogenization, as well as adverse economic impacts (Lacerda et al., 2013).

The introduction and transfer of invasive alien species among continents, regions and nations has often had significant impacts on the recipient aquatic and terrestrial ecosystems (Hulme, 2006). Introduced species severely threaten some native flora and fauna communities, either directly or by modifying ecosystem processes and functions (Reaveley et al., 2009). The introduction of non-native plants or animals has usually been deliberately facilitated by humans, and although most introduced species do not deleteriously affect ecosystems, a small proportion become invasive. The non-native species not only exhibit higher reproductive success but their superior foraging ability and large colony size may lead to competitive displacement of natives (Hulme, 2006). Traits that enable introduced species to persist or spread successfully include high fecundity, good defense mechanisms, high survival rates, adaptability and a lack of natural enemies, such as predators and diseases. Introduced species that become invasive can have catastrophic effects on native biodiversity assets and ecological processes by altering nutrient levels, hydrological cycles, fire regimes and community composition, including the removal of keystone species (Reaveley et al., 2009).

Large-scale introductions of exotic fish species are a relatively recent phenomenon. Most introductions were made for aquaculture purpose, for recreational fisheries and to improve wild stocks. A large part of introductions were either accidental through escape from aquaculture ponds and aquaria, or for reasons that are unknown. In tropical America, the earliest introductions were of sport fish followed by extra-tropical carps for culture. For Northeast Brazil, the impact of piscivorous fish was well-documented with relation to the changes in the reservoir fisheries. Local, Amazonian and exotic species of fish were introduced and the fish yields in reservoirs are substantial. Tilapias contribute about 30% of the catch in reservoirs. Piscivorous fishes have caused a negative impact on indigenous fish but the role of exotic sport fish, and extra-tropical herbivores and omnivores, remain altogether negligible. Although, the temporal scale is an important factor to be considered in evaluating the impact of species transfers, generally, it is considered that these operations may have consequences in regard to fish interactions, environment and habitat modifications, genetic deterioration, introduction of new diseases and socioeconomic context (Lazard & Dabbadie, 2002).

Just as the role of parasites has had late recognition in community ecology, it also took some time to be considered in invasion ecology (Lacerda et al., 2013). Keane & Crawley (2002) present the 'enemy release hypothesis' (ERH) which explains the success of introduced species using three predictions: (1) specialist enemies (parasites, pathogens or predators) will be absent from the new region; (2) specialist enemies of native hosts will rarely switch to exotic invasive species and (3) generalist enemies will have a smaller impact on exotic species than on natives.

In addition, Williams & Williams Jr. (1994) argue that: 1) Repeated introduction of fish from a native region predisposes the new environment to support parasites with complex life cycles; 2) Complex life cycles may be completed by introducing a single key species; 3) Even unsuccessful attempts to introduce fish may succeed in introducing their parasites; 4) Exotic parasites often change their behaviors and host preferences immediately in new environmental conditions; 5) Exotic parasites often cause serious disease in native fish or fish with no prior exposure and defenses against them, conversely, native parasites can cause serious diseases in exotic fish; 6) Established exotic parasites that maintain strict host specificity can be eradicated by eliminating their host. More generalized parasites can only be eradicated in the early stages of introduction; and 7) Aquaculture management practices and treatments for disease problems actually select for the most hardy and adaptive parasite species. Often fish with these parasites are then sent into new areas for aquacultural purposes and exotic parasites that become established among fish having few parasites will eventually change their host and attachment-site preferences to take advantage of these new niches. Ultimately, the new environmental system will drive the speciation of parasites.

10.5 Fish Farming in Natural Environments

The intensive culturing of fish in cages, such as floating net cages is widespread around the world, and is indicated as one of the major methods for intense fish production in the tropics. Over the last decade, cage aquaculture in Brazil has expanded in an expressive way, generating substantive income and an increase in the production and exportation of fish (Gorlach-Lira et al., 2013). Fish farming activities have been increasing and becoming an important source of protein for human consumption. Brazil has a great potential for the development of aquaculture, due to its vast territory and favorable climate conditions, which is also internationally acknowledged (Pavanelli et al., 2008). According to the Food and Agriculture Organization (FAO), the commodity that has the greatest interest for aquaculture production is fish. In 2008, the production of fish was responsible for 55% of world aquaculture production, and activity in 2009 contributed about 28.8 million tons of fish, generating in the order of 40.5 billion dollars. In Brazil, according to the Ministério da Pesca e Aquicultura report, aquaculture production in 2009 was 394,340 tons; there were plans to double that number by 2015. This production in Brazil is considered low, even though the country has large spatial water availability, with about 20% of global freshwater, and space for implementation of new fisheries projects (Ueda et al., 2013).

In hydroelectric plant reservoirs, the inclusion of these systems may be a new source of impact on the ichthyofauna of reservoirs (Brandão et al., 2012).

The environmental impact of waste (fish faeces, uneaten food, bacterial biomass) from the fish culture industry, notably from cage fish farms, is an increasing issue of concern around the world. The intensive fish culture in cages can lead to the eutrophisation of water bodies and to the emergence of deleterious effects on the water quality, such as the blooms of toxic cyanobacteria that are harmful for wildlife and humans (Gorlach-Lira et al., 2013). The problem is that this activity produces a considerable amount of unused food, and other products which are often lost to the aquatic environment. Furthermore, the food ingested by cage fish farming that is eliminated in the form of excreta (feces and metabolites), are also utilized by resident fish fauna occupying areas close to fish farming cages. Thus, part of this food is not fully exploited by fish in the farming cage and is, therefore, lost to the aquatic environment and can be used by the local biota. The availability of food may cause changes in the potential of aquatic organisms through trophic interactions. Additionally, the input of fish food into the ecosystem may confer greater dietary advantage to opportunistic native fish species in relation to non-opportunistic species. The increasing inclusion of net cages in littoral areas of large rivers can lead to modification of these important spaces that serve as protection, feeding and breeding areas for many species of fish and other organisms, and, consequently, change the dynamics of the natural environment. For this reason, the technological development and negative potential impacts caused by fish farming have been assessed in various regions of the world across marine and freshwater environments (Brandão et al., 2013).

Intensive cultivation is widely used. As a consequence, fish are subjected to chronic stress, caused by confinement, a situation that starts to facilitate the transmission of pathogens, as there is a lack of space for the fish to develop (Ueda et al., 2013). The fish live in balance with the parasites, but this balance can be broken, mainly by environmental disturbances, among which changes in the water quality have a relevant role, as well as inadequate management and high stocking densities of fish (Pantoja et al., 2012).

It is necessary to mention that infectious, parasitic and nutritional diseases cause serious health and economic problems for fish farmers. The exact values of the economic losses caused by these agents in fish farms in Brazil are not known, but it is known that there is great potential to cause significant mortality in farmed fish. Fish are likely to be parasitized by various groups: Amoebae, Flagellata, Apicomplexa, Ciliophora, Myxozoa, Monogenea, Digenea, Cestoda, Nematoda, Acanthocephala, Hirudinea, Crustacea and Mollusca (Ueda et al., 2013).

Therefore, in systems of intensive culture, infections caused by protozoan and metazoan are a frequent problem. Protozoan parasites are common in farmed fish and can cause economic losses. Metazoan are parasites that can cause gill infections, damage to eyes and internal organs, starvation, inflammation of the swim bladder, and inhibited oxygen exchange across gill lamella. They provide portals of entry for bacteria in fish. Therefore, these parasites can be limiting factors for the development of fish farms as they contribute to the low growth of fish and occurrence of diseases, reducing profitability and increasing the costs of production due to treatments. Thus, epidemiological studies in fish farms are important for adapting the management techniques and providing sanitary guidelines (Pantoja et al., 2012).

According to Nowak (2007), the rapid development of fish culture in cages has been associated with an emergence of parasitic diseases in native fish populations. In these systems, cages protect the cultivated fish against predators, but free water flow allows parasites to be transmitted to wild fish which causes serious environmental problems such as the decline of wild populations. Another impact observed in areas used for cage farms is the attraction of piscivorous birds, fish and mollusks (Ramos et al., 2013). These authors evaluated the influence of cage farming on *Austrodiplostomum compactum* (Lutz, 1928) Dubois, 1970 metacercariae infections in fish *Plagioscion squamosissimus*

(Heckel, 1840) specimens caught in an area close to the fish farm and an area that is not affected by the fish farm from the Chavantes reservoir, São Paulo State, Brazil, and their results showed a significant increase in the rate of infection of *A. compactum* metacercariae in eyes of *P. squamosissimus* at sites used for cage farming and argued that the possible causes of this increase may be related to the organic enrichment of the sediment and water because areas close to fish farms receive much of the generated effluent (fish faeces and unconsumed rations). Additionally, the high availability of fish, either cultivated or local, in these areas acts as an attractive factor for piscivorous birds. Together, with increased numbers of molluscs, fish and piscivorous birds, all of which are hosts involved in the life cycle of *A. compactum*, could lead to increased transmission of *A. compactum* to definitive and intermediate hosts. Furthermore, species of the genus *Austrodiplostomum*, including adults and metacercariae, can infect various host species from different animal groups (Eiras, 1994). Thus, the increase in metacercariae in areas used for cage farming as a result of the increased numbers of their intermediate and definitive hosts, together with the low host specificity of this parasite, may increase the possibility of infection in other fish species (Ramos et al., 2013).

An additional problem is the use of pesticides in agriculture. Specifically, we highlight herbicides - biological agents or chemicals that act by killing or suppressing the development of weeds (Moura et al., 2008). Herbicides represent a large share of the total pesticides marketed worldwide. Unfortunately, many of these molecules are likely to contaminate water resources through a high potential shift in soil profile (leaching), high persistence in soil, moderate water solubility and moderate adsorption of the organic matter present in the soil. In studies examining their impact on water resources, it is known that various formulations can cause a decrease in water quality used in public water supply, as well as behavioral and deep metabolic changes in fish. Cases of high fish mortality caused by pesticides are rare, but rather fish become subjected to long periods of stress, caused by sublethal concentrations. The sublethal concentrations in the long term, can promote greater deleterious effects for the survival of fish species than the lethal concentrations. This is due to changes in behaviors, affecting predator x prey relationships, position in school, food habits and reproductive success (Murty, 1988). All these changes can generate stress that makes fish more susceptible to attack by parasites.

10.6 Final Considerations

Recent decades have seen a great increase in the number of studies related to aquatic organism parasites, especially parasites with hosts that are important for cultivation and commercialization. Many problems such as accelerated deforestation, unplanned urban development and point source pollution, which involve conservation of freshwater resources in developing countries, should be considered. Rivers in agricultural and urban landscapes are affected by numerous natural and anthropogenic disturbances. These disturbances may affect resource quality and availability, physical and chemical conditions, ecological integrity, as well as disrupt ecosystem processes and biotic structures and may affect fish assemblages and their parasites. The parasitology of fish has been included in numerous studies, and in this chapter we presented some of these studies that have current relevance. Given the immense biodiversity found in Brazil, the issues related to parasites uncover an additional dimension which expands this biodiversity further.

Bibliography

Boxshall, G.A., & Montú, M. (1997). Copepods parasitic on Brazilian coastal fishes: a handbook. Nauplius, 5, 1-225.

Brandão, H., Lobón-Cerviá, J., Ramos, I.P., et al. (2012). Influence of a cage farming on the population of the fish species *Apareiodon affinis* (Steindachner, 1879) in the Chavantes reservoir, Paranapanema River SP/PR, Brazil. Acta Limnologica Brasiliensia, 24, 438-448.

Brandão, H., Nobile, A.B., Souto, A.C., et al. (2003). Influence of cage fish farming on the diet and biological attributes of Galeocharax knerii in the Chavantes Reservoir, Brazil. Boletim do Instituto de Pesca, 39, 157-167.

Buckup, P.A., Menezes, N.A., & Ghazzi, M.S. (2007). Catálogo das espécies de peixes de água doce do Brasil. Rio de Janeiro: Museu Nacional.

Casatti, L. (2004). Ichthyofauna of two streams (silted and reference) in the Upper Paraná river basin, Southeastern Brazil. Brazilian Journal of Biology, 64, 757-765.

Cohen, S.C., Justo, M.C.N., & Kohn, A. (2013). South American Monogenoidea parasites of fishes, amphibians and reptiles. Rio de Janeiro: Oficina de Livros.

Diamant, A., Banet, A., Paperna, I., et al. (1999). The use of fish metabolic, pathological and parasitological indices in pollution monitoring. II. The Red Sea and Mediterranean. Helgoland Marine Research, 53, 195-208.

Eiras, J.C. (1994). Elementos de ictioparasitologia. Porto: Fundação Engenheiro Antônio de Almeida.

Eiras, J.C., Takemoto, R.M., & Pavanelli, G.C. (2010). Diversidade dos parasitas de peixes de água doce do Brasil. Maringá: Cliche Tec Editora.

Gómez, A., & Nichols, E. (2013). Neglected wild life: Parasitic biodiversity as a conservation target. International Journal for Parasitology: Parasites and Wildlife, 2, 222-227.

Gorlach-Lira, K., Pacheco, C., Carvalho L.C.T., et al. (2013). The influence of fish culture in floating net cages on microbial indicators of water quality. Brazilian Journal of Biology, 73, 457-463.

Horwitz, P., & Wilcox, B.A. (2005). Parasites, ecosystems and sustainability: an ecological and complex systems perspective. International Journal for Parasitology, 35, 725-732.

Hulme, P.E. (2006). Beyond control: wider implications for the management of biological invasions. Journal of Applied Ecology, 43, 835-847.

Keane, R.M., & Crawley, M.J. (2002). Exotic plant invasions and the enemy release hypothesis. Trends in Ecology and Evolution, 17, 164-170.

King, K.C., McLaughlin, J.D., Boily, M., Marcogliese, D.J. (2010). Effects of agricultural landscape and pesticides on parasitism in native bullfrogs. Biological Conservation, 143, 302-310.

Kohn, A., & Cohen, S.C. (1998). South American Monogenea – list of species, hosts and geographical distribution. International Journal for Parasitology, 28, 1517-1554.

Kohn, A., Fernandes, B. M.M., & Cohen, S.C. (2007). South American trematodes of parasites of fishes. Rio de Janeiro: Fundação Oswaldo Cruz.

Kohn, A., & Paiva, M.P. (2000). Fishes parasitized by Monogenea in South America. In G. Salgado-Maldonado, A.N. Garcia-Aldrete, & V.M. Vidal (Eds.), Metazoan parasites in the neotropics: a systematic and ecological perspective (p.25-60). México: Universidade Nacional Autonma (UNAM).

Lacerda, A.C.F., Takemoto, R.M., Poulin, R., & Pavanelli, G.C. (2013). Parasites of the fish Cichla piquiti (Cichlidae) in native and invaded Brazilian basins: release not from the enemy, but from its effects. Parasitology Research, 112, 279-288.

Landsberg, J.H., Blakesley, B.A., Reese, R.O. et al. (1998). Parasites of fish as indicators of environmental stress. Environmental Monitoring and Assessment, 51, 211-232.

Luque, J.L., Aguiar, J.C., Vieira, F.M., et al. (2011). Checklist of Nematoda associated with the fishes of Brazil. Zootaxa, 3082, 1-88.

Luque, J.L., & Poulin, R. (2007). Metazoan parasite species richness in Neotropical fishes: Hotspots and the geography of biodiversity. Parasitology, 134, 865-878.

Luque, J.L., & Tavares, L.E.R. (2007). Checklist of Copepoda associated with fishes from Brazil. Zootaxa, 1579, 1-39.

Luque, J.L., Vieira, F.M., Takemoto, R.M., et al. (2013). Checklist of Crustacea parasitizing fishes from Brazil. Check List, 9, 1449-1470.

Marcogliese, D.J. & Price, J. (1997). The paradox of parasites. Global Biodiversity, 7, 7-15.

Moravec, F. (1998). Nematodes of Freshwater Fishes of the Neotropical Region. Praga: Academia.

Moura, M.A.M., Franco, D.A.S., Matallo, M.B. (2008). Impacto de Herbicidas sobre os Recursos Hídricos. Tecnologia e Inovação Agropecuária, 1, 142-51.

Muniz-Pereira, L.C., Vieira, F.M., & Luque, J.L. (2009). Checklist of helminth parasites of threatened vertebrate species from Brazil. Zootaxa, 2123, 1-45.

Murty, A.S. (1988). Toxicity of pesticides to fish. Flórida: CRC Press.

Nachev, M. (2010). Bioindication capacity of fish parasites for the assessment of water quality in the Danube River. Germany: Universität Duisburg-Essen.

Nowak, B.F. (2007). Parasitic diseases in marine cage culture – an example of experimental evolution of parasites? International Journal for Parasitology, 37, 581-588.

Pantoja, W.M.F., Fishing Engineer, L.N., Fishing Engineer, M.D., et al. (2012). Protozoan and metazoan parasites of Nile tilapia *Oreochromis niloticus* cultured in Brazil. Revista MVZ Córdoba, 17, 2812-2819.

Pavanelli, G.C., Eiras, J.C., & Takemoto, R.M. (2008). *Doenças de peixes profilaxia, diagnóstico e tratamento*. Maringá: Eduem.

Poulin, R., & Morand, S. (2004). Parasite Biodiversity. Washington: Smithsonian Books.

Ramos, I.P., Franceschini, L., Zica, E.O., et al. (2013). The influence of cage farming on infection of the corvine fish *Plagioscion squamosissimus* (Perciformes: Sciaenidae) with metacercariae of *Austrodiplostomum compactum* (Digenea: Diplostomidae) from the Chavantes reservoir, São Paulo State, Brazil. Journal of Helminthol*ogy*, 30, 1-7.

Reavely, K., Bettink, K., & Valentine, L. (2009). Impacts of introduced species on biodiversity. In B.A. Wilson & L.E. Valentine (Eds.), Biodiversity values and threatening processes of the Gnangara groundwater system (p.1-51). Australia: Gnangara Sustainability Strategy and the Department of Environment and Conservation.

Rego, A.A. (1983-1984). Sinopse dos pentastomídeos da região neotropical. Garcia de Orta, Série Zoológica, 11, 45-56.

Sthapak, M. (2013). Impact of Tourism on Environment. Research Link, 117, 26-28.

Sures, B. (2001). The use of parasites as bioindicators of heavy metals in aquatic ecosystems: a review. Aquatic Ecology, 35, 245-255.

Sures, B. (2005). Effects of pollution on parasites, and use of parasites in pollution monitoring, In K. Rohe (Ed.), Marine Parasitology (p.421-425). Collingwood: CSIRO Publishing.

Sures, B. (2008). Environmental Parasitology. Interactions between parasites and pollutants in the aquatic environment. Parasite, 15, 434-438.

Thatcher, V.E. (1991) Amazon fish parasites. Amazoniana, 11, 263-572.

Thatcher, V.E. (1993). Trematódeos Neotropicais. Manaus: Instituto Nacional de Pesquisas da Amazônia.

Thatcher, V.E. (2006). Amazon fish parasites. Sofia and Moscow: Pensoft Publishers.

Travassos, L., Artigas, P., & Pereira, C. (1928). Fauna helminthologica dos peixes de água doce do Brasil. Archivos do Instituto Biológico de *São Paulo, 1*, 5-68.

Travassos, L., Freitas, J.F.T., & Kohn, A. (1969). Trematódeos do Brasil. Rio de Janeiro: Memórias do Instituto Oswaldo Cruz.

Ueda, B.H., Karling, L.C., Takemoto, R.M., et al. (2013). Parasites of the freshwater fish trade in Brazil: science metric study. Pesquisa Veterin*ária Brasileira, 33*, 851-854.

Viana, L. (1924). Tentativa de catalogação das espécies brasileiras de trematódeos. Mem*órias do Instituto Oswaldo Cruz,* 17, 95-227.

Vicente, J.J. & Pinto, R.M. (1999). Nematóides do Brasil. Nematóides de peixes. Atualização: 1985-1998. Revista Brasileira de Zoologia, 16, 561-610.

Vicente, J.J., Rodrigues, H.O., & Gomes, D.C. (1985). Nematóides do Brasil. 1ª parte: Nematóides de peixes. Atas da Sociedade de Biologia do Rio de Janeiro, 25, 1-79.

Williams, L.B. & Williams Jr. E.H. (1994). Parasites of Puerto Rican Freshwater Sport Fishes. Puerto Rico: Sportfish Disease Project. Department of Marine Sciences. University.

Young, P.S. (1998). Catalogue of crustacean of Brazil. Rio de Janeiro: Museu Nacional.

Zampella, R.A. & Bunnell, J.F. (1998). Use of reference-site fish assemblages to assess aquatic degradation in pinelands streams. Ecological Applications, 8, 645-658.

Alex Augusto de Abreu Bovo, Katia Maria P. M. B. Ferraz, Luciano M. Verdade, José Roberto Moreira

11 Capybaras (*Hydrochoerus hydrochaeris*) in Anthropogenic Environments: Challenges and Conflicts

Abstract: Capybaras are the largest living rodents, found only in the Neotropics. The species is expanding their range as agricultural deforestation occurs, especially in Southeastern Brazil, Paraguay and Colombia. Major conflicts between capybaras and humans have been described, especially in the context of Brazilian spotted fever (BSF), road kill and crop damage. Even though capybaras are just one of many possible tick-borne hosts, the species has been blamed by the public as the main agent responsible for its occurrence. In addition, capybaras are among the species most frequently involved in road accidents causing extensive damage including human casualties due to its large body size and social behavior. Lastly, the increase of capybara populations in agricultural landscapes may lead to crop damage. Often such conflicts are not well managed, bringing up only short-term solutions that are inefficient in detecting the real problem. Instead of a plague, capybaras should be considered as a great natural resource still lacking the proper public policy to be effectively managed in Brazil.

11.1 Capybaras' Natural History

The capybara, *Hydrochoerus hydrochaeris* Linnaeus, 1766, is the largest living rodent with maximum dimensions of 1.30 m in length and 70 kg in weight (Ojasti, 1973; Ferraz et al., 2005; Lord, 2009). This species is widespread in Brazil and other countries of South America east of the Andes (Lord, 2009). The capybara family, Hydrochoeridae, includes another asympatric species, *Hydrochoerus isthmius* Goldman, 1912, which is spatially limited to Panama, Colombia and Venezuela (Moreira et al., 2013; Vucetich et al., 2013).

Capybaras are closely associated with aquatic environments, shrub vegetation and grasslands (Ojasti, 1973; Alho et al., 1987a; Herrera & Macdonald, 1989; Moreira & Macdonald, 1997). However, their primary use of water is for thermoregulation (Lord, 2009), reproduction (Azcárate, 1980; Macdonald, 1981) and refuge (Macdonald, 1981). Grasslands are preferred for grazing (Alho et al., 1987a; Macdonald, 1981), while shrub forest is more used for protection against predators, as well as for breeding sites (Herrera & Macdonald, 1989). Several studies have been carried out with the species in two biomes: the Llanos, in Venezuela (Cordero & Ojasti, 1981; Macdonald, 1981; Herrera & Macdonald, 1989) and the Pantanal, on the border between Brazil,

Paraguay and Bolivia (Schaller & Crawshaw, 1981; Alho et al., 1987a,b). These areas are flat savannas with ponds, isolated patches of woodland and seasonally flooded grassland, with two well-defined seasons: wet and dry. However, capybaras can also use other habitats and resources, generally having a stable territory which always includes available water bodies (Alho & Rondon, 1987), seasonally flooded grassland and dry land (Herrera & Macdonald, 1989).

Capybaras are grazing rodents with an utterly herbivorous diet, consisting mainly of grasses and aquatic plants (Ojasti, 1973; Escobar & Gonzalez-Jimenez, 1976; Herrera & Macdonald, 1989). The species shows different grazing patterns dependent on environmental variables. In general, individuals are more selective during the rainy season when food availability increases (Quintana et al., 1994; Barreto & Herrera, 1998; Herrera & Macdonald, 1989; Quintana et al., 1998). Diet consists primarily of *Hymenachne amplexicaulis*, *Leersia hexandra*, *Paratheria prostrata*, *Reimarochloa acuta* and some species of Cyperaceae in natural environments (Escobar & González-Jiménez, 1976; Barreto & Herrera, 1998; Quintana et al., 1994; Quintana et al., 1998). Activity patterns are strongly influenced by the ambient temperature. Herds usually stay close to the water or in the forest during the warmest period of the day and graze during periods of milder temperature (Alho et al., 1987a; Herrera & Macdonald, 1989; Lord, 2009).

An interesting aspect of capybara behavior is their social structure, forming groups with strong relationships between members (Ojasti, 1973; Herrera & Macdonald, 1987). The mean group size is about 15 individuals in natural areas (Alho & Rondon, 1987; Alho et al., 1987b; Herrera, 2013), varying from four to 40 individuals according to the region and season (Alho & Rondon, 1987; Alho et al., 1987a; Lord, 2009). During the dry season, food and water availability becomes more limited; so larger aggregations are formed by individuals of two or more groups, which tend to keep their unit, with no interaction between individuals of different groups (Herrera & Macdonald, 1987). On the other hand, groups tend to split during the wet season, when resources are more abundant (Ojasti, 1973; Herrera, 2013). There is a relatively stable but markedly linear hierarchy among males (Herrera & Macdonald, 1993), followed by adult females with their offspring of different ages. Subordinate males tend to stay at the periphery of the herd (Lord, 2009).

11.2 Capybaras in Anthropogenic Environments

Capybaras are expanding their range as agricultural deforestation occurs, especially in more developed regions of Southeastern Brazil (Moreira & Macdonald, 1997; Verdade & Ferraz, 2006; Ferraz et al., 2007), Paraguay (Campos-Krauer & Wisely, 2011) and Colombia (Aldana-Domingues et al., 2002). Such expansion is primarily determined by food availability and natural predator decline (Ross, 1998; Ferraz et al., 2003; Ferraz et al., 2007; Caso et al., 2008; Queirogas et al., 2012; Verdade et al., 2012). Croplands and anthropogenic wetlands, which provide an extensive source of food for capybaras, can have even higher carrying capacity than pristine wetlands (Tab. 11.1). In such circumstances, hunting (even illegal) can be the only effective population control. However, urban areas including university *campi*, recreational parks and suburban development offer protection against hunters and predators and plenty of food resources for the species, which can result in conflicts with humans (see section 11.2.2).

Tab. 11.1: Population densities of capybaras in distinct environments.

Location	Environment	Density (ind./ha)	Source
Taim Ecological Station. RS. Brazil	Natural	5.87	Garcias & Bager (2009)
Brazilian Pantanal	Natural	0.13	Schaller & Crawshaw (1981)
Brazilian Pantanal	Natural	0.07	Alho et al. (1987b)
Venezuelan Llano	Natural – Forest	1.84	Cordero & Ojasti (1981)
Venezuelan Llano	Natural – Savanna	2.06	Cordero & Ojasti (1981)
São Paulo State	Anthropogenic landscapes	0.82	Vargas et al. (2007)
São Paulo State	Anthropogenic landscapes	1.24	Verdade & Ferraz (2006)
São Paulo State	Anthropogenic landscapes	3.28	A.A.A. Bovo, São Paulo University, pers. obs.

11.2.1 Adaptation

Land use change frequently results in behavioral changes in wild species. Adaptive processes can occur in variable spatial and temporal scales, along a single or multiple number of generations (Rosalino et al., 2014). The former is usually based on phenotypic plasticity (West-Eberhard, 2003), and might be considered as "acclimation" (e.g. Watling & Press, 2000; Brown, 2001), whereas the later involves genetic changes to stronger selective pressures, usually called contemporary or rapid evolution (e.g. O'Steen et al., 2002; Sasaki et al., 2009). The evolution process in unsteady environments has been studied for decades (Simpson, 1944; Levins, 1968; Paterniani, 1969). However, evolutionary concepts have been commonly ignored in the fields of conservation biology and wildlife management. Processes associated with acclimation are likely more relevant to long-lived species. Alternatively, contemporary or rapid evolution is more relevant to species with small body size and short generation length (Rosalino et al., 2014).

Capybaras are relatively long-lived animals, but their generation length is similar to that of small rodents (Moreira & Macdonald, 1996; Eisenberg & Redford, 1999). The species thus present high phenotypic plasticity and genetic variability as possible responses to environmental change, which may result in their remarkable success in anthropogenic areas.

Even though capybaras prefer grass species from wetlands, they have a plastic diet being able to include other food sources in disturbed environments (Ferraz et al., 2003; Verdade & Ferraz, 2006; Barreto & Quintana, 2013). Several crops, like corn, rice, soybean and sugarcane can be consumed by the species, which occasionally results in conflict with humans (Bilenca & Kravetz, 1995; Ferraz et al., 2007). In addition, the species can show behavioral changes when colonizing man-made habitat (e.g. variation in herd size and activity patterns) (Verdade, 1996).

Groups of more than 50 individuals have been living in an agricultural landscape (approx. 40 ha, Fig. 11.1) for almost two decades, resulting in conflicts with humans even on a university campus in Piracicaba, state of São Paulo, in Southeastern Brazil (Verdade & Ferraz, 2013). Under such circumstances, a non-intentional process of gradual domestication may occur resulting in decreasing vigilant behavior by the species in the presence of humans.

Fig. 11.1: Capybara herd in Piracicaba, state of São Paulo, in Southeastern Brazil.

Spatial patterns of capybaras in such circumstances (rural and peri-urban areas) are different compared to those exhibited in natural environments (Fig. 11.2). In Pantanal and Venezuelan Llanos, the species' use of space is related to the variation in water, pasture and dry land availability (Alho et al., 1987b; Macdonald, 1981). Herds follow persistent water bodies during the dry season when intermittent ponds get dry, which results in the approximation of distinct groups (Alho et al., 1987b; Macdonald, 1981). On the other hand, patterns of land use change and productive cycles of crops in agricultural landscapes are determined by crop management and harvest (e.g. annual or perennial). In such circumstance, the use of space can be determined by the "productive pulse" of biomass in the landscape matrix (i.e., the agroecosystem), more than by seasonal variation of the water level (Verdade et al., 2014b). In urban areas, like Paranoá Lake (an artificial reservoir in Brasília, D.C., Brazil), the proximity to human habitations (including fenced properties by the lakeshore) reduced the available habitat for the species. In such circumstances, the species' home range was reduced (Moreira et al., 2001) and vigilance behavior enhanced (Moreira & Macdonald, 1997).

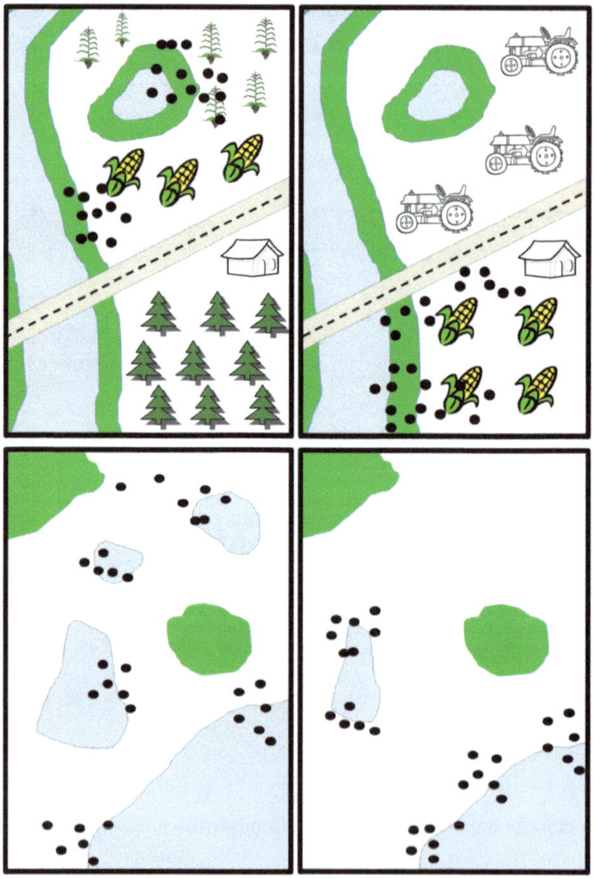

Fig. 11.2: Capybaras' use of space in anthropic (above) and natural (below) environments. Small black: individual capybaras. Larger light grey spots: water bodies. Larger dark grey spots: forest patches.

11.2.2 Conflicts with Humans

Major conflicts between capybaras and humans have been described in the last decade in Southeastern Brazil. The main causes of such conflicts are the association of capybaras with the Brazilian spotted fever, road kill events and crop damage.

a) Capybaras, ticks and the Brazilian spotted fever

Conflicts involving humans and capybaras have become frequent in anthropic environments in Southeastern Brazil, especially concerning the Brazilian spotted fever (BSF), since capybaras play an important role in the disease dynamics. Capybaras are hosts of the tick *Amblyomma cajennense*, which is a vector of *Rickettsia rickettsii*, the bacteria that transmits BSF to humans (Labruna et al., 2001; Labruna et al., 2004). Opossums (*Didelphis* sp.) and horses (*Equus caballus*) are, respectively, amplifiers and primary hosts of this pathogen (Labruna, 2013b).

BSF is treated as one of the most dangerous *rickettsiosis* in the world, with death occurring 40% of the time for the 557 cases reported in the state of São Paulo (official data between 1985 and 2013 from the São Paulo State Health Office, available at http://www.cve.saude.sp.gov.br/). A total of 34 cases were confirmed in the Piracicaba region between 2007 and 2012, leading to 16 deaths (Angerami et al., 2013). There are also confirmed cases in other regions of Brazil (Angerami et al., 2013; Labruna, 2013b).

The disease transmission depends on the time of contact between the tick and a person, requiring at least four to six hours for transmission to occur (Labruna, 2013a). Symptoms arise 2 to 14 days after transmission (Angerami et al., 2013), leading to fatality when the infection is not treated.

Even though capybaras are just a part of the disease cycle as one of the many possible tick hosts, the species has been blamed by the public as the main factor responsible for its occurrence (Ribeiro et al., 2010; Labruna, 2013b). The tick is capable of transovarial transmission; however, a host is required to prolong the tick reproduction cycle for many generations (Labruna, 2013a,b). Therefore, some crucial steps are necessary to decrease infections in humans. In addition, the monitoring of environmental conditions to reduce primary host populations, thus controlling tick populations, is crucial. It is also critical that clinical diagnoses be improved to avoid more casualties, as early treatment can dramatically reduce death risk (Angerami et al., 2013). For example, the patients who died of BSF in the Piracicaba region, had access to few doctors who failed in diagnosing the disease during its first stages (Verdade & Ferraz, 2013).

b) Road kill events

Roads and highways are extensively distributed throughout Southeastern Brazil, with frequent occurrences of accidents involving wild species (e.g. Cáceres et al., 2010; Bager & Fontoura, 2013). Capybaras are among the species most frequently involved in road accidents. Due to the species' large body size and social behavior, such accidents tend to cause extensive damage and casualties (Pinowski, 2005; Huijser et al., 2009;

Cáceres et al., 2010; Souza et al., 2010; Dornas et al., 2012; Bager & Fontoura, 2013). For example, a seven-year study examining a highway in the state of São Paulo found that capybaras accounted for 462 (28.4%) out of 1627 road kill events (Huijser et al., 2013). Fences and under- and above-ground passages should be implemented in critical road and highway areas in order to decrease such events (Huijser et al., 2013).

c) Crop damage

Capybara population increase in agricultural landscapes may lead to crop damage. Sugarcane and corn plantations are usually damaged by capybaras by herbivory (Fig. 11.3). Crop damage is commonly related to areas close to water bodies (Alho et al., 1987a; Azcárate, 1980; Bueno et al., 2013; Herrera & Macdonald, 1989; Macdonald, 1981) or forested areas (Ferraz et al., 2003), frequently associated with high-density populations (Verdade & Ferraz, 2006). In addition, damage to trees has also been detected in areas under revegetation (Pedro Brancalion[1], personal communication) (Fig. 11.4). In such circumstances, capybaras not only eat plants, but they also physically destroy them when marking territory. Damage to ornamental plants like heliconias, palm trees and bamboos in urban parks and gardens have also been described (Moreira et al., 2001).

Fig. 11.3: Damage caused by capybaras in sugarcane crop in Piracicaba, state of São Paulo, Brazil.

1 Departamento de Ciências Florestais, Escola Superior de Agricultura 'Luiz de Queiroz', Universidade de São Paulo

Fig. 11.4: Damage caused by capybaras in forest restoration area.

11.3 Final Considerations

Capybaras have been associated with the conflicts mentioned above, mostly due to misinformation from the media and even from some academic sectors (Verdade & Ferraz, 2013). Frequently, such conflicts are not well managed, bringing up only short-term solutions that are inefficient in detecting the real problem. However, this species has possibly the best potential to change the current conservative philosophy of wildlife management in Brazil (Verdade, 2004). The species status regionally varies from damaging to economic, demanding distinct management practices including monitoring (Verdade et al., 2014a). It is common, abundant and fast-growing enough to allow management mistakes, usual during the learning process (Verdade, 2004). For these reasons capybaras should be considered as a great natural resource (Moreira et al., 2013) still lacking – like others in Brazil – the proper public policy to be effectively managed. In order to do so, future studies should prioritize causes and consequences of conflicts between capybaras and humans, taking into consideration the many dimensions of both species.

Acknowledgments: We would like to thank the Wildlife Ecology, Management and Conservation Lab (LEMaC). We also thank the Forest Science Department, Luiz de Queiroz College of Agriculture, University of São Paulo (ESALQ/USP) and the Forestry Science and

Research Institute (IPEF). Part of this study was granted by the Biota Program / FAPESP. LMV has a productivity fellowship of CNPq (Proc. No. 309468/2011-6). KMPMBF has a productivity fellowship of CNPq (Proc. No. 308503/2014-7).

Bibliography

Aldana-Dominguéz, J., Forero-M., J., Betancur, J., et al. (2002). Dinâmica y estructura de la población de chigüiros (*Hydrochaeris hydrochaeris*: Rodentia, Hydrochaeridae) de Caño Limón, Arauca, Colombia. Caldasia, 24, 445–458.

Alho, C.J., & Rondon, N.L. (1987). Habitats, population densities, and social structure of capybaras (*Hydrochaeris hydrochaeris*, Rodentia) in the Pantanal, Brazil. Revista Brasileira de Zoologia, 4 (2), 139-149.

Alho, C.J.R., Campos, Z.M.S., & Gonçalves, H.C. (1987b). Ecologia de capivara (Hydrochaeris hydrochaeris, Rodentia) do Pantanal:-I Hábitats, densidades e tamanho de grupo. Revista Brasileira de Biologia, 47 (1/2), 87-97.

Alho, C.J.R., Campos, Z.M.S., & Gonçalves, H.C. (1987a). Ecologia de capivara (Hydrochaeris hydrochaeris, Rodentia) do Pantanal:-II Atividade, sazonalidade, uso do espaço e manejo. Revista Brasileira de Biologia, 47 (1/2), 99-110.

Angerami, R.N., Silva. M.V., Santos, F.C.P. et al. (2013). Febre Maculosa Brasileira: aspectos clínicos, epidemiológicos, diagnósticos e terapêuticos. In A.M. Meira, J.A. Monti, K.M.P.M.B. Ferraz et al. (Eds.), Febre Maculosa: Dinâmica da Doença, Hospedeiros e Vetores (p.32-51). Piracicaba, SP: ESALQ.

Azcárate, T. (1980). Sociobiologia y manejo del capibara (*Hydrochoerus hydrochaeris*). Doñana Acta Vertebrata, 7 (6), 1–228.

Bager, A., & Fontoura, V. (2013). Evaluation of the effectiveness of a wildlife roadkill mitigation system in wetland habitat. Ecological Engineering, 53, 31-38.

Barretto, A.G.D.O.P., Sparovek, G., & Giannotti, M. (2006). Atlas rural de Piracicaba. Instituto de Pesquisas e Estudos Florestais-IPEF.

Barreto, G.R., & Herrera, E.A. (1998). Foraging patterns of capybaras in a seasonally flooded savanna of Venezuela. Journal of Tropical Ecology, 14 (1), 87-98.

Barreto, G.R., & Quintana, R.D. (2013). Foraging strategies and feeding habits of capybaras. In J.R. Moreira, K.M.P.M.B. Ferraz, E.A. Herrera et al. (Eds.), Capybara: Biology, use and conservation of an Exceptional Neotropical Species (p.83-96). New York: Springer.

Bilenca, D.N., & Kravetz, F.O. (1995). Daños a maíz por roedores en la región pampeana (Argentina) y un plan para su control. Vida Silvestre Neotropical, 4, 51-57.

Brown, J.S. (2001). Fit of form and function, diversity of life, and procession of life as an evolutionary game. In S.H. Orzack & E. Sober (Eds.) Adaptationism and optimality (p.114-160). Cambridge: Cambridge University Press.

Bueno, C., Faustino, M.T., & Freitas, S. (2013). Influence of landscape characteristics on capybara road-kill on highway br-040, Southeastern Brazil. Oecologia Australis, 17 (2), 320-327.

Cáceres, N.C., Hannibal, W., Freitas, D.R. et al. (2010). Mammal occurrence and roadkill in two adjacent ecoregions (Atlantic Forest and Cerrado) in south-western Brazil. Zoologia (Curitiba), 27 (5), 709-717.

Campos-Krauer, J.M., & Wisely, S.M. (2011). Deforestation and cattle ranching drive rapid range expansion of capybara in the Gran Chaco ecosystem. Global Change Biology, 17 (1) 206-218.

Caso, A., Lopez-Gonzalez, C., Payan, E. et al. (2008). *Puma concolor*. In IUCN 2012. IUCN Red List of Threatened Species. Version 2012.2. <www.iucnredlist.org>. Downloaded on 15 February 2013.

Cordero, G.A. & Ojasti, J. 1989, Comparison of capybara populations of open and forested habitats. Journal of Wildlife Management, 45 (1), 267-271.

Dornas, R.A.P., Kindel, A., Bager, A. et al. (2012). Avaliação da mortalidade de vertebrados em rodovias no Brasil. In A. Bager. (Ed.) Ecologia de estradas: tendências e pesquisas (p.139-152). Lavras, MG: Editora UFLA.

Eisenberg, J.F., & Redford, K.H. (Eds.). (1999). Mammals of the Neotropics, Volume 2. Chicago: University of Chicago Press.

Escobar, A., & González-Jiménez, E. (1976). Estudio de la competencia alimenticia de los herbívoros mayores del Lhano inundable com especial al chiguire (*Hydrochaeris hydrochaeris*). Agronomía Tropical, 26 (3) 215-227.

Ferraz, K.M.P.M.B., Lechevalier, M.A., Couto, H.T.Z. et al. (2003). Damage caused by capybaras in a corn field. Scientia Agricola, 60 (1), 191-194.

Ferraz, K.M.P.M.B., Ferraz, S.F.B., Moreira, J.R. et al. (2007). Capybara (*Hydrochoerus hydrochaeris*) distribution in agroecosystems: a cross-scale habitat analysis. Journal of Biogeography, 34 (2), 223-230.

Ferraz, K.M.P.M.B., Bonach, K., & Verdade, L.M. (2005). Relationship between body mass and body length in capybaras (*Hydrochoerus hydrochaeris*). Biota Neotropica, 5 (1), 197-200.

Garcias, F.M., & Bager, A. (2009). Estrutura populacional de capivaras na Estação Ecológica do Taim, Brazil, RS. Ciência Rural, 39 (8), 2441-2447.

Herrera, E.A., & Macdonald, D.W. (1987). Group stability and structure of a capybara population. Symposium of the Zoological Society of London, 58, 115-130

Herrera, E.A., & Macdonald, D.W. (1989). Resource utilization and territoriality group-living capybaras (*Hydrochoerus hydrochaeris*). Journal of Animal Ecology, 58, 667-679.

Herrera, E.A., & Macdonald, D.W. (1993). Aggression, dominance, and mating success among capybara males (*Hydrochaeris hydrochaeris*). Behavioral Ecology, 4 (2), 114-119.

Herrera, E.A. (2013). Capybara social behavior and use of space: patterns and processes. In J.R. Moreira, K.M.P.M.B. Ferraz, E.A. Herrera et al. (Eds.), Capybara: Biology, use and conservation of an Exceptional Neotropical Species (p.195-207) New York: Springer.

Huijser, M.P., Duffield, J.W., Clevenger, A.P.et al. (2009). Cost-benefit analyses of mitigation measures aimed at reducing collisions with large ungulates in the United States and Canada: A decision support tool. Ecology and Society, 14 (2).

Huijser, M.P., Abra, F.D., & Duffield, J.W. (2013). Mammal road mortality and cost–benefit analyses of mitigation measures aimed at reducing collisions with capybara (*Hydrochoerus hydrochaeris*) in São Paulo state, Brazil. Oecologia Australis, 17 (1), 129-146.

Labruna, M.B., Kerber, C.E., Ferreira, F.et al. (2001). Risk factors to tick infestations and their occurrence on horses in the state of São Paulo, Brazil. Veterinary Parasitology, 97 (1), 1-14.

Labruna, M.B., Whitworth, T., Horta, M.C. et al. (2004). Rickettsia species infecting *Amblyomma cooperi* ticks from an area in the state of Sao Paulo, Brazil, where Brazilian spotted fever is endemic. Journal of Clinical Microbiology, 42 (1), 90-98.

Labruna, M.B. (2013a). Brazilian spotted fever: the role of capybaras. In J.R. Moreira, K.M.P.M.B. Ferraz, E.A. Herrera et al. (Eds.), Capybara: Biology, use and conservation of an Exceptional Neotropical Species (p.371-383). New York: Springer.

Labruna, M.B. (2013b). Epidemiologia da Febre Maculosa no estado de São Paulo. In M. Meira, J.A. Monti, K.M.P.M.B. Ferraz et al. (Eds.), Febre Maculosa: Dinâmica da Doença, Hospedeiros e Vetores (p.32-51). Piracicaba, SP: ESALQ.

Levins, R. (1968). Evolution in changing environments: some theoretical explorations (No. 2). Princeton: Princeton University Press.

Lord, R.D. (2009). Capybaras: a natural history of the world's largest rodent. Baltimore, Maryland: Johns Hopkins University Press.

Macdonald, D.W. (1981). Dwindling resources and the social behaviour of capybaras (*Hydrochoerus hydrochaeris*) (Mammalia). Journal of Zoology, 194 (3), 371-391.

Moreira, J.R., & Macdonald, D.W. (1996). Capybara use and conservation in South America. In V.J. Taylor & N. Dunstone (Eds.). The exploitation of mammal populations (p. 88-101). Dordrecht: Springer Science & Business Media

Moreira, J.R., & Macdonald, D.W. (1997). Técnicas de manejo de capivaras e outros grandes roedores na Amazônia. In C. Valladares-Pádua & R.E. Bodmer (Eds.), Manejo e Conservação de Vida Silvestre no Brasil (p.196-213). Brasília: Conselho Nacional de Desenvolvimento Científico e Tecnológico.

Moreira, J.R., Pinha, P.R.S., & Cunha, H.J. (2001) Capivaras do Lago Paranoá. In F.O. Fonseca (Ed.), Olhares sobre o Lago Paranoá (p.141-147). Brasília: Secretaria de Meio Ambiente e Recursos Hídricos.

Moreira, J.R., Alvarez, M.R., Tarifa, T. et al. (2013). Taxonomy, natural history and distribution of the capybara. In J.R. Moreira, K.M.P.M.B. Ferraz, E.A. Herrera et al. (Eds.), Capybara: Biology, use and conservation of an Exceptional Neotropical Species (p.3-37). Springer New York.

O'Steen, S., Cullum, A.J., & Bennett, A.F. (2002). Rapid evolution of escape ability in Trinidadian guppies (*Poecilia reticulata)*. Evolution, 56 (4), 776-784.

Ojasti, J. (1973). Estudio biológico del chigüire o capibara. Caracas: Fondo Nacional de Investigaciones Agropecuarias.

Ojasti, J. (1991). Human exploitation of capybara. In J.G. Robinson & K.H. Redford (Eds.), Neotropical wildlife use and conservation (p.236-252). Chicago: University of Chicago Press.

Paterniani, E. (1969). Selection for reproductive isolation between two populations of maize, *Zea mays* L. Evolution, 23 (4), 534-547.

Pinowski, J. (2005). Roadkills of vertebrates in Venezuela. Revista Brasileira de Zoologia, 22 (1). 191-196.

Queirogas, V.L., Del Claro, K., Nascimento, A.R.T. et al. (2012). Capybaras and ticks in the urban areas of Uberlândia, Minas Gerais, Brazil: ecological aspects for the epidemiology of tick-borne diseases. Experimental and Applied Acarology, 57 (1), 75-82.

Quintana, R.D., Monge, S., & Malvárez, A.I. (1994). Feeding habits of capybara (*Hydrochaeris hydrochaeris*) in afforestation areas of the Lower Delta of the Parana River, Argentina. Mammalia, 58 (4), 569-580.

Quintana, R.D., Monge, S., & Malvárez, A.I. (1998). Feeding patterns of capybara *Hydrochaeris hydrochaeris* (Rodentia, Hydrochaeridae) and cattle in the non-insular area of the Lower Delta of the Paraná River, Argentina. Mammalia, 62 (1), 37-52.

Ribeiro, K.T., da Silva Rocha, G.F., Saraiva et al. (2010). Das capivaras e carrapatos a uma proposta de comunicação e manejo no parque nacional da serra do cipó para redução de riscos à saúde. Oecologia Australis, 14 (3), 668-685.

Rosalino, L.M., Lyra-Jorge, M.C. & Verdade, L.M. (2014). Adaptation and evolution in changing environments. In L.M. Verdade, M.C. Lyra-Jorge & C.I. Piña (Eds.). Applied ecology and human dimensions on biological conservation (p.53-71). Heidelberg: Germany: Springer-Verlag.

Ross, J.P. (1998). Crocodiles. Status Survey and Conservation Action Plan. 2nd Edition. Gland: IUCN.

Sasaki, K., Fox, S.F., & Duvall, D. (2009). Rapid Evolution in the Wild: Changes in Body Size, Life-History Traits, and Behavior in Hunted Populations of the Japanese Mamushi Snake. Conservation Biology, 23 (1), 93-102.

Schaller, G.B., & Crawshaw, P.G. (1981). Social organization in a capybara population. Säugetierkundliche Mitteilungen, 29 (1).

Simpson, G.G. (1944). Tempo and mode in evolution. New York: Columbia University Press.

Souza, S.A., Lucca, A.L.T., Dickfeldt, E.P. et al. (2010). Impactos de atropelamentos de animais silvestres no trecho da rodovia SP-215 confrontante ao Parque Estadual de Porto Ferreira–Porto Ferreira, SP. Revista do Instituto Florestal, 22 (2), 315-323.

Vargas, F.C., Vargas, S.C., Moro, M.E.G. et al. (2007). Monitoramento populacional de capivaras (*Hydrochaeris hydrochaeris* Linnaeus, 1766) em Pirassununga, SP, Brasil. Ciência Rural, 37, 1104-1108.
Verdade, L.M. (1996). The influence of hunting pressure on the social behavior of vertebrates. Revista Brasileira de Biologia, 56 (1), 1-13.
Verdade, L.M. (2004). A exploração da fauna silvestre no Brasil: jacarés, sistemas e recursos humanos. Biota Neotropica, 4 (2), 1-12.
Verdade, L.M., & Ferraz, K.M.P.M.B. (2006). Capybaras in an anthropogenic habitat in southeastern Brazil. Brazilian Journal of Biology, 66 (1B), 371-378.
Verdade, L.M. & Ferraz, K.M.P.M.B. (2013). Capivaras de Piracicaba: o bom, o mau e o feio. In A.M. Meira, J.A. Monti, K.M.P.M.B. Ferraz et al. (Eds.), Febre Maculosa: Dinâmica da Doença, Hospedeiros e Vetores (p.143-161). Piracicaba, SP: ESALQ.
Verdade, L. M., Gheler-Costa, C., Penteado, M. et al. (2012). The Impacts of Sugarcane Expansion on Wildlife in the State of São Paulo, Brazil. Journal of Sustainable Bioenergy Systems, 2, 138.
Verdade, L.M, Lyra-Jorge, M.C. & Piña, C.I. (2014a). Redirections in conservation biology. In L.M. Verdade, M.C. Lyra-Jorge & C.I. Piña (Eds.), Applied ecology and human dimensions on biological conservation (p.3-17). Heidelberg, Germany: Springer-Verlag.
Verdade, L.M., Penteado, M., Gheler-Costa, C.et al. (2014b). The conservation value of agricultural landscapes. In L.M. Verdade, M.C. Lyra-Jorge & C.I. Piña (Eds.). Applied ecology and human dimensions on biological conservation (p.91-102). Springer-Verlag, Heidelberg, Germany.
Vucetich, M.G., Deschamps, C.M., & Pérez, M.E. (2013). Paleontology, evolution and systematics of capybara. In J.R. Moreira, K.M.P.M.B. Ferraz, E.A. Herrera et al. (Eds.), Capybara: Biology, use and conservation of an Exceptional Neotropical Species (p.39-59). Springer New York.
Watling, J.R., & Press, M.C. (2000). Light heterogeneity in tropical rain forests: photosynthetic responses and their ecological consequences. In M.J. Hutchings, L.A. John & A.J.A. Stewart (Eds.) The ecological consequences of environmental heterogeneity (p.131-154). Oxford: Blackwell Science.
West-Eberhard, M.J. (2003). Developmental plasticity and evolution. Oxford: Oxford University Press.

Karina L. Silva-Brandão, Danilo Bandini Ribeiro, Mariana Monteiro de Brito, Alberto José Arab Olavarrieta, André V. L. Freitas

12 The Influence of Agricultural and Forest Landscapes on the Diversity of Lepidoptera

Abstract: Lepidoptera is one of the most diverse groups of insects in the neotropics. It has been thoroughly used as model in several areas of Biology, with special interest in studies of evolution and interactions with larval host plants. Although a huge body of information is available for host plant use for lepidopterans, the influence of crop plants in Brazilian agroecosystems conditions on the evolution and diversification of populations is still reduced. This chapter presents current available data on genetic diversity of moths of economic interest in Brazilian conditions, with special discussion on the role of cultivated and wild host plants surrounding crops systems affecting the genetic structure of populations. The role of host plants, including important crops in Brazil, will also be presented in the discussion on Ecological Speciation in Lepidoptera, with information about what it is been done on this subject in Brazilian agroecosystems aiming the application on Integrated Pest Management. How the information gathered with model pest insects on Ecological Speciation due to host plant use can be used to understand speciation mechanisms on wild herbivorous insects will also be discussed. The third main discussion will focus on how the surrounding vegetation formed by contact zone between pastures and secondary forests influence the structure of butterfly assemblages and determine richness and abundance of groups, and how these studies can offer broader understanding of patterns of distribution of butterflies in fragmented landscapes.

12.1 Introduction

Intensification of agriculture is one of the main drivers of biodiversity declines worldwide (Tilman et al., 2002; Benton et al., 2003; Tscharntke et al., 2007a). The main environmental impacts of agriculture come from the conversion of natural ecosystems into a fragmented landscape composed by remnants of native vegetation surrounded by a matrix of crops, forestry, pastures and/or vegetation in different stages of succession (Saunders et al., 1991). However, besides fragmentation and obvious habitat loss, connectivity among the remaining habitats becomes impaired by agricultural matrix, highly impermeable for several organisms. Moreover, the widespread use of agricultural pesticides and fertilizers has profound effects on organisms and water bodies (Pimentel & Edwards, 1982; Logan, 1993; Paoletti et al., 1995; Groenendijk et al., 2002). Conversely, agroecosystems may also provide environmental benefits and important ecosystem services, such as pollination and biological control (Tscharntke et al., 2007b), also providing a greater diversity of conditions and resources to native organisms. As a result, they end up sheltering a larger number of species per unit area compared to more homogeneous environments (Magurran, 2004). Additionally, adoption of "conservation agriculture" measures can highly improve local biodiversity in agroecosystems (Paoletti et al., 1992; Hobbs et al., 2008).

In Brazil, vast areas of tropical forests have been converted into farmland in the last decades (Nepstad et al., 2006, Gibbs et al., 2010). As an example, from 1999 to 2011 agricultural areas grew from 530,000 to 680,000 km^2, especially due to large scale farming of commodity crops (such as soybean, sugarcane and corn) (Lapola et al., 2013). However, besides being one of the largest agrarian countries in the world, there is a general lack of information on the effects of agricultural landscapes on biodiversity in Brazil.

In this way, a good approach to understand the effects of agricultural landscapes on biodiversity is by better understanding the effects of habitat fragmentation on different biological groups. Studies focusing on habitat fragmentation and its effects on biodiversity have been attentive to both patch and landscape levels for several taxonomic groups, mainly insects (Duelli & Obrist, 2003; Bianchi et al., 2006; Hendrickx et al., 2007; Attwood et al., 2008). Among the insect groups, Lepidoptera (butterflies and moths) are excellent models to investigated such effects, for the following reasons: 1) they are very specific concerning their local habitat requirements, such as food resources and larval host-plants (Munguira et al., 2009), 2) many species are influenced by habitat degradation (Dover & Settele, 2009; Hanski, 1999), 3) several species may exploit resources from the surrounding matrix (Dennis & Hardy, 2007), and 4) local assemblages respond to modification on vegetation structure (Dover & Settele, 2009; Ribeiro et al., 2012).

In the present chapter we discuss two complementary approaches and levels of study with high potential to help to understand the role of agricultural and forest landscapes in shaping lepidopteran communities. First, we describe **genetic diversity** studies of polyphagous pest lepidopterans and how alternative crops and wild plants influence genetic and reproductive isolation in the field. Second, we provide examples of studies showing the effects of farmland and forested landscapes on the **community diversity** of frugivorous (fruit-feeding) butterflies.

12.2 Genetic Diversity of Lepidopteran Pests in Brazil

12.2.1 Lepidopterans-larval Host Plant Interactions and their Role in Genetic and Reproductive Incompatibility

With few exceptions, caterpillars are mainly herbivorous, that is, they feed on a broad variety of plant tissues. Most of them feed on external tissues of superior plants, a derived habit within the Order Lepidoptera, whose ancestors were endophagous that used to feed internally as miners on plant tissues (Menken et al., 2010). Although barely present in butterflies (exceptions are some species in the family Lycaenidae and species within *Dynamine* (Nymphalidae: Biblidinae), that feed partially inside *Dalechampia* (Euphorbiaceae) sprouts (Leite et al., 2014)), the endophagous habit is broadly found in many groups of moths, mainly microlepidopterans (Powell, 1980).

Herbivorous caterpillars however do not feed indiscriminately on any species of host plant (Ehrlich & Raven, 1964), and most of them show restrictions related to host plant use, mainly due to plants secondary compounds, which distinguish many species and superior taxa (Jaenike, 1990; Futuyma et al., 1993; Bernays, 1998; Schoonhoven et al., 2008). In this way, herbivorous species can be **monophagous** and feed on a few related host plant species; otherwise, **oligophagous** or **polyphagous** species can be associated to different host plant species within the same or distinct plant families, respectively (Schoonhoven et al., 2008).

Populations of polyphagous insects associated with different host plants through their geographical range can become genetically isolated in distinct lineages (Pashley, 1986), mainly due to ecological specialization followed by reduced gene flow and speciation processes (Funk, 1998; Dres & Mallet, 2002; Nosil et al., 2008). The resulting lineages could later become differentiated host races or biotypes (Emelianov et al., 1995; Groman & Pellmyr, 2000; Martel et al., 2003; Machado et al., 2008) due to the increased reproductive isolation (Dres & Mallet, 2002).

The characterization of insect lineages associated to crops is of considerable importance to answering both basic (e.g. speciation) and applied (e.g. pest management) questions. First, the biological information obtained for one lineage may not be applied to the other, compromising the strategies of pest control. Furthermore, the quantification of intra and interpopulation divergence among pest populations is pivotal to its management as the incorrect identification of species complexes can result in an erroneous employment of control techniques (Rosen, 1986; Paterson, 1991; Mills & Kean, 2010). For example, highly divergent populations of pest insects can differently damage their hosts (Rugman-Jones et al., 2010), and can respond differentially to control methods (Bickford et al., 2007).

Molecular markers have been broadly applied to evaluate the increasing reproductive isolation among insect populations, and to investigate species boundaries (Sperling & Hickey, 1994; Porreta et al., 2007; Dumas et al., 2015). Those markers can indicate if the reproductive isolation estimated by the lack of gene flow among individuals occupying different host plants is due to: 1) genetic drift acting distinctly in each host, or 2) **divergent selection** (i.e. selection arising from environmental differences or ecological interactions that bear polymorphic characters to evolve in contrasting directions on two populations (Nosil et al., 2009)) related to host plant use (Funk, 1998; Dres & Mallet, 2002). **Ecological speciation** studies (Rundle & Nosil, 2005; Matsubayashi et al., 2010) focus on the many ways that divergent selection can cause reproductive isolation among populations (Rundle & Nosil, 2005; Funk & Nosil, 2008). For polyphagous insects, genetic divergence and reproductive isolation are expected to be stronger between sympatric populations using different host plants than between populations using the same host because ecological divergence is indicative of divergent selection (Funk, 1998; Nosil et al., 2009).

12.2.2 Case Studies on Polyphagous Species

Population genetics studies of agricultural lepidopteran pests are indispensable to estimate genetic and reproductive isolation by the amount of gene flow among populations, in order to evaluate ecological factors shaping their present distribution (Scott et al., 2005; Endersby et al., 2006). Brazilian conditions comprise a complex scenario to evaluate the role of alternative larval host plants in the genetic structure of insect populations for several reasons. Among these we can highlight the vast human modified areas for intensive agriculture and livestock production, and more recently, forestry (Hirakurim et al., 2012); and the extraordinary natural biodiversity, coupled with high diversity of biomes and ecosystems. Abundant crop fields of the main Brazilian commodities provide food all year-round for polyphagous pest species, which allows them to maintain large persistent populations at regional levels. Simply, two possible mechanisms of genetic structure can emerge from that model: 1) populations are allowed to spread its range to all available areas containing preferred and alternative hosts, where they are permitted to evolve in isolation from other populations; 2) populations of pest species are seasonally subject to pest control practices and every season they go through a "bottleneck" that randomly sweep genetic variability. In this scenario, both divergent selection and random drift can cause genetic structuring in those populations. An additional local source of genetic variability may be composed of populations from neighbor wild host plants (Blanco et al., 2008), although those populations are hardly sampled in any study of population genetics of pest insects, and its influence is usually assumed.

12.2.2.1 *Spodoptera frugiperda*

The fall armyworm *Spodoptera frugiperda* (J.E. Smith) (Noctuidae) is the classic lepidopteran example of insect with well characterized host strains. *S. frugiperda* is a polyphagous pest species which cause damage in corn (*Zea mays* L.), rice (*Oryza sativa* L.), cotton (*Gossypium hirsutum* L.), sorghum (*Sorghum bicolor* (L.) Moench), and other crops in Brazil and worldwide. Numerous evidences suggest the existence of host strains, or genetic groups (Groot et al., 2010), in *S. frugiperda* populations (Busato et al., 2004; Nagoshi et al., 2007; Juarez et al., 2014; Dumas et al., 2015). Caterpillars collected in corn, sorghum and cotton represent typically the **corn strain**, while individuals from rice and several other pasture grasses are usually recognized as the **rice strain** (Pashley, 1986). Corn and rice strains are morphologically indistinguishable, but they show several ecological, genetic and physiological discriminant characters (Nagoshi & Meagher, 2008; Groot et al., 2010). Two main possible mechanisms of reproductive isolation are present: differences in female pheromones composition (Groot et al., 2008) and nocturnal timing of mating activity (Pashley et al., 1992). Hybridization between strains suggests incipient speciation (Groot et al., 2010), and although reproductive isolation is incomplete in the two lineages (Pashley, 1993), hybrids show lower fertility (Busato et al., 2008). The role of host plants to the reproductive isolation between *S. frugiperda* strains is unknown (Groot et al., 2010). Busato et al. (2004) used AFLP patterns to investigate the genetic diversity of populations from adjacent areas of corn and irrigated rice in south Brazil and found genetic variation associated to host plants on those localities. Groot et al. (2010) suggest several approaches to better understand how much host plants shape the evolution and isolation of strains in *S. frugiperda*. Certainly a broad population genetics study using *S. frugiperda* as model coupled with techniques of population genomics (Luikart et al., 2003; Ekblom & Galindo, 2011; Silva-Brandão et al., 2015b), and based on both neutral and non-neutral molecular markers (Kirk & Freeland, 2011), could provide an opportunity of finally investigate ecological aspects on the nature and amount of gene flow among Brazilian populations. That approach might contribute to the characterization of genes involved in the process of specialization due to host plant use, and could gather information on the interaction between ecological and reproductive isolation and their role in the process of speciation in herbivorous insects.

12.2.2.2 *Heliothis virescens*

The tobacco budworm *Heliothis virescens* (F.) (Noctuidae) is a polyphagous species described feeding on 14 families of host plants (Waldvogel & Gould, 1990), which could possibly lead to a high diversity related to host plant specialization. The species is an important pest of cotton in the American continent and in Brazil (Degrande, 1998), where it also has recently become an important pest of soybean (*Glycine max* (L.) Merr.) (Tomquelski & Maruyama, 2009). To investigate the possible impact on IPM strategies for controlling this pest, Albernaz et al. (2012) and Domingues et al. (2012) developed population genetics studies with *H. virescens* populations sampled from the main cotton- and soybean-producing areas in Brazil. Using mitochondrial DNA neutral markers, Albernaz et al. (2012) found unique haplotypes at specific localities (Fig. 1 A), but they found no significant genetic structure associated to host plant. The authors hypothesized however that the origin of rare haplotypes may be the migration of individuals from wild host plants found nearby crop fields every season; those rare haplotypes would be different in each sampled region due to the natural flora found in each Brazilian biome.

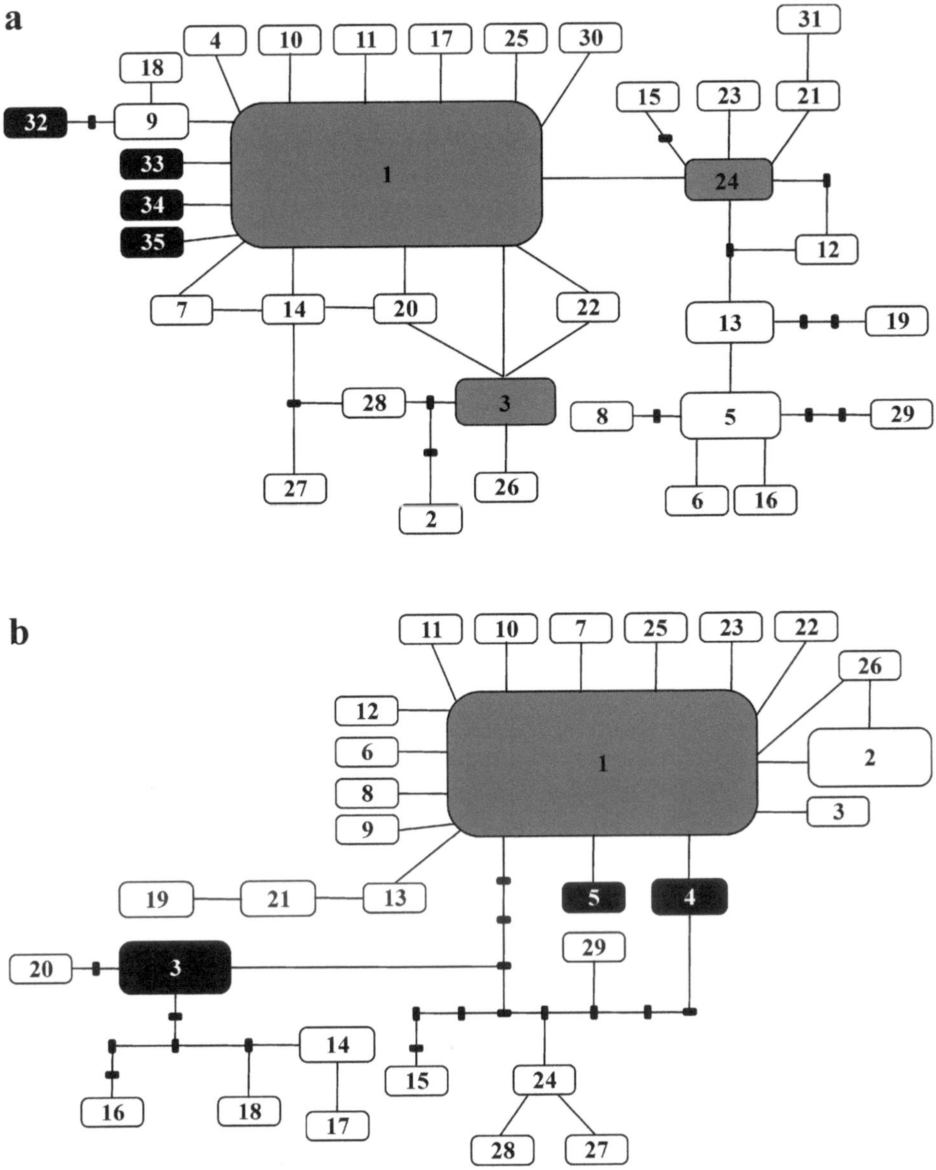

Fig. 12.1: Haplotypes networks of A) *Heliothis virescens*. White: haplotypes recorded only in populations from cotton, Black: haplotypes recorded only in populations from soybean, Grey: haplotypes recorded in cotton and soybean populations (modified from Albernaz et al., 2012); B) *Diatraea saccharalis*. White: haplotypes recorded only in populations from sugarcane, Black: haplotypes recorded only in populations from corn, Grey: haplotypes recorded in sugarcane and corn populations (modified from Silva-Brandão et al., 2015a).

12.2.2.3 *Diatraea saccharalis*

The sugarcane borer *Diatraea saccharalis* (F.) (Crambidae) feeds on several cultivated grasses, and is a pest of sugarcane (*Saccharum officinarum* L.), corn, rice, and sorghum (Dyar & Heinrich, 1928; Box, 1931; Myers, 1935; Long & Hensley, 1972; Pashley et al., 1990). The species is the main pest of sugarcane in Brazil (Pinto et al., 2006), and it is also turning into an important pest of corn following the expansion of Brazilian agriculture to the Savannah region (Cruz, 2007). A population genetic study focusing on the overall corn- and sugarcane-producing areas in Brazil suggested population structure associated to the crop plant from where individuals were sampled (Silva-Brandão et al., 2015a) (Fig. 1 B). The authors went further and raised the hypothesis of a possible influence of wild and alternative hosts as intake of individuals of the sugarcane borer to crop fields, considering that endemic species of *Saccharum* are indeed available nearby crop fields (Vianna et al., 2006). Similarly, they advocated that dispersal from cultivated rice- and sorghum-fields should certainly be estimated to better understand host plant dynamics for this important pest species in Brazil.

12.3 Diversity of Butterflies in Agricultural and Forest Ecosystems

12.3.1 Fruit-feeding Butterflies

Butterflies can be divided in two groups according to adult feeding habit: **fruit-feeding**, that acquired most of their nutritional requirements from rotten fruits, carcasses and plant sap, and **nectar-feeding**, that use flower nectar as their main nutritional source (DeVries, 1987). The fruit-feeding butterflies are represented, exclusively, by the Nymphalidae subfamilies Satyrinae (in the neotropics, three tribes are more commonly reported to be attracted to bait traps: Satyrini, Morphini and Brassolini; the Haeterini can be also captured by using horizontal bait traps), Biblidinae, Charaxinae and also some Nymphalinae (a non-monophyletic group of genera previously treated as tribe Coeini) (Wahlberg et al., 2009).

Fruit-feeding butterfly assemblages are clearly affected by land use (Barlow et al., 2007; Ribeiro et al., 2012; Ribeiro & Freitas 2012, Brito, 2013). Although these modifications rarely reflects in changes in species richness, they are evident in species composition, with some species increasing in abundance while other decreases with habitat change. These differences were documented in all studies comparing areas with contrasting levels of human disturbance, including logged versus unlogged areas (Barlow et al., 2007; Koh, 2007; Ribeiro & Freitas, 2012), continuous versus fragmented landscapes (Uehara-Prado et al., 2007), and primary versus secondary forests (Barlow et al., 2007; Uehara-Prado et al., 2009). Fruit-feeding butterfly assemblages are also affected by the surrounding landscape (Barlow et al., 2007, Ribeiro et al., 2012), as detailed in the section 12.4, and have also been successfully used as model organisms in studies of diversity in agroecosystems worldwide (Horner-Devine et al., 2003; Mas & Dietsch, 2003; Dolia et al., 2008).

The magnitude of effects on fruit-feeding butterflies is dependent on the intensity of land use modifications. In tropical forests, butterfly fauna usually shows a clear vertical stratification, with different species occurring exclusively in canopy and understory (DeVries et al., 1999; Ribeiro & Freitas, 2012). Intense forest modifications like **traditional logging** (the process of felling and extracting timber from forests without previous planning) can disrupt this pattern (DeVries, 1988; Schulze et al., 2001; Fermon et al., 2005), while **Reduced Impact Logging** (the use of several approaches to reduce damage to non-target trees and operational risks in timber extraction) apparently does not affect it (Ribeiro & Freitas, 2012). The understory fauna is usually more affected by changes in forest structure because it changes microclimatic conditions due to the increase of sunlight in the forest ground, reducing humidity and raising temperature. Another important consequence of those changes in forest structure affecting butterflies is the increase of pioneer plants and the reduction of some shadow love plants that modify the availability of host plant for butterfly larvae with consequent changes in butterfly assemblage composition.

In the opposite situation, forested habitats are mostly or completely converted in open habitats, as is the case of annual crops (e.g. corn, wheat (*Triticum* spp L.), and soybean) and even perennial cultures (e.g. coffee, *Coffea arabica* L.). In these now highly fragmented landscapes, conditions inside the small forest remnants can be usually far from ideal to harbor viable populations of forest specialists or species with poor dispersal abilities (Hanski, 1999; Steffan-Dewenter & Tscharntke, 2000). Accordingly, a recent study in a fragmented landscape in the Atlantic forest in Alfenas region, Minas Gerais State (southeastern Brazil) found a significant relationship between the degree of isolation of forest fragments, the shape of the fragment and the percentage of surrounding matrix with the functional composition of fruit-feeding butterfly assemblages. Surveys were conducted in nine sites containing forest fragments of semi-deciduous seasonal Atlantic rainforest and adjacent matrices of pastures and plantations of coffee and sugarcane. This study showed that abundance of butterflies in both habitats was equivalent; species composition however differed significantly between the forest fragments and the surrounding matrix. Species richness however was higher in forest fragments (based on rarefaction analysis).

The above results suggest that both richness and abundance of fruit-feeding butterflies can be predicted by the landscape structure of the study site (Tab. 12.1). Richness and abundance of forest species had a positive relationship with the proximity and size of surrounding forest fragments, while species that may exploit the matrix were particularly sensitive to the shape of the forest fragment and the percentage of surrounding matrix. Additionally, the species favored by more intensive agriculture in this study were typically mobile users or have grasses as host plants (Devries, 1987). It seems that generalists without habitat preference are able to compensate for a loss of their natural forest habitat by moving into the agricultural matrix. In contrast, forest specialists are less likely to move across adjacent forest fragments, therefore, the positive relationship between forest specialists and isolation may reflect their limited dispersal ability.

Tab. 12.1: Effect of the landscape structure on the richness and abundance of the ecological groups of fruit-feeding butterflies in a fragmented landscape in Southeastern Brazil. Values inside parentheses indicate the regression coefficients of the valid general linear models (Brito et al., 2014).

Ecological Groups	**Factors affecting abundance**		
Forest specialists	PROX (5.517)		
Matrix specialists	Sugarcane (1.746)	Pastures (5.167)	AWM (666.7)
Generalists	Pastures (1.667)	Coffee (0.443)	
Total forests assemblage	PROX (14.13)		
Total matrix assemblage	Pastures (7.323)	AWM (939.0)	Natural Cover (-1.915)
Generalists	NNA (8.206)	Pastures (0.646)	
Total forests assemblage	PROX (3.186)		
Total matrix assemblage	AWM (170.1)	Pastures (1.403)	

Natural cover = percentage of natural forest; Coffee = percentage of coffee plantations; Sugarcane = percentage of sugarcane plantations; Pastures = percentage of pastures; NNA = Nearest Neighbor Area (ha); PROX = Proximity index; AWM = forest fragment shape complexity

Microhabitat generalist insects have been shown to have more mobility among forest fragments than habitat specialists (Rosch et al., 2013), and responses to isolation could differ between matrix-tolerant species and forest specialists. Therefore, for butterflies, tolerance to the matrix may be an important criterion to predict species occurrence on fragmented landscapes. In addition, the availability of resources in the matrix is a key factor determining the presence of fragment-dependent species in fragmented landscapes (e.g. Jokimaki & Huhta, 1996)

The connectivity between fragments is an important feature of the landscape to specialist species. Therefore maintaining the connectivity among forests fragments is important for the conservation of forest specialists, and modifying the matrix in order to increase its permeability to dispersing butterflies may be an easier approach to restore connectivity than creating/restoring ecological corridors (Ricketts, 2001). For the conservation and maintenance of biodiversity management recommendations should take into account the characteristics of the ecological groups and not only focus on all of the species. To do this, it is important to understand which local and landscape factors are necessary for the conservation of habitat specialists.

12.4 Management of Agricultural Systems and Biodiversity

As previously stated, fruit feeding butterfly assemblages are affected not only by modifications in their habitats but also by what happens in the surrounding landscape (Ribeiro et al., 2012). The way butterflies use landscape is probably affected by their mobility, with more vagile butterflies exploring bigger areas than more resident species. This vagility is correlated with some species traits, like size, and long distance dispersal is very rare even in naturally fragmented landscapes (Marini–Filho & Martins, 2010). Thus, changes in the immediate vicinity vegetation (within a radius of 100-200 m) are more prone to affect local butterfly assemblages than modifications occurring in larger scales (1000-2000 m) (Ribeiro et al., 2012).

These effects are usually noted in the abundance and not in richness, and have been shown to be taxon dependent. Accordingly, for some butterfly groups whose larvae feeds mainly on monocots, the proportion of pasture lands on 100 and 200 m of radius are the most important landscape features explaining their abundances (Ribeiro et al., 2012), possibly reflecting the availability of adult and larval resources in these areas of contact between forests and pastures. Conversely, other butterfly groups showed more ambiguous responses to vegetation cover, and could probably respond to other landscape features not measured in that study.

The use of pesticides and transgenic cultures in the surrounding landscapes are also deleterious for butterflies but data about the extension of this damage are still absent. As a general pattern, we could suggest that more intensely managed landscapes have a greater potential to affect fruit-feeding butterflies assemblages than less intense ones.

12.5 Final Considerations and Future Directions

As aforementioned, though being one of the largest agrarian countries in the world, there is a general lack of information about biological diversity in agrarian systems in Brazil, and also about the effects of farmland management on its neighboring habitats. Regarding Lepidoptera, this gap is even greater, and most of the available information consists of preliminary lists of species for some few agricultural areas, published as abstracts in scientific meetings, preventing any kind of broader comparative studies.

Contrariwise, two lines of research have contributed considerably to bridge this gap: 1) Detailed studies on the genetic diversity of lepidopteran pests in agricultural habitats have increased our knowledge of how different types of agricultural management can affect many aspects of the populations of these pest species, and 2) concerning forested habitats, a growing body of information has served as the basis for an initial understanding of the main effects of different types of disturbance in the composition of local butterfly assemblages. In this chapter, we present recent results of these two lines of research in Brazil as an initial attempt to reconcile two approaches that have been separate for many years.

Even so, studies focusing on lepidopteran biodiversity in agrarian systems are needed in Brazil, and the following research themes are considered priorities:

1. Basic inventories in different kind of agroecossystems in all Brazilian biomes, with publication of lists of species (qualitative and/or quantitative), which will serve as basis for future studies;
2. Comparative studies of lepidopteran assemblages in different agricultural habitats with different crops and/or management systems (including organic farming);
3. Studies focusing on the influence of the agricultural matrix on the taxonomic and genetic composition of lepidopteran assemblages in areas adjacent to the agricultural habitats;
4. Long-term monitoring (of both populations, assemblages and/or genetic diversity) of Lepidoptera in agricultural habitats;
5. Based on the above information, promote management strategies, which would match production and conservation of biodiversity in tropical agroecosystems, the "conservation agriculture" (see FAO CA web site: http://www.fao.org/ag/ca/1a.html).

Acknowledgements: We thank Alexandre M. Dos Santos (PIBICT/FAPEMIG) and Eduardo Loureiro Abreu who helped throughout the field work. Part of this research was financially supported by a grant from Coordenação de Aperfeiçoamento de Pessoal de Nível Superior to MMB.

KLSB thanks "Fundação de Amparo à Pesquisa do Estado de São Paulo" (Fapesp Process # 2012/16266-7). AVLF thanks the Brazilian Research Council – CNPq (fellowship 302585/2011-7), and the National Science Foundation (DEB-1256742). This publication is part of the RedeLep "Rede Nacional de Pesquisa e Conservação de Lepidópteros" SISBIOTA-Brasil/CNPq (563332/2010-7), and of the BIOTA-FAPESP Program (Process # 2011/50225-3).

Bibliography

Albernaz, K.C., Silva-Brandão, K.L., Fresia, P., et al. (2012). Genetic variability and demographic history of *Heliothis virescens* (Lepidoptera: Noctuidae) populations from Brazil inferred by mtDNA sequences. Bulletin of Entomological Research, 102, 333-343.

Andrade, V.M., & Queiroz, J.M. (2007). Bases ecológicas para a conservação da biodiversidade: diversidade de espécies e interações de borboletas em agroecossistema de produção orgânica de alimentos. Anais do VIII Congresso de Ecologia do Brasil, 1-2.

Attwood, S.J., Maron, M., House, A.P.N., et al. (2008). Do arthropod assemblages display globally consistent responses to intensified agricultural land use and management? Global Ecology and Biogeography, 17, 585–599.

Barlow, J., Overal, W.L., Araujo, I.S., et al. (2007). The value of primary, secondary and plantation forests for fruit-feeding butterflies in the Brazilian Amazon. Jounal of Applied Ecology, 44, 1001-1012.

Benton, T.G., Vickery, J.A. & Wilson, J.D. (2003). Farmland biodiversity: is habitat heterogeneity the key? Trends in Ecology and Evolution, 18, 182–188.

Bernays, E.A. (1998). Evolution of feeding behavior in insect herbivores - success seen as different ways to eat without being eaten. Bioscience, 48, 35-44.

Bianchi, F.J.J.A., Booij, C.J.H., Tscharntke, T. (2006). Sustainable pest regulation in agricultural landscapes: a review on landscape composition, biodiversity and natural pest control. Proceedings of the Royal Society B, 273, 1715–1727.

Bickford, D., Lohman, D.J., Sodhi, N.S., et al. (2007). Cryptic species as a window on diversity and conservation. Trends in Ecology and Evolution, 22, 148-155.

Blanco, C.A., Teran-Vargas, A.P., Abel, C.A., et al. (2008). Plant host effect on the development of *Heliothis virescens* F. (Lepidoptera: Noctuidae). Environmental Entomology, 37, 1538-1547.

Box, H.E. (1931). The crambine genera *Diatraea* and *Xanthopherne* (Lep., Pyral.). Bulletin of Entomological Research, 22, 1-49.

Brito, M.M. (2013). A influência de diferentes tipos de matriz na guilda de borboletas frugívoras (Insecta: Lepidoptera) em fragmentos de mata atlântica. Brazil: Universidade Federal de Alfenas, UNIFAL-MG.

Busato, G.R., Grutzmacher, A.D., de Oliveira, A.C., et al. (2004). Analysis of the molecular structure and diversity of *Spodoptera frugiperda* (JE Smith) (Lepidoptera : Noctuidae) populations associated to the corn and rice crops in Rio Grande do Sul State, Brazil. Neotropical Entomology, 33, 709-716.

Busato, G.R., Loeck, A.E., Garcia, M.S., et al. (2008). Compatibilidade reprodutiva entre os biótipos “milho” e “arroz” de *Spodoptera frugiperda* (J.E. Smith) (Lepidoptera: Noctuidae). Revista Brasileira de Agrociência, 14, 273-278.

Cruz, I. (2007). A broca da cana-de-açúcar, *Diatraea saccharalis*, em milho, no Brasil. Circular Técnica, 90, 12. EMBRAPA/CNPMS.

Degrande, P.E. (1998). Guia prático de controle das pragas do algodoeiro. Dourados, UFM: 60.

Dennis, R.L.H., & Hardy, P.B. (2007). Support for mending the matrix: resource seeking by butterflies in apparent non-resource zones. Journal of Insect Conservation, 11, 157–168.

Dumas, P., Barbut, J., Le Ru, B., et al. (2015). Phylogenetic molecular species delimitations unravel potential new species in the pest genus *Spodoptera* Guenee, 1852 (Lepidoptera, Noctuidae). Plos One, 10, e0122407.

DeVries, P.J. (1987). The butterflies of Costa Rica and their natural history: Papilionidae, Pieridae, and Nymphalidae. Princeton: Princeton University Press.

DeVries, P.J. (1988). Stratification of fruit-feeding nymphalid butterflies in a Costa Rican rainforest. Journal of Research on the Lepidoptera 26, 98-108.

DeVries, P.J., Walla, T.R., & Greeney, H.F. (1999). Species diversity in spatial and temporal dimensions of fruit-feeding butterfly community in an Ecuadorian rainforest. Biological Journal of the Linnean Society, 68, 333-353.

Dolia, J., Devy, M.S., Aravind, N.A., et al. (2008). Adult butterfly communities in coffee plantations around a protected area in the Western Ghats, India. Animal Conservation, 11, 26-34.

Domingues, F.A., Silva-Brandão, K.L., Abreu, A.G., et al. (2012). Genetic structure and gene flow among Brazilian populations of *Heliothis virescens* (Lepidoptera: Noctuidae). Journal of Economic Entomology, 105, 2136-2146.

Dover, J., & Settele, J. (2009). The influences of landscape structure on butterfly distribution and movement: a review. Journal of Insect Conservation, 13, 3–27.

Duelli, P., & Obrist, M.K. (2003). Regional biodiversity in an agricultural landscape: the contribution of seminatural habitat islands. Basic and Applied Ecology, 4, 129–138.

Dres, M., & Mallet, J. (2002). Host races in plant-feeding insects and their importance in sympatric speciation. Philosophical Transactions of the Royal Society of London Series B-Biological Sciences, 357, 471-492.

Dyar, H.G., & Heinrich, C. (1928). The American moths of the genus *Diatraea* and allies. Proceedings of the United States National Museum, 71, 1-48.

Ehrlich, P.R., & Raven, P.H. (1964). Butterflies and plants: a study in coevolution. Evolution, 18, 586-608.

Ekblom, R., & Galindo, J. (2011). Applications of next generation sequencing in molecular ecology of non-model organisms. Heredity, 107, 1-15.

Emelianov, I., Mallet, J. & Baltensweiler, W. (1995). Genetic differentiation in *Zeiraphera diniana* (Lepidoptera, Tortricidae, the Larch Budmoth) - polymorphism, host races or sibling species. Heredity, 75, 416-424.

Endersby, N.M., McKechnie, S.W., Ridland, P.M., et al. (2006). Microsatellites reveal a lack of structure in Australian populations of the diamondback moth, *Plutella xylostella* (L.). Molecular Ecology, 15, 107-118.

Fermon, H., Waltert, M., Vane-Wright, R.I., et al. (2005). Forest use and vertical stratification in fruit-feeding butterflies of Sulawesi, Indonesia: impacts for conservation. Biodiversity and Conservation 14, 333-350.

Funk, D.J. (1998). Isolating a role for natural selection in speciation: Host adaptation and sexual isolation in *Neochlamisus bebbianae* leaf beetles. Evolution, 52, 1744-1759.

Funk, D.J., & Nosil, P. (2008). Comparative analyses of ecological speciation. In K.J. Tilmon (Ed.), Specialization, Speciation, and Radiation (p.117-135.) Berkeley and Los Angeles: University of California Press.

Futuyma, D.J., Keese, M.C., & Scheffer, S.J. (1993). Genetic constraints and the phylogeny of insect-plant associations - responses of *Ophraella communa* (Coleoptera, Chrysomelidae) to host plants of its congeners. Evolution, 47, 888-905.

Gibbs, H.K., Ruesch, A.S., Achard, F., et al. (2010). Tropical forests were the primary sources of new agricultural land in the 1980s and 1990. Proceendings of the National Academy of Science USA, 107, 16732-16737.

Groenendijk, D., van Mannekes, M., & van den Berg, M. (2002). Butterflies and insecticides: a review and risk analysis of modern Dutch practice. Proceedings of the Section Experimental and Applied Entomology of the Netherlands Entomological Society (N.E.V.), 13, 29-34.

Groman, J.D., & Pellmyr, O. (2000). Rapid evolution and specialization following host colonization in a yucca moth. Journal of Evolutionary Biology, 13, 223-236.

Groot, A.T., Marr, M., Heckel, D.G., et al. (2010). The roles and interactions of reproductive isolation mechanisms in fall armyworm (Lepidoptera: Noctuidae) host strains. Ecological Entomology, 35, 105-118.

Groot, A.T., Marr, M., Schofl, G., et al. (2008). Host strain specific sex pheromone variation in *Spodoptera frugiperda*. Frontiers in Zoology, 5, DOI:10.1186/1742-9994-1185-1120.

Hanski, I. (1999). Metapopulation ecology. New York: Oxford University Press.

Hendrickx, F., Maelfait, J.P., van Wingerden, et al. (2007). How landscape structure, land-use intensity and habitat diversity affect components of total arthropod diversity in agricultural landscapes? Journal of Applied Ecology, 44, 340–351.

Hirakurim, M. H., Debiasi, H., Procópio, S. O., et al. (2012). Sistemas de Produção: conceitos e deflnlções no contexto agrícola. Londrlna, PR, Embrapa: Embrapa Soja,

Hobbs, P.R., Sayre K., & Gupta R. (2008). The role of conservation agriculture in sustainable agriculture. Philosophical Transactions of the Royal Society, series B., 363, 543-555.

Horner-Devine, M.C.; Daily, G.C.; Ehrlich, P.R.; & Boggs, C.L. 2003. Countryside biogeography of tropical butterflies. Conservation Biology, 17 (1), 168–177.

Jaenike, J. (1990). Host specialization in phytophagous insects. Annual Review of Ecology and Systematics, 21, 243-273.
Jokimaki, J., & Huhta, E. (1996). Effects of landscape matrix and habitat structure on a bird community in northern Finland: a multi-scale approach. Ornis Fennica, 73, 97–113.
Juarez, M.L., Schofl, G., Vera, M.T., et al. (2014). Population structure of *Spodoptera frugiperda* maize and rice host forms in South America: are they host strains? Entomologia Experimentalis et Applicata, 152, 182-199.
Kirk, H., & Freeland, J.R. (2011). Applications and implications of neutral versus non-neutral markers in molecular ecology. International Journal of Molecular Sciences, 12, 3966-3988.
Koh, L.P. (2007). Impacts of land use change on South-east Asian forest butterflies: a review. Journal of Applied Ecology, 44, 703-713.
Lapola, D.M., Martinelli, L.A., Peres, C.A., et al. (2013). Pervasive transition of the Brazilian land-use system. Nature Climate Change, 4, 27-35.
Leite, L.A.R., Freitas, A.V.L., Barbosa, E.P., et al. (2014). Immature stages of nine species of genus *Dynamine* Hübner, [1819]: morphology and natural history (Lepidoptera: Nymphalidae, Biblidinae). SHILAP Revista de Lepidopterología, 42, 27-55.
Logan, T.J. (1993). Agricultural best management practices for water pollution control: current issues. Agriculture, Ecosystems and Environment, 46, 223-231.
Long, W.H., & Hensley, S.D. (1972). Insect pests of sugar cane. Annual Review of Entomology, 17, 149-176.
Luikart, G., England, P.R., Tallmon, D., et al. (2003). The power and promise of population genomics: From genotyping to genome typing. Nature Reviews Genetics, 4, 981-994.
Machado, V., Wunder, M., Baldissera, V.D., et al. (2008). Molecular characterization of host strains of *Spodoptera frugiperda* (Lepidoptera: Noctuidae) in Southern Brazil. Annals of the Entomological Society of America, 101, 619-626.
Magurran, A.E. (2004). Measuring biological diversity. Oxford: Blackwell Science.
Marini–Filho, O.J., & Martins, R.P. (2010). Nymphalid butterfly dispersal among forest fragments at Serra da Canastra National Park, Brazil. Journal of Insect Conservation, 14, 401-411.
Martel, C., Réjasse, A., Rousset, F., et al. (2003). Host-plant-associated genetic differentiation in Northern French populations of the European corn borer. Heredity, 90, 141–149.
Mas, A.H., & Dietsch, T.V. (2003). An index of management intensity for coffee agroecosystems to evaluate butterfly species richness. Ecological Applications, 13, 1491-1501.
Matsubayashi, K.W., Ohshima, I., & Nosil, P. (2010). Ecological speciation in phytophagous insects. Entomologia Experimentalis et Applicata, 134, 1-27.
Menken, S.B.J., Boomsma, J.J., & van Nieukerken, E.J. (2010). Large-scale evolutionary patterns of host plant associations in the Lepidoptera. Evolution, 64, 1098-1119.
Mills, N.J., & Kean, J.M. (2010). Behavioral studies, molecular approaches, and modeling: Methodological contributions to biological control success. Biological Control, 52, 255-262.
Munguira, M.L., García-Barros, E., & Martín Cano, J., (2009). Butterfly herbivory and larval ecology. In J. Settele, T. Shreeve, M. Konvicka, H. Van Dyck (Eds.), Ecology of Butterflies in Europe. Cambridge: Cambridge University Press.
Myers, J.G. (1935). The ecological distribution of some South American grass and sugar-cane borers (*Diatraea* spp., Lep., Pyralidae). Bulletin of Entomological Research, 26, 335-342.
Nagoshi, R.N., & Meagher, R.L. (2008). Review of fall armyworm (Lepidoptera: Noctuidae) genetic complexity and migration. Florida Entomologist, 91, 546-554.
Nagoshi, R.N., Silvie, P., & Meagher, R.L. (2007). Comparison of haplotype frequencies differentiate fall armyworm (Lepidoptera: Noctuidae) corn-strain populations from Florida and Brazil. Journal of Economic Entomology, 100, 954-961.
Nepstad, D.C., Stickler, C.M., & Almeida, O.T. (2006). Globalization of the Amazon soy and beef industries: opportunities for conservation. Conservation Biology, 20, 1595-1603.

Nosil, P., Egan, S.P., & Funk, D.J. (2008). Heterogeneous genomic differentiation between walking-stick ecotypes: "Isolation by adaptation" and multiple roles for divergent selection. Evolution, 62, 316-336.

Nosil, P., Harmon, L.J., & Seehausen, O. (2009). Ecological explanations for (incomplete) speciation. Trends in Ecology and Evolution, 24, 145-156.

Paoletti, M.G., Pimentel, D., Stinner, B.R., et al. (1992). Agroecosystem biodiversity: matching production and conservation biology. Agricultural, Ecosystems and Environment 40, 2-23.

Paoletti, M.G., Schweigl, U., & Favretto, M.R. (1995). Soil macroinvertebrates, heavy metals and organochlorines in low and high input apple orchards and a coppiced woodland. Pedobiologia, 39, 20-33.

Pashley, D.P. (1986). Host-associated genetic differentiation in fall armyworm (Lepidoptera: Noctuidae): a sibling species complex? Annals of the Entomological Society of America, 79, 898-904.

Pashley, D.P. (1993). Causes of host-associated variation in insect herbivores: an example from fall armyworm. In K.C. Kim & B.A. McPheron (Eds.), Evolution of insect pests: patterns of variation (p.351-359.) New York: John Wiley and Sons, Inc.

Pashley, D.P., Hammond, A.M., & Hardy, T.N. (1992). Reproductive isolating mechanisms in fall armyworm host strains (Lepidoptera, Noctuidae). Annals of the Entomological Society of America, 85, 400-405.

Pashley, D.P., Hardy, T.N., Hammond, A.M., et al. (1990). Genetic evidence for sibling species within the sugarcane borer (Lepidoptera, Pyralidae). Annals of the Entomological Society of America, 83, 1048-1053.

Paterson, H.E.H. (1991). The recognition of cryptic species among economically important insects. In M.P. Zalucki (Ed.), *Heliothis:* Research methods and prospects (p.1-10) New York: Springer-Verlag.

Pimentel, D., & Edwards, C.A. (1982). Pesticides and ecosystems. BioScience, 32, 595-600.

Pinto, A.S., Cano, M.A.V., & Santos, E.M. (2006). A broca-da-cana, *Diatraea saccharalis*. Boletim Técnico Biocontrol 1. In A.S. Pinto (Ed.), Controle de Pragas da Cana-de-açúcar (p.15-20). Sertãozinho: Biocontrol.

Porreta, D., Canestrelli, D., Bellini, R., et al. (2007). Improving insect pest management through population genetic data: a case study of the mosquito *Ochlerotatus caspius* (Pallas). Journal of Applied Ecology, 44, 682-691.

Powell, J.A. (1980). Evolution of larval food preferences in microlepidoptera. Annual Review of Entomology, 25, 133-159.

Ribeiro, D.B., Batista, R., Prado, P.I., et al. (2012). The importance of small scales to the fruit-feeding butterfly assemblages in a fragmented landscape. Biodiversity and Conservation, 21, 811-827.

Ribeiro, D.B., & Freitas, A.V.L. (2012). The effect of Reduced-Impact Logging on Fruit-feeding butterflies in Central Amazon, Brazil. Journal of Insect Conservation, 16, 733-744.

Ribeiro, D.B., Batista, R., Prado, P.I., et al. (2012). The importance of small scales to the fruit-feeding butterfly assemblages in a fragmented landscape. Biodiversity and Conservation, 21, 811-827.

Ricketts, N.T. (2001). The matrix matters: effective isolation in fragmented landscapes. The American Naturalist, 158, 87-99.

Rosch, V., Tscharntke, T., Scherber, C., et al. (2013). Landscape composition, connectivity and fragment size drive effects of grassland fragmentation on insect communities. Journal of Applled Ecology, 50, 387–394.

Rosen, D. (1986). The role of taxonomy in effective biological control programs. Agriculture Ecosystems and Environment, 15, 121-129.

Rugman-Jones, P.F., Hoddle, M.S., & Stouthamer, R. (2010). Nuclear-mitochondrial barcoding exposes the global pest Western Flower Thrips (Thysanoptera: Thripidae) as two sympatric cryptic species in its native California. Journal of Economic Entomology, 103, 877-886.

Rundle, H.D., & Nosil, P. (2005). Ecological speciation. Ecology Letters, 8, 336-352.
Saunders, D.A., Hobbs, R.J., & Margules, C.R. (1991). Biological consequences of ecosystem fragmentation: a review. Conservation Biology, 5, 18–32.
Schoonhoven, L.M., van Loon, J.J.A., & Dicke, M. (2008). Insect-Plant Biology. New York: Oxford University Press.
Schulze, C.H., Linsenmaier, K.E., & Fiedler, K. (2001). Understorey versus canopy: patterns of vertical stratification and diversity among Lepidoptera in a Bornean rain forest. Plant Ecology, 153, 133-152.
Scott, K.D., Lawrence, N., Lange, C.L., et al. (2005). Assessing moth migration and population structuring in *Helicoverpa armigera* (Lepidoptera: Noctuidae) at the regional scale: Example from the Darling Downs, Australia. Journal of Economic Entomology, 98, 2210-2219.
Silva-Brandão, K.L., Santos, T.V., Consoli, F.L., et al. (2015a). Genetic diversity and structure of Brazilian populations of *Diatraea saccharalis* (Lepidoptera: Crambidae): Implications for pest management. Journal of Economic Entomology, 108, 307-316.
Silva-Brandão, K.L., Silva, O.A.B.N.E., Brandão, M.M., et al. (2015b). Genotyping-by-sequencing approach indicates geographic distance as the main factor affecting genetic structure and gene flow in Brazilian populations of *Grapholita molesta* (Lepidoptera, Tortricidae). Evolutionary Applications, 8, 476-485.
Sperling, F.A.H., & Hickey, D.A. (1994). Mitochondrial-DNA sequence variation in the spruce budworm species complex (Choristoneura, Lepidoptera). Molecular Biology and Evolution, 11, 656-665.
Steffan-Dewenter, I., & Tscharntke, T., (2000). Butterfly community structure in fragmented habitats. Ecology Letters, 3, 449–456.
Tilman, D., Cassman, K.G., Matson, P.A., et al. (2002). Agricultural sustainability and intensive production practices. Nature, 418, 671–677.
Tomquelski, G.V., & Maruyama, L.C.T. (2009). Lagarta-da-maçã em soja. Revista Cultivar, 117, 20-22.
Tscharntke, T., Tylianakis, J. M., Wade, M.R., et al. (2007a). Insect conservation in agricultural landscapes. In A.J.A. Stewart, T.R. New & O.T. Lewis (Eds.), Insect Conservation Biology - Proceedings of the Royal Entomological Society's 23rd Symposium (p.383–404). Wallingford, UK: CAB International.
Tscharntke, T., Bommarco, R., Clough, Y., et al. (2007b). Conservation biological control and enemy diversity on a landscape scale. Biological Control, 43, 294–309.
Uehara-Prado, M., Brown, K.S., & Freitas, A.V.L. (2007). Species richness, composition and abundance of fruitfeeding butterflies in the Brazilian Atlantic forest: comparison between a fragmented and a continuous landscape. Global Ecology and Biogeography, 16, 43–54.
Vianna, J.A., Bonde, R.K., Caballero, S., et al. (2006). Phylogeography, phylogeny and hybridization in trichechid sirenians: implications for manatee conservation. Molecular Ecology, 15, 433-447.
Wahlberg, N., Leneveu, J., Kodandaramaiah, U., et al. (2009). Nymphalid butterflies diversify following near demise at the cretaceous/tertiary boundary. Proceedings of the Royal Society B, 276, 4295-4302.
Waldvogel, M., & Gould, F. (1990). Variation in oviposition preference of *Heliothis virescens* in relation to macroevolutionary patterns of Heliothine host range. Evolution, 44, 1326-1337.

Pedro Gerhard, Luciano M. Verdade

13 Stream Fish Diversity in an Agricultural Landscape of Southeastern Brazil

Abstract: The Corumbataí watershed is a typical agricultural landscape of Southeastern Brazil, an important area for sugarcane production, while pasture fields for cattle production are also a relevant agroecosystem. This study was performed to understand how stream fish diversity related to land-use (LU). A Geographic Information System was used to split the watershed into smaller catchments and classify each according to its dominant LU in Sugarcane, Pasture or native Forest remnants categories, both at the catchment (*DC*) or riparian network scale (*DB*). Sixty catchments were assigned to these groups, classified by *DC* or *DB*. Stream reaches 150 meters long were selected for sampling fish by electrofishing, each one sampled twice during the rainy and the dry seasons of 2003 and 2004. Analysis focused on the comparison of species observed (*Sobs*), abundance data (individuals in species) and species occurrences in the Sugarcane, Pasture and Forest groups, grouped by *DC* and *DB*. Kruskal-Wallis non-parametric tests, Analysis of Similarities and Similarities Percentages were used for the comparisons. Results showed that, in general, *DC* is a poor predictor of *Sobs*. while *DB* impacted total *Sobs* only for the Rainy season sample. Species occurrences were associated with Sugarcane and Pasture areas, especially by the *DB* predictor: a different set species occurred or prevailed on stream reaches bordered by Sugarcane or Pasture *versus* Forest. Species occurrences were interpreted to be linked to species life traits, so that these species can be considered as indicators of the impacts of poorly managed agroecosystems in relation to lotic environments.

13.1 Introduction

Fish adapted for living in stream (or headwater) ecosystems are characterized by short life span, small body sizes and early sexual maturity; a host of life traits which favors rapid colonization after severe disturbances, characteristic of upstream reaches (Schlosser, 1991). High rates of local extinctions and recolonizations are expected due to the dendritic nature of the stream channel network, which is highly hierarchical and spatially structured across landscapes (Fagan, 2002). Headwater stream ecosystems display diverse assemblages in which fish can be assigned into: (1) true headwaters specialists, (2) generalists that use both upstream and downstream river reaches and (3) downstream (riverine) species that use headwaters for spawning and nursery areas (Meyer et al., 2007).

Anthropogenic activities on small catchments determine disturbances at spatial and temporal extends beyond those that stream fishes would be able to cope.

The small ratio of aquatic ecosystem to its drainage area places severe impacts to the former, disrupting the local dominant control processes (Gomi et al., 2002) in nonlinear, multi-scale, time and space contingent manners (Allan, 2004). Both riparian and catchment-scale land use affects a particular stream reach morphology, water quality, fish distribution and density (Roth et al., 1996; Vondracek et al., 2005). Stream networks may also be impacted by headwaters 'shrinking' due to deforestation and poor soil conservation practices (Faria & Marques, 1999); a process that will lead to a diminishing area of available fluvial habitat. An integrated response of fish fauna to the disruption of the fluvial continuum is that of biotic homogenization (Scott & Helfman, 2001), a process of colonization of generalist species onto upstream areas that, under anthropogenic impact, presents structures and functions of downstream reach stream ecosystems.

Brazil is one of the leading agricultural economies of the world, and some of its areas have undergone agricultural intensification for more than two hundred years (Dean, 1996). The State of São Paulo, on the Southeastern portion of Brazil is one of the oldest colonization fronts of the country, but widespread deforestation motivated by the coffee plantation system started only on the second half of the XIX century (Victor et al., 2005). Despite a relatively well documented stream fish fauna, most studies developed on the country have focused on species autoecology or general stream limnology, but with little attention to the relations between agroecosystems and aquatic environments. Corumbataí river basin is characteristic of the land use history of the State of São Paulo, including early coffee plantations, and subsequent development of pasture for livestock production, sugarcane and *Eucalyptus* plantations. The objective of this study is to understand the effects of land use on the diversity of stream fish assemblages, considering two landscape metrics defined by the spatial scales: whole catchment and riparian zone.

13.2 Study Area

Corumbataí river is an affluent to the Piracicaba river, located on the East-Central region of State of São Paulo, Brazil, approximately between parallels 22°04'46"S and 22°41'28"S and meridians 47°26'23"W and 47°56'15"W (Fig. 13.1). Climate is subtropical, with a dry winter (from April to September) and a rainy summer (from October to March), type *Cwa* (in Köppen's category). Geology is mostly associated with Paleozoic, Mesozoic and Cenozoic basalts, sandstones and limestones. The watershed develops on the Middle Tietê zone, Peripheral Depression of the State of São Paulo; while its headwaters dissect a terrain characterized by *cuestas*, mid and low sections drain gentle hilly terrains. The Corumbataí watershed has an area of 170,775.6 hectares, and develops from 1,050 meters high at its headwaters to 470 meters at its mouth (Viana et al., 2002). The municipality of Rio Claro, with around 190,000 inhabitants, is entirely located in its area. However, most of the watershed is agricultural, with

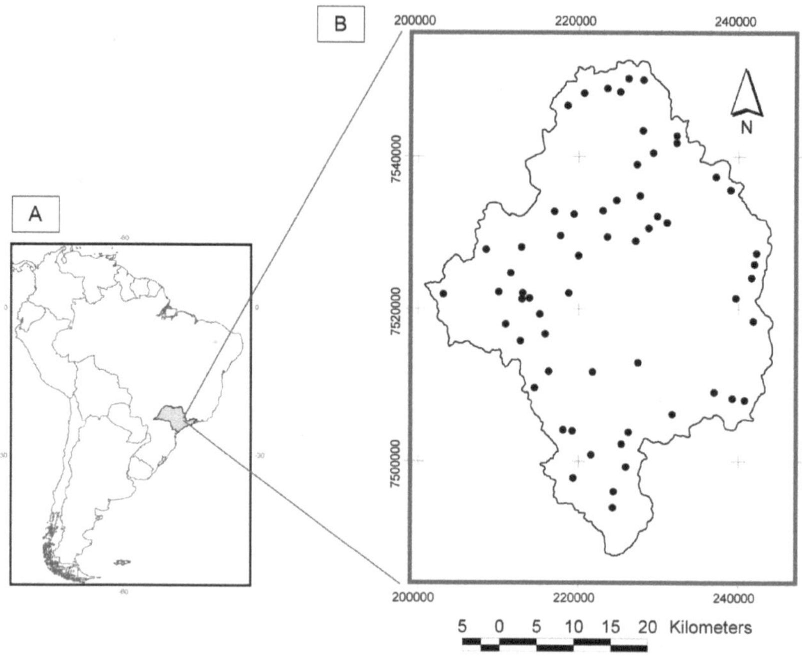

Fig. 13.1: Location of the study area. A) Location in South America and Brazil, highlighting the State of São Paulo. B) Corumbataí Watershed, showing the 60 stream fish sampling sites (black dots). The large sampling void (center to western portion of the watershed) is the municipality of Rio Claro.

sugarcane monoculture (26%) and pastures (44%), occupying most of its area. Native forest fragments accounted for 12% (Valente & Vettorazzi, 2008). The region has been occupied extensively for at least 200 years (Dean, 1977) and intensive and widespread deforestation probably started about 100 years ago (Victor et al., 2005). Over the last few years, many of the municipalities of central and western portions of the State of São Paulo have seen an increase in sugarcane plantations (Azevedo, 2013; Rudorff et al., 2010; Sparovek et al., 2008).

13.3 Methods

13.3.1 Study Design and Sampling Site Selection

This work started with the construction of a GIS for the for the Corumbataí River watershed. Two main geographic digital data were employed. The first data is a land use (LU) map based on SPOT-4 and LANDSAT-5 satellite images for the year 1999 (Valente & Vettorazzi, 2003). Second, a Digital Terrain Model (DTM) was obtained from the GIS database of “Projeto Corumbataí” (2001). Both had a pixel resolution of 20 meters. A line vector hydrographic map is included in the database.

Catchment delineation employed the software SWAT2000 (Luzio et al., 2002) based on the DTM and the hydrography. SWAT2000 employs a raster analysis over the terrain data, guided by the hydrography vector lines, in order to model a stream network and a segmentation of the watershed areas into smaller units, catchments. A 'seed' value was chosen for the SWAT2000 delineate drainage areas corresponding to streams of orders from 1 to 3 (Strahler stream ordering). After removing software errors and inter-basins (those that receive stream-water from headwaters catchments), the process resulted in 209 hydrological independent units.

An overlay operation between the LU map and the catchment boundaries allowed for the calculation of LU classes proportions for each of the 209 hydrological units. These were accounted for direct percentage dominance in each of the land use classes: Sugarcane, Pasture and Forest. For each catchment channel network, it was also calculated a stream buffer zone of 30 m from each stream margin. The value of 30 m was chosen after the Brazilian Forest Code (Federal Act n° 7803/89) that, by then, had demanded that all water bodies on the studied category (streams with width up to 10 meters) should be protected by such buffer areas, termed as Areas of Permanent Preservation (APP). LU was also computed for these APPs, and each catchment was classified according to the dominant classes, i.e. Sugarcane, Pasture and Forest. Throughout this study, each catchment should be classified for its dominant upland (total area) land use class (*DC* predictor) or the dominant land use class in its APP (*DB*).

A field work campaign was conducted in order to check for each catchment land use category (*DC* and *DB*), sampling accessibility and other constraints, such as proximity to urban areas, presence of paved roads, point-source pollution from poultry farms, among others. A total of 130 catchments were assessed, and the validity of the LU mapping was checked, owing to the time lapse between map production (year 1999) to the field campaign (years 2003-2004).

After reconnaissance of the potential sampling catchments, considering the criteria described above, the study aimed at an optimal distribution of sites along the Corumbataí watershed, among its main land use classes. Final selection defined a group of 60 catchments that, for the *DC* predictor included: 11 Forest catchments (11 out 13 delineated, two presented sampling constraints for both *DC* and *DB*); 19 sugarcane catchments (of which, by *DB* criterion 2 are of sugarcane, 11 are Pasture and 6 Forest); 30 Pasture catchments (*DB* criterion: 16 Pasture, 14 Forest). Sums for the *DB* predictor were, thus: 2 Sugarcane, 27 Pasture, and 31 Forest.

13.3.2 Habitat and Fish Data Collection

All field work was carried out along a 150 meter stream reach at the downstream end of selected catchments. At the 150 m selected sampling reach, LU classes were inspected by visual recognition up to a distance of 30 m from each stream margin. This allowed for a match between LU maps produced and the local conditions of selected sites. Sites that presented LU classes other than Sugarcane, Pasture or Forest along channel network or uplands were discarded. Sampling reaches were always placed at least 50 m upstream to roads crossing the stream channel and at least 150 m upstream from a confluence between the sampled channel into a larger river (order $n + 1$). A collection of physical structure both at the riparian, reach and instream channel scales, together with stream-water samples, were collected at each site, but were not included in this study.

Fish collection was performed along four field campaigns, along the Rainy and Dry seasons of years 2003 and 2004. All fish were collected by an electrofishing AC apparatus, on the 15 m stream reach, blocked at both ends with nets of mesh size 5 mm. The protocol included a three-pass, downstream, removal method performed with a two-person crew. Fish stunned in dip-nets were sacrificed to a lethal solution of anesthesia, and then transferred to 10% formaldehyde solution. After three days the specimens were preserved in a 70% ethanol solution. Identification occurred at the laboratory, aided with specific literature (e.g. Menezes et al., 2007; Reis et al., 2003). Voucher specimens were deposited at the Museu de Zoologia da Universidade de São Paulo (MZUSP).

13.3.3 Analytical Methodology

Species observed (S_{obs}) in each sample (Rainy and Dry seasons), and those observed in each group of streams of predictors (*DC* and *DB*), were used as an estimate of number of species (S_{est}) by the extrapolation procedure (Colwell et al., 2012). The method also calculates Confidence Limits for each extrapolation step, which can be inspected for differences among groups at a given level. The lack of overlap of two Confidence Limit bands is an overly conservative indication of differences of S_{est} among groups, but can be used as a simple criterion of a statistical difference. Extrapolations beyond two to three times that of the original sample size tend to produce high variances of S_{est} (Colwell et al., 2012). The software EstimateS 9.1.0 (Colwell, 2013) did the analysis of the S_{est} and its Confidence Limits using an analytical solution and a bootstrap estimator, respectively. Then, in this work, the dataset for Rainy and Dry seasons were extrapolated to 3 times that of actual sample size, a number that approximates to the population of streams of orders 1-3 in Corumbataí watershed. For comparison among streams grouped by the *DC* and *DB* predictors, *Sest* was calculated for a "reference sample" that equaled the number of streams of the larger sample in each comparison. So, the number of sampling units was not higher than 1.5 times that of actual sample size within groups.

Abundance data (individuals in species) collected in each stream and season was used to compute richness, diversity and evenness indexes. Indexes are scanned for normal distribution properties. All indexes were tested for correlations in order to study which aspects of sample diversity were informative, using the Kendall's tau correlation coefficient. Pairs of indexes that presented high correlations ($p < 0.05$) were considered for removal on the following analysis. Selected indexes were tested for significant differences between the samples taken during Rainy and Dry seasons with the paired Wilcoxon non-parametric test. Abundance data were converted to a square matrix of dissimilarities among sites employing the Chord's distance, considered an appropriate and unbiased metric for ecological studies (Legendre & Legendre, 2012). The Mantel's permutation test assessed the correlation between the dissimilarity matrices of the Rainy and Dry season samples. The Kruskal-Wallis non-parametric test compared diversity indexes medians among the Sugarcane, Pasture and Forest groups of streams, ordered by the predictors *DC* and *DB*. When appropriate, tests were followed by Mann-Whitney pairwise tests, with Bonferroni-corrected p values for multiple comparisons. The Analysis of Similarities (ANOSIM, Clarke, 1993) tested for significant differences in species abundances among groups. The Similarities Percentages (SIMPER, Clarke, 1993) procedure followed ANOSIM, focusing on the statistically significant differences found. SIMPER allowed for the detection of species that determined the largest differences among tested groups, a form of indicator species analysis. The inspection of unique, duplicates and common occurrences in each group of streams (Sugarcane, Pasture, Forest on the predictors *DC* and *DB*) was also deemed a criterion for indicator species of those groups. These analyses were performed with the software PAST 2.17c (Hammer et al., 2001).

13.4 Results

13.4.1 Stream Fish Richness

The 118 samples presented 49 fish species for 52,942 individuals. Estimates of total species richness by sample extrapolation run from the Rainy season sample (m_R = 60 streams) and Dry season sample (m_D = 58 streams) revealed that the number of observed species (S_{obs}) approximated the estimated (S_{est}) for three times (m_{extr}) the sample size of m_R and m_D (Tab. 13.1). The percentage of observed species in relation to the estimated number of species ($S_{obs}/S_{est} \times 100$) is 97.76 and 95.12 for Rainy and Dry season samples, respectively. Considering the upper 95% Confidence Interval for the S_{est} at m_{extr}, the total number of species attains a maximum of 55 species. Estimates for the two seasons did not vary, and confidence intervals present large overlaps. It can be concluded that the total stream fish richness for streams of orders 1 to 3 in Corumbataí watershed is around 49 to 55 species.

Tab. 13.1: Number of streams (*m*) sampled on two seasons (Rainy and Dry samples) for stream fish species (S_{obs}) and estimates by extrapolation of sample size. S_{est} = Number of estimated species for $m_{R,D}$ and m_{extr} sample sizes. m_{extr} = sample sizes extrapolated over the *m* sampled. $-CL_{(95\%)}$ and $+CL_{(95\%)}$ = lower and upper Confidence Limits with 95% of probability.

Sample	m	S_{obs}	$S_{est\ (m)}$	$-CL_{(95\%)}$	$+CL_{(95\%)}$	m_{extr}	$S_{est\ (mextr)}$	$-CL_{(95\%)}$	$+CL_{(95\%)}$
Rainy	60	48	48.00	45.96	50.04	180	49.10	45.73	52.47
Dry	58	47	47.00	44.01	49.99	174	49.41	43.96	54.87

13.4.2 Stream Fish Diversity

A number of diversity indexes were calculated for the two samples. The inspection of the square matrix of Kendall's tau coefficients revealed that all computed indexes are highly correlated. The total number of observed species (S_{obs}) was chosen for the subsequent analysis, as no other aspect of diversity is informed by indexes that include species abundances, including total individuals abundance per site. The use of S_{obs} also parallels the estimates of number of species by extrapolation (S_{est}), as used in the analysis below.

The number of species collected in each stream varied from 1 to 26, from 1 to 5,114 individuals. Rainy season sampling units revealed 1 to 26 species (median 12) while the Dry season sample presented 1 to 23 (median 10). Total abundance varied from 7 to 1,426 (median 374.0) and 1 to 5,114 (median 349.5) individuals for the Rainy and Dry samples. No outliers were found for species richness, but for total fish abundance three sites were detected, one in the Rainy and two in the Dry season sample. These outliers, larger than four times the sample median, were removed for the following analysis. Species richness observed on the Rainy season sample was significantly higher than for the Dry season (paired Wilcoxon test $W = 966.5$, $p < 0.001$, 55 d.f.). Species collected only in the Rainy season sample were *Apareiodon piracicabae* (Characiformes, Parodontidae) and *Rineloricaria latirostris* (Siluriformes: Loricariidae: Loricariinae). *Planaltina britskii* (Characiformes: Characidae: Glandulocaudinae) was collected only on the Dry season sample. No differences were found between total individual abundances per site sampled in each season ($W = 842.0$, $p = 0.5460$, 53 d.f.). A Mantel permutation test comparing the species abundance matrices for Rainy and Dry season samples revealed that datasets had similar structures (Mantel test on the abundance data converted to Chord's distance with 9,999 permutations, $n = 58$ sites, Pearson correlation $r = 0.6362$, $p = 0.0001$).

13.4.3 Stream Fish Diversity and LU

Land use was related to the observed number of species in each sampling period (Rainy and Dry seasons) employing the predictors *DC* (Dominant Use) and *DB* (Dominant

Buffer). Results for the *DC* predictor, with the groups Sugarcane, Pasture and Forest, revealed no statistical significance either for the Rainy season sample (Kruskal-Wallis non-parametric test, tied-corrected, Hc = 0.2458, p = 0.8843, 59 d.f.) or the Dry season sample (Hc = 0.5987, p = 0.7413, 55 d.f.). The *DB* tested for the groups Pasture and Forest presented statistical significance for the Rainy season sample (Hc = 4.4200, p = 0.0355, 59 d.f.) but not for the Dry season sample (Hc = 2.3470, p = 0.1256, 55 d.f.). The Rainy season sample for *DB* presented high S_{obs} for the Pasture group.

For the estimation of the number of species (S_{est}) of each LU predictor (*DC* or *DB*), the number of samples to extrapolate - the reference sample - was considered as the larger number of sampling units among the predictor levels (Tab. 13.2). For predictor *DC* on the Rainy season sample the reference sample was 30 sampling units, for the Dry season sample, 28 units. For the *DB* predictor these reference samples were 31 and 30 units. These figures, summed along the levels of each comparison (Rainy or Dry *DC*, Rainy or Dry *DB*), adds up nearly to 90 samples, which is around 1.5 times the actual sample sizes of 58 (Dry) and 60 (Rainy). At these extrapolation levels, *Sobs* was between 72.6 to 90.6% of S_{est} for *DC* and 97.0 to 97.8% of S_{est} for *DB*. Comparisons for S_{est} and its correspondent 95% Confidence Intervals for each group, predictor and season revealed that these estimates largely overlap (Tab. 13.2). This is an overall indication that neither *DC* nor *DB* were good predictors of the estimated number of species extrapolated to equal sampling sizes.

Tab. 13.2: Number of streams (*n*) sampled grouped by *DC* and *DB* predictors on two seasons (Rainy and Dry samples) for stream fish species (S_{obs}) and estimates by extrapolation of sample size. S_{est} = Number of estimated species for *m* and m_{extr} sample sizes. m_{extr} = sample sizes extrapolated over the *m* sampled. $-CL_{(95\%)}$ and $+CL_{(95\%)}$ = lower and upper Confidence Limits with 95% of probability.

Predictor	Season	Group	m	S_{obs}	$S_{est\,(m)}$	$-CL_{(95\%)}$	$+CL_{(95\%)}$	$m_{(ref)}$	$S_{est\,(ref)}$	$-CI_{(95\%)}$	$+CI_{(95\%)}$
DC	**Rainy**	**Sugarcane**	19	42	42.00	36.87	47.13	30	46.33	39.74	52.92
		Pasture	30	46	46.00	41.78	50.22	30	46.00	41.78	50.22
		Forest	11	35	35.00	30.19	39.81	30	41.76	32.45	51.07
	Dry	**Sugarcane**	19	38	38.00	32.58	43.42	28	41.93	35.34	48.52
		Pasture	28	43	43.00	38.34	47.66	28	43.00	38.34	47.66
		Forest	11	33	33.00	26.43	39.57	28	45.44	31.99	58.88
DB	**Rainy**	**Pasture**	27	47	47.00	41.44	52.56	31	48.47	42.67	54.27
		Forest	31	45	45.00	41.30	48.70	31	45.00	41.30	48.70
	Dry	**Pasture**	26	44	44.00	40.51	47.49	30	45.00	41.34	48.65
		Forest	30	42	42.00	37.65	46.35	30	42.00	37.65	46.35

The analysis of similarities (ANOSIM) among levels of *DC* and *DB* resulted in significant results only for the later (Tab. 13.3). ANOSIM tests, run post-hoc for pairwise comparisons among groups on the *DB* predictor, revealed more significant results for Rainy season sample than for the Dry season sample. Catchments where *DB* is classed as Forest differed from Sugarcane and Pasture on the Rainy season, the last comparison holds also for the Dry season sample (Tab. 13.4).

Tab. 13.3: Analysis of similarities (ANOSIM) results for comparisons among groups (*Sugarcane*, *Pasture* and *Forest*) of two predictors (*DC*, *DB*) for two samples of stream fish assemblages on the Rainy ($m = 60$ streams) and Dry ($m = 58$ streams) seasons.

Predictor	DC		DB	
Season	**Rainy**	**Dry**	**Rainy**	**Dry**
Mean rank within	855.2	807.2	805.1	746.7
Mean rank between	903.6	838.4	954.4	895.5
r	0.0546	0.0378	0.1687	0.1801
p	0.0710	0.1552	0.0001	0.0002

Tab. 13.4: Results for pairwise ANOSIM tests for the *DB* predictor on two samples of stream fish assemblages (Rainy and Dry seasons). The upper right triangle of the matrix presents the ANOSIM statistic *R*, while the lower left triangle presents the Bonferroni-corrected p values of the comparisons.

Season	Rainy			Dry		
Groups	**Sugarcane**	**Pasture**	**Forest**	**Sugarcane**	**Pasture**	**Forest**
Sugarcane	-	-0.036	0.392	-	0.286	0.498
Pasture	1.000	-	0.163	0.089	-	0.149
Forest	0.038	0.001	-	0.060	0.002	-

13.4.4 Indicator Species in LU Catchment Groups

Figure 13.2 shows the unique, duplicate and common species as related to occurrences on the *DC* and *DB* groups (data pooled for Rainy and Dry seasons samples). Some species were unique for the *DC* or *DB* predictors. *Steindachnerina insculpta* (Characiformes: Curimatidae) and *Planaltina britskii* only in pasture dominated catchments (*DC*) or stream networks bordered with pastures (*DB*). Most differences in duplicates occurrences between *DC* and *DB* can be related to the fact that the *DC* grouping includes streams with different riparian LU classes. For instance, most sugarcane *DC* streams possessed forests along the catchment channel network. This explains why *Hyphessobrycon eques* (Characiformes: Characidae: "Tetragonopterinae") was a duplicate for the *DC* Sugarcane × Forest, but a unique in forested *DB* sites. *Corydoras*

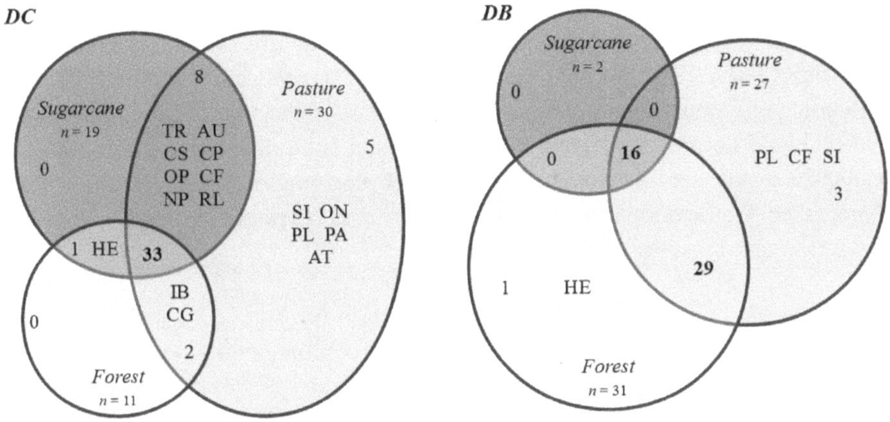

Key to species:

AT = *Astyanax sp.*	IB = *Imparfinis borodini*
AU = *Australoheros facetus*	NP = *Neoplecostomus paranaensis*
BI = *Bryconamericus iheringii*	ON = *Oreochromis niloticus*
CF = *Corydoras flaveolus*	OP = *Oligosarcus pintoi*
CG = *Characidium gomesi*	PA = *Piabina argentea*
CP = *Characidium sp.*	PL = *Planaltina britskii*
CS = *Cyphocarax modestus*	SI = *Steindachnerina insculpta*
HE = *Hyphessobrycon eques*	TR = *Tilapia rendalli*

Fig. 13.2: Venn-diagram for unique, duplicate and common species as related to the predictors DC and DB. Number of sampling units in each group = n. Numbers indicate unique and duplicate and, when marked in bold, common and duplicate species in major groups.

flaveolus (Siluriformes: Callichthyidae) were only found on catchments in pasture dominated stream networks buffers (*DB*), but occurred in sugarcane dominated catchments (*DC*) that were classified as Pasture on the *DB* grouping. The large amount of duplicates on the *DB* Pasture × Forest may reveal further heterogeneity of LU along riparian buffers.

The Similarity Percentage (SIMPER) analysis was run only for those significant ANOSIM results, those that grouped stream samples by the *DB* predictor. Differences in species abundance on the comparison of groups Sugarcane × Pasture (Rainy and Dry season samples) and Sugarcane × Forest (Rainy season sample) were considered. Only those species contributing to more than 5% similarity between groups were selected for the results interpretation. This criterion resulted on a list of nine species, in which seven of these are present in all considered comparisons (Tab. 13.5). Sugarcane dominated stream network buffers occurred only twice. Thus, conclusions drawn from those comparisons with these groups, although significant, should be regarded with caution. However, as for the number of species overlapping in the Sugarcane and Pasture groups (Fig. 13.2), these differences in abundances (Sugarcane × Pasture) are related to rivers buffered with forests, in opposition to those where anthropic use is dominant. Stream networks bordered by riparian forests present four out of nine species in which abundances are larger than the other groups: *Trichomycterus* sp. (Siluriformes: Trichomycteridae), *Imparfinis mirini* (Siluriformes: Heptapteridae), *Corumbataia cuestae* (Siluriformes: Loricariidae: Hypoptopomatinae) and *Phalloceros caudimaculatus* (Cyprinodontiformes: Poeciliidae). The disturbed groups Sugarcane and Pasture present major abundances of *Astyanax paranae, Astyanax altiparanae, Bryconamericus iheringii* (Characiformes: Characidae: "Tetragonopterinae"), *Poecilia reticulata* (Cyprinodontiformes: Poeciliidae) and *Geophagus brasiliensis* (Perciformes: Cichlidae).

Tab. 13.5: Results of the Similarity Percentage analysis (SIMPER) for pairwise comparisons of stream fish assemblage samples taken on the Rainy and Dry seasons. The percentage contribution to the overall dissimilarity between groups (Contr.) is presented for each taxon*, together with the mean abundance of the species in each group. Bold numbers highlight the most abundant species in each pairwise comparison.

	DB - Rainy season				**DB – Rainy season**				**DB - Dry season**			
	Taxon	**Contr.**	**Sugarcane**	**Forest**	**Taxon**	**Contr.**	**Pasture**	**Forest**	**Taxon**	**Contr.**	**Pasture**	**Forest**
1	AP	25.47	**46.00**	17.90	TS	17.23	42.10	**69.60**	TS	22.16	35.60	**99.70**
2	TS	17.21	0.00	**69.60**	IM	13.95	57.30	**84.70**	IM	15.86	86.60	**92.10**
3	IM	11.12	10.00	**84.70**	CO	10.04	24.10	**64.60**	PR	11.64	**41.90**	12.20
4	CO	10.46	0.00	**64.60**	AA	8.77	**50.70**	19.20	CO	8.15	18.70	**60.80**
5	PR	8.27	**39.50**	9.29	PC	8.32	19.60	**36.20**	PC	7.61	28.80	**31.90**
6	PC	7.19	29.00	**36.20**	AP	6.97	**27.50**	17.90	AP	7.17	**25.40**	24.50
7	AA	6.67	**33.50**	19.20	PR	5.14	**31.70**	9.29	AA	7.10	**42.90**	10.70
8	BI	5.39	**31.00**	8.81					GP	5.37	**29.70**	3.17

*AP = *Astyanax paranae*, AA = *Astyanax altiparanae*, BI = *Bryconamericus iheringii*, TS = *Trichomycterus* sp., IM = *Imparfinis mirini*, CO = *Corumbataia cuestae*, PC = *Phalloceros caudimaculatus*, PR = *Poecilia reticulata*, GP = *Geophagus brasiliensis*.

13.5 Discussion

The number of stream fish species collected (49) was nearly equal to the estimated total number of species (49-55), indicating that the 60 streams sample included most of the diversity of ecological conditions found in streams of orders 1 to 3 in Corumbataí watershed. These are streams that should present a set of headwater conditions, such as fast-moving waters, relatively cold, clear and highly oxygenated waters and channels presenting shallow habitats. Riverine conditions are mostly opposed to headwaters, and may include increased local primary productivity and availability of coarse and fine organic debris (Allan & Castillo, 2007). While some of the collected species are typically headwater species, such as *Astyanax paranae*, *Characidium gomesi* (Characiformes: Crenuchidae) and *Neoplecostomus paranaensis*, (Siluriformes: Loricariidae: Neoplecostominae) others, such as *Serrapinnus* spp. (Characiformes: Characidae: Cheirodontinae), *Steindachnerina insculpta*, *Hypostomus ancistroides* (Siluriformes: Loricariidae: Hypostominae) and *Geophagus brasiliensis*, may be ascribed to riverine conditions. The headwaters species listed are insectivores or specialized peryphython grazers from torrential habitats (Braga et al., 2008; Ferreira et al., 2012; Rondinelli et al., 2011). Riverine species are of detritivore, algivore and omnivore trophic guilds, respectively (Casatti et al., 2006; 2008). The observation of such differential stream habitat occupancy may be devised from trophic guilds (Zeni & Casatti, 2013), where those species typical of larger fluvial bodies are, then, found in headwater streams. The mechanism proposed for such changes is that small streams draining agricultural fields, especially those devoid of riparian forests, imitate river conditions and allows for the colonization of riverine species into upper reaches of the stream network (Scott & Helfman, 2001). The relatively high number of species found in this study should be regarded as a sum of a set of specialized headwater species (v. Meyer et al., 2007, for a description of life history attributes associated to this fauna) and one of riverine, tolerant widespread species. Corumbataí stream fish fauna may be regarded as a product of biotic homogenization through the longitudinal continuum, probably due to an extensive and long history of agricultural land use change (Scott & Helfman, 2001; Burcher et al., 2008; Maloney et al., 2008).

A large range of total number of observed species and individuals was found in this study in both sampling seasons, which is an indication of local instream conditions variability. Sampled reaches varied from highly degraded, for instance, presenting heavy sand sedimentation, coupled with high daily water temperature fluctuations, to near pristine reaches presenting high diversity of habitat units and bordered by well developed, stabilizing riparian forests (Paula et al., 2011). These extremes, however, cannot be easily assigned to land use categories, either at the catchment or riparian scales. Reach morphology, i.e. local conditions, dictates at least part of fish richness and assemblage structure (Nerbonne & Vondracek, 2001; Inoue & Nunokawa, 2002). The same applies for total fish abundance: some of the sampled reaches that presented high abundances (outliers removed from some analysis) were locations where large pools (large total habitat volume), in well insulated areas, determine high local primary productivity, rather than large abundances of individuals (see Matthews, 1998, for a general discussion on stream productivity, pool size and stream fish richness).

The analysis on the number of observed (S_{obs}) and estimated (S_{est}) species presented different results, as S_{obs} indicates a significant difference between riparian stream channel networks bordered by pasture or forest for the sample taken on the rainy season. However, comparing results of observed species bears the problem of comparing different rates of accumulation of sampled species due to habitat heterogeneities (Krebs, 1998; Colwell et al., 2012). For instance, sugarcane dominated catchments may determine simplification of the stream reach morphology, which will host a smaller number of species than streams draining forested catchments, that present distinctive and specialized habitats (higher within-group beta diversity). The species rarefaction procedure circumvents this problem, but limits the analysis to a comparison to the sample size of the smallest group. This can be further resolved by the extrapolation method described by Colwell et al. (2012), a better quantitative and logic quantitative solution, and should be preferred. Thus, a conservative interpretation of S_{obs} and S_{est} results is to choose for the lack of difference within stream groups, either for land use predictors or seasons. Despite being practical and integrate information at the assemblage level, studies pointed against the usage of (assemblage level) indexes for the analysis of biota diversity among habitats (Magurran & McGill, 2011). Part of the response of species level to different habitat stressors could be blurred by species-specific tolerances and degradation processes that may operate at a variety of spatial and temporal scales. That may explain this no-difference scenario in species richness observed in this study: headwaters degradation, longitudinal homogenization of stream fish occurrences, both operating at a long term scale (over 100 years). This probably resulted in a fish assemblage composed mostly of tolerant, widespread species, some of which may have adapted to particular degraded environmental conditions.

The present study allowed for the identification of species that are uniquely or mostly associated with combinations of land use at the whole catchment or at the riparian stream network buffer.

Unique species were associated with the forest-agriculture gradient, some species, as noted above, related to riverine conditions (e.g. detritus feeders such as *Steindachnerina insculpta*, aquatic insectivores such as *Piabina argentea* - Characiformes: Characidae: "Tetragonopterinae"-, *Corydoras flaveolus*) and the exotic (allochthonous) species *Oreochromis niloticus* (Perciformes: Cichlidae). *Tilapia rendalli*, another exotic Cichlidae species also occurred as a duplicate only in agriculture dominated catchments. This exotic species, the 'tilapias', are reported as pests in many river systems of the world, including the Americas (see Canonico et al., 2005, for general discussion of tilapias introductions worldwide; Sanches et al., 2012, for a discussion on behavioral mechanisms explaining *O. niloticus* success upon a Neotropical Cichlidae species and CABI, 2014a, for general information on the species introduction and its environmental harms). The duplicate species also fall into the forest to agriculture gradient: many can relate to detritus or autochthonous feeding resources (see Vannote et al., 1980, for a consideration on trophic groups and resource basis along a river longitudinal continuum) and the rationale on the faunal homogenization also applies here (see Vilella et al., 2002; Casatti et al., 2003; Casatti & Castro, 2006; Ferreira et al., 2012; Zeni & Casatti, 2013, for information on diet, feeding behavior and guilds of the species cited in this study). Two species that presented a limited occurrence in this dataset (as duplicates in pasture and forest catchments), *C. gomesi* and *Imparfinis borodini* (Siluriformes, Heptapteridae), probably are habitat specialists, collected only at torrential habitats and boulder-cobble substrata on 'run' habitat units. Results obtained from the analysis of similarities also highlighted the importance of forest on riparian corridors on fish abundances, as a distinct group of species could be defined for forest and non-forest groups.

The pair of Poeciliidae species, *Phalloceros caudimaculatus* and *Poecilia reticulata*, is emblematic of such differences. *P. reticulata* is the third exotic species recorded on this study, widely introduced across freshwater ecosystems of the world. Its wide thermal tolerance, high fecundity and vicious reproductive behavior allows for the colonization of harsh, polluted environments, limiting the successor occupying habitats inappropriate for close (taxonomic and functional) relatives such as *P. caudimaculatus* (Chung, 2001; Valero et al., 2008). An overview of the biology of *P. reticulata* is presented in CABI (2014b). *P. caudimaculatus*, on the other hand, was collected only from mildly to well preserved stream reaches, bordered by native forests. The other species, separated according to streams bordered with forest or agricultural environments, have insect-based diets, but particular feeding behaviors may explain their significant differences in abundance. For instance, *Trichomycterus* sp., *Imparfinis mirini*, both Siluriformes, base their diets on insects associated with the stream bottom, mostly when structured with rocky substrate. Other species, such as *Astyanax altiparanae*, *A. paranae* and *Bryconamericus iheringii* are nektonic species that either relies on a wide variety of aquatic and terrestrial insects or on other food items such as algae and plant debris (Vilella et al., 2008; Ferreira et al., 2012). *Corumbataia cuestae*, a miniature armored catfish, is iconic to the studied region, named after the municipality of Corumbataí and after the *cuestas* relief found on the watershed (Britski, 1997). It is associated to fast-flowing waters, rocky substrate and the riparian inputs of medium to large woody debris; conditions found on forest remnants occurring on the steep slopes of the watershed *cuestas*. Thus, species occurrence and abundances across the agriculture impact gradient upon streams can be mostly interpreted in the light of longitudinal habitat homogenization and species life traits.

13.6 Final Considerations

This study showed that catchment dominant land use is a poor predictor of fish species richness. Land use along a 30 m riparian buffer impacts total species richness, but evidence is not unequivocal, and limited to the Rainy season sample. On the other hand, common agriculture production systems in Corumbataí watershed place impacts on species occurrence of stream fish assemblages. Differences are clearer and larger for the riparian buffer land use. Different species occur on stream reaches bordered by contrasting land use classes such as sugarcane plantations or pastures *versus* forest remnants. A group of nine fish species presented difference abundances to stream networks bordered by pastures or forest remnants, which can be linked to species life traits. These species can be considered as indicators of the impacts of poorly managed agricultural production systems, but further studies on (thermal, nutrient) pollution tolerances and general species life traits are needed. The study also points that preserving riparian corridors across small catchments is a necessary condition for the maintenance of fish species that are habitat specialists.

Acknowledgements: This study was funded by Fundação de Amparo à Pesquisa do Estado de São Paulo (FAPESP), grant number 01/13251-4, and a doctoral scholarship for Pedro Gerhard, number 00/14284-0. Luciano M. Verdade has a Research Productivity Fellowship from the Conselho nacional de Desenvolvimento Científico e Tecnológico (CNPq). The study benefited from field and laboratory facilities of the Escola Superior de Agricultura Luiz de Queiroz (University of São Paulo). Edwin Ortega, José Eurico P. Cyrino, Carlos Alberto Vettorazzi, Naércio de Aquino Menezes, Flávio C. T. Lima, Roberta de Oliveira Averna Valente, Sílvio F. B. Ferraz, Kátia M. P. M. B. Ferraz, Anderson Ferreira, Lilian Casatti and Francisco Langeani helped on earlier versions of this work. Gabriel Brejão, Felipe Rosseti de Paula, Maurício Tassoni Filho, Henrique Sawakuchi, Ricardo Brassalotti, Gustavo Libardi and Bruno Cavichioli assisted on field work.

Bibliography

Allan, J.D. (2004). Landscapes and Riverscapes: The Influence of Land Use on Stream Ecosystems. Annual Review of Ecology, Evolution, and Systematics, 35(1), 257–284.

Allan, J.D., & Castillo, M.M. (2007). Stream Ecology: Structure and function of running waters. Dordrecht: Springer.

Azevedo, T. (2013). Efeito da expansão do cultivo de cana-de-açúcar na composição da paisagem do estado de São Paulo. Universidade de São Paulo. Available at: http://www.teses.usp.br/teses/disponiveis/41/41134/tde-15102013-140418/en.php [Accessed September 23, 2015]

Braga, F.M., Gomiero, L.M., & Souza, U.P. (2008). Aspectos da reprodução e alimentação de Neoplecostomus microps (Loricariidae, Neoplecostominae) na microbacia do Ribeirão Grande, serra da Mantiqueira oriental (Estado de São Paulo). Acta Scientiarum (Biological Sciences), 30 (4), 455–463.

Britski, H.A. (1997). Descrição de um novo gênero de Hypoptopomatinae, com duas espécies novas (Siluriformes, Loricariidae). Papéis Avulsos de Zoologia, 40 (15), 231–255.

Burcher, C.L., McTammany, M.E., Benfield, E.F., et al. (2008). Fish assemblage responses to forest cover. Environmental Management, 41 (3), 336–46.

CABI (2014a). *Oreochromis niloticus*. In: Invasive Species Compendium. Wallingford: CAB International, 39. Available at: http://www.cabi.org/isc. [Accessed September 23, 2015]

CABI (2014b). *Poecilia reticulata*. In: Invasive Species Compendium. Wallingford: CAB International, 25. Available at: http://www.cabi.org/isc. [Accessed September 23, 2015]

Canonico, G.C., Arthington, A., McCrary, J.K., et al. (2005). The effects of introduced tilapias on native biodiversity. Aquatic Conservation: Marine and Freshwater Ecosystems, 15 (5), 463–483.
Casatti, L., & Castro, R.M.C. (2006). Testing the ecomorphological hypothesis in a headwater riffles fish assemblage of the rio São Francisco, southeastern Brazil. Neotropical Ichthyology, 4 (2), 203–214.
Casatti, L., Ferreira, C.P., & Langeani, F. (2008). A fish-based biotic integrity index for assessment of lowland streams in southeastern Brazil. Hydrobiologia, 623 (1), 173–189.
Casatti, L., Langeani, F., Silva, A.M., et al. (2006). Stream fish, water and habitat quality in a pasture dominated basin, southeastern Brazil. Brazilian Journal of Biology (Revista Brasileira de Biologia), 66 (2B), 681–96.
Casatti, L., Mendes, H.F., & Ferreira, K.M. (2003). Aquatic macrophytes as feeding site for small fishes in the Rosana Reservoir, Paranapanema River, southeastern Brazil. Brazilian Journal of Biology (Revista Brasileira de Biologia), 63 (2), 213–22.
Chung, K.S. (2001). Critical thermal maxima and acclimation rate of the tropical guppy Poecilla reticulata. Hydrobiologia, 462 (1-3), 253–257.
Clarke, K.R. (1993). Non-parametric multivariate analyses of changes in community structure. Austral Ecology, 18 (1), 117–143.
Colwell, R.K. (2013). EstimateS 9.1.0 User's Guide. Available at: http://viceroy.eeb.uconn.edu/estimates/EstimateSPages/EstSUsersGuide/EstimateSUsersGuide.htm. [Accessed September 23, 2015]
Colwell, R.K., Chao, A., Gotelli, N.J., et al. (2012). Models and estimators linking individual-based and sample-based rarefaction, extrapolation and comparison of assemblages. Journal of Plant Ecology, 5 (1), 3–21.
Dean, W. (1977). Rio Claro: um sistema brasileiro de grande lavoura (1820-1920). Rio de Janeiro: Paz & Terra.
Dean, W. (1996). A Ferro e Fogo: A história e a devastação da Mata Atlântica brasileira. São Paulo: Companhia da Letras.
Fagan, W.F. (2002). Connectivity, fragmentation, and extinction risk in dendritic metapopulations. Ecology, 83 (12), 3243–3249.
Faria, A.P., & Marques, J.S. (1999). O desaparecimento de pequenos rios brasileiros. Ciência Hoje, 25 (146), 56–61.
Ferreira, A., Gerhard, P., & Cyrino, J.E.P. (2012). Diet of Astyanax paranae (Characidae) in streams with different riparian land covers in the Passa-Cinco River basin, southeastern Brazil. Iheringia. Série Zoologia, 102 (1), 80–87.
Gomi, T., Sidle, R.C., & Richardson, J.S. (2002). Understanding Processes and Downstream Linkages of Headwater Systems. BioScience, 52 (10), 906–916.
Hammer, Ø., Harper, D.A.T., & Ryan, P.D. (2001). PAST : Paleontological Statistics Software Package for Education and Data Analysis. Paleontologia Electronica, 4 (1), 1–9.
Inoue, M., & Nunokawa, M. (2002). Effects of longitudinal variations in stream habitat structure on fish abundance: an analysis based on subunit-scale habitat classification. Freshwater Biology, 47 (9), 1594–1607.
Krebs, C.J. (1998). Ecological Methodology. Second edition. Menlo Park: Benjamin Cummings.
Legendre, P., & Legendre, L.F.G. (2012). Numerical Ecology. Third edition. Amsterdam: Elsevier.
Luzio, M. Di, Srinivasan, R., Arnold, J.G., et al. (2002). ArcView Interface for SWAT2000. Temple: Texas Water Resources Institute.
Magurran, A.E., & McGill, B.J. (2011). Biological Diversity: Frontiers in Measurement and Assessment. Oxford: Oxford University Press.

Maloney, K.O., Feminella, J.W., Mitchell, R.M., et al. (2008). Landuse legacies and small streams: identifying relationships between historical land use and contemporary stream conditions. Journal of the North American Benthological Society, 27 (2), 280–294.

Matthews, W.J. (1998). Patterns in Freshwater Fish Ecology. Dordrecht: Springer.

Menezes, N.A., Weitzman, S.H., Weitzman, M.J., et al. (2007). Peixes de Água Doce da Mata Atlântica. São Paulo: Museu de Zoologia da Universidade de São Paulo.

Meyer, J.L. Strayer, D.L., Wallace, J.B., et al. (2007). The Contribution of Headwater Streams to Biodiversity in River Networks. Journal of the American Water Resources Association, 43 (1), 86–103.

Nerbonne, B.A., & Vondracek, B. (2001). Effects of Local Land Use on Physical Habitat, Benthic Macroinvertebrates, and Fish in the Whitewater River, Minnesota, USA. Environmental Management, 28 (1), 87–99.

Paula, F.R., Ferraz, S.F.B., Gerhard, P., et al. (2011). Large woody debris input and its influence on channel structure in agricultural lands of Southeast Brazil. Environmental Management, 48 (4), 750–63.

Reis, R.E., Ferraris, C.J., & Kullander, S.O. (2003). Check List of the Freshwater Fishes of South and Central America. Porto Alegre: EDIPUCRS.

Roth, N.E., Allan, J.D., & Erickson, D.L. (1996). Landscape influences on stream biotic integrity assessed at multiple spatial scales. Landscape Ecology, 11 (3), 141–156.

Rudorff, B.F.T., Aguiar, D.A., Silva, W.F., et al. (2010). Studies on the Rapid Expansion of Sugarcane for Ethanol Production in São Paulo State (Brazil) Using Landsat Data. Remote Sensing, 2 (4), 1057–1076.

Sanches, F.H.C., Miyai, C.A., Costa, T.M., et al. (2012). Aggressiveness overcomes body-size effects in fights staged between invasive and native fish species with overlapping niches. PloS One, 7 (1), 1-4 (e29746).

Schlosser, I.J. (1991). Stream Fish Ecology: A Landscape Perspective. BioScience, 41 (10), 704–712.

Scott, M.C., & Helfman, G.S. (2001). Native Invasions, Homogenization, and the Mismeasure of Integrity of Fish Assemblages. Fisheries, 26 (11), 6–15.

Sparovek, G., Barretto, A., Berndes, G., et al. (2008). Environmental, land-use and economic implications of Brazilian sugarcane expansion 1996–2006. Mitigation and Adaptation Strategies for Global Change, 14 (3), 285–298.

Valente, R.O.A., & Vettorazzi, C.A. (2002). Análise da estrutura da paisagem na Bacia do Rio Corumbataí, SP. Scientia Forestalis, 62, 114–129.

Valente, R.O.A., & Vettorazzi, C.A. (2008). Definition of priority areas for forest conservation through the ordered weighted averaging method. Forest Ecology and Management, 256 (6), 1408–1417.

Valero, A., Macías-Garcia, C., & Magurran, A.E. (2008). Heterospecific harassment of native endangered fishes by invasive guppies in Mexico. Biology letters, 4 (2), 149–52.

Vannote, R.L., Minshall, G.,W., Cummins, K.W., et al. (1980). The River Continuum Concept. Canadian Journal of Fisheries and Aquatic Sciences, 37 (1), 130–137.

Viana, V., Vettorazzi, C.A., & Zakia, M. (2002). Plano diretor para conservação dos recursos hídricos por meio da recuperação e da conservação da cobertura florestal da bacia do rio Corumbataí. Piracicaba, IPEF.

Victor, M.A.M., Cavalli, A.C., Guillaumon, J.R., et al. (2005). Cem anos de devastação: revisitada 30 anos. Brasília: Ministério do Meio Ambiente.

Vilella, F.S., Becker, F.G., & Hartz, S.M. (2002). Diet of Astyanax species (Teleostei, Characidae) in an Atlantic Forest River in Southern Brazil. Brazilian Archives of Biology and Technology, 45 (2), 223–232.

Vondracek, B., Blann, K.L., Cox, C.B., et al. (2005). Land use, spatial scale, and stream systems: lessons from an agricultural region. Environmental Management, 36 (6), 775–791.
Zeni, J.O., & Casatti, L. (2013). The influence of habitat homogenization on the trophic structure of fish fauna in tropical streams. Hydrobiologia, 726 (1), 259–270.

Part III: **Case Studies**

Paula C. Lopes, Thiago S. Marques, Paula S. Martin, Carla Gheler-Costa, Luciano M. Verdade

14 Patterns of Leaf-litter Amphibian Diversity in a Silvicultural Landscape of Southeastern Brazil

Abstract: Amphibians are susceptible to environmental changes due to a combination of morphological, physiological and behavioral characteristics adapted to specific environmental conditions. Few studies focus on the patterns of amphibian diversity in disturbed habitats by human action. However, recent studies with other taxa suggest that agricultural landscapes can support considerable biodiversity. The main goal of this study was to evaluate the spatial-temporal distribution and abundance patterns of leaf-litter amphibians inhabiting a silvicultural landscape. This study was conducted at Três Lagoas and Arca farms, located in the Upper Paranapanema basin, state of São Paulo, in southeastern Brazil. The data collection was carried out from August 2007 to July 2009 in monthly sampling campaigns. A grid of sampling units ($n = 30$) formed by pitfall traps were distributed over *Eucalyptus* plantations ($n = 18$), native vegetation ($n = 7$) and abandoned pastures ($n = 5$). A total of 1047 leaf-litter individuals from 18 species were captured in the study area. The leaf-litter amphibian assemblage was composed predominantly by *Physalaemus cuvieri* (58.6%), *Physalaemus nattereri* (13.1%), and *P. marmoratus* (13.0%). There was a significant difference in species richness and abundance between environments, surprisingly with the lowest species richness and abundance in the native vegetation. However, there was no relationship between species richness and relative abundance considering the distance of water bodies or the native vegetation. The results of this study suggest that silvicultural landscapes may have some conservation value. In order to improve it, it is necessary to maintain conservation areas within the landscape, as well as to stimulate innovation in silvicultural wildlife-friendly management practices.

14.1 Introduction

The expansion of socioeconomic activities has proven to be a major driver of the alteration and destruction of natural habitats (Rocha et al., 2006). This process caused extensive impacts on biodiversity in different regions of the world due to the replacement of native vegetation by agricultural crops and livestock areas (Diamond, 1997; Noris, 2008). These agricultural systems are spatially homogeneous (e.g. monoculture of sugarcane and *Eucalyptus*), but temporally heterogeneous due to the production cycle. The management of these systems may result in profound changes in the wildlife by modifying the spatial distribution of resources (Eterovick & Sazima, 2000; Silvano et al., 2003). These processes can compromise gene flow and dispersal frequency resulting in population decline (Feder & Burggren, 1992; Young et al., 2000).

The socioeconomic development of the state of São Paulo in Southeastern Brazil reflects the global trend of natural resources utilization. The economic development model adopted by the state caused profound changes in its original vegetation, gradually replaced by agroecosystems (Victor et al., 2005). Such process resulted in the current critical environmental scenario, with only 17.5% of the native vegetation remaining (IF, 2009).

The state of São Paulo has approximately 1.2 Mha covered by forest plantations that correspond to the second largest in Brazil (ABRAF, 2013). Most of this area was formerly occupied by low productivity pastures (Vianna et al., 2007). In turn, environmental laws due to the relatively well established certification process (i.e. Forest Stweardship Council – FSC) in this sector provides the implementation of Areas of Permanent Protection (APP) and Legal Reserve (LR). Such change on the landscape structure can have direct and indirect consequences on local biodiversity.

A priori, amphibians can be considered good indicators of environmental change as they generally have relatively small home range and most depend on the availability of aquatic environments for reproduction and terrestrial environments for foraging, aestivation, migration and dispersion (Stebbins & Cohen, 1995; Zug et al., 2001; Pough et al., 2004). Habitat fragmentation, deforestation and water contamination by agrochemicals have been considered the main causes of global decline of amphibians (Gray et al., 2004; Young et al., 2004). Many species are also sensitive to changes in the vegetation structure near water bodies (Jim, 1980; Valan, 2002; Renken et al., 2004) and their dispersal patterns are intrinsically related to habitat quality (Duelman & Trueb, 1994; Peltzer et al., 2003). However, tolerant amphibians have been recorded in anthropic environments (e.g. Demaynadier & Hunter, 1998; Campos & Vaz-Silva, 2010; Machado et al., 2011).

The main goal of this study was to evaluate spatial-temporal variation in diversity of leaf-litter amphibians inhabiting a silvicultural landscape during the first three years (0-3 year) of *Eucalyptus* plantations. The following null-hypotheses have been tested: 1) There is no difference in leaf-litter amphibians diversity among the different components of silvicultural landscapes (*Eucalyptus* plantations, abandoned pastures and native vegetation remnants); 2) There is no relation between leaf-litter amphibians diversity and their distance to the nearest border of native vegetation inside the matrix (*Eucalyptus* plantations); 3) There is no time variation in leaf-litter amphibians diversity in silvicultural landscapes; and, 4) There is no intraspecific difference in body length in leaf-litter amphibians in relation to the silvicultural landscape components.

14.2 Material and Methods

14.2.1 Study Area

This study was carried out in two farms, Três Lagoas (23°22'0"-23°20'41"S/48°28'0"-48°27'57"W) and Arca (23°20'0"-23°18'51"S/48°27'30"-48°28'20"W), located in the Alto Paranapanema watershed, in the state of São Paulo, in Southeastern Brazil. The regional climate is classified as subtropical (Cwa) according to the Köppen system with an average monthly temperature of 20.6°C (ranging from 14.1°C in June 2008 to 24.3°C in March 2009) and average monthly cumulative rain of 113 mm (ranging from 0 mm in July 2008 to 314 mm in January 2009) during the study period (data from Estação Experimental de Ciências Florestais de Itatinga - Escola Superior de Agricultura "Luiz de Queiroz", Universidade de São Paulo).

The two farms together have a total area of 4333 ha (Três Lagoas: 3242 ha and Arca: 1123 ha) composed by a mosaic of *Eucalyptus* plantations (as the landscape matrix), fragments of native vegetation and abandoned pastures. In the past, livestock production was the major activity; however, between 2006 and 2008, much of the pastures were converted into *Eucalyptus* plantations. The remaining area was abandoned for natural revegetation, according to the Brazilian environmental law.

Local *Eucalyptus* plantations are composed by *Eucalyptus grandis*, *E. urophila* and their hybrid *E. urograndis*. Native vegetation is composed by fragments of Semideciduous Forest (dominated by *Cariniana estrelensis*, *Cedrela fissilis*, *Coccoloba mollis*, *Crysophyllum marginatum*, *Gallesia integrifolia*, *Lonchocarpus muehlbergianus*, *Prunus sellowii* and *Machaerium villosum*), Cerrado *sensu stricto* (dominated by *Annona cacans*, *A. coriacea*, *Aspidosperma tomentosa*, *Anadenanthera falcata*, *Dalbergia miscolobium*, *Eryotheca gracilipes*, *Caryocar brasiliense*, *Dimorphandra mollis*, *Stryphnodendron adstringens* and *Ouratea spectabilis*) and Cerradão (dominated by *Tapirira guianensis*, *A. crassiflora*, *Xylopia brasiliensis*, *A. australe*, *Gochnatia polymorpha*, *Terminalia brasiliensis*, *Miconia chartacea*, *Pera glabrata*, *Amaioua guianenis* and *Vochisia tucanorum*). Abandoned pastures are dominated by *Brachyarya* syn. *Urochloa* spp. (Athayde, 2013).

14.2.2 Methodology

14.2.2.1 Sampling Methodology

The study design is based on the methodology proposed by the Biodiversity Research Program (PPBio) where sampling units form a grid with nodules distant one kilometer from each other. This methodology is appropriate for long-term ecological research because it allows rapid biota inventories (Magnusson et al., 2005). In the present study, 30 sampling units (located in the grid nodules) were implemented taking into account the proportion of the three landscape elements: *Eucalyptus* plantations (n = 18), native vegetation (n = 7) and abandoned pasture (n = 5).

The amphibian captures were conducted with drift fence pitfall traps (Corn, 1994; Block et al., 1998; Cechin & Martins, 2000; Dixo & Verdade, 2006; Bernarde & Macedo, 2008). Each sample unit was composed by four 100 L buckets buried to the surface of the ground connected by guide fence (plastic net) with 80 cm high (10 cm buried) in shape of "Y" with 15 m long segments. Buckets had small holes for rainwater drainage and remained capped while not in use. Small mammal studies were also carried out simultaneously (Martin et al., 2012; Rosalino et al., 2013).

Data collection occurred in 23 monthly campaigns from August 2007 to July 2009 (skipping November 2008 due to logistical impediments). The buckets remained open for two consecutive nights in each campaign and were checked every morning. Captured animals were identified, measured (body length), sexed and released at the capture site. Vouchers of amphibian species were deposited in the collection "Célio F. B. Haddad" (CFBH) of Instituto de Biociências - Universidade Estadual Paulista - Campus Rio Claro, under appropriate license from the Instituto Chico Mendes de Conservação da Biodiversidade – ICMBio (License n° 12623-1 and 12623-2). Species identification followed the Brazilian Society of Herpetology (Segalla et al., 2012).

14.2.2.2 Analytical Methodology

The sampling effort sufficiency to characterize the leaf-litter amphibian diversity was evaluated for each landscape unit and for the whole study area by species incidence curves (adapted from Colwell & Coddington, 1994) where the time-series sigmoid model bootstrapped (1000 randomizations by EstimateS software - Colwell, 2004) asymptote was considered the estimated number of species. The sampling effort (*SE*) of pitfall traps methodology was calculated according to the formula:

$$SE = C \times D \times U \times PF$$

where C is the number of sampling campaigns, D the number of sampling days, U the number of sampling units and PF the number of pit-fall stations. The capture success (CS) was calculated according to the formula:

$$CS = (T / SE) \times 100$$

where T is the total number of individuals captured and SE is the sampling effort. Abundance index (AI) was calculated according to the formula:

$$AB = T / SE.$$

Prior to statistical analysis we tested data distribution and homoscedasticity, respectively by Anderson-Darling and Levene's tests. One-way analysis of variance (ANOVA) (Zar, 2010) and Kruskal-Wallis test (Siegel, 1956) were used to respectively test whether leaf-litter amphibian species richness and relative abundance differ among the three distinct landscape units (*Eucalyptus* plantations, abandoned pastures and native vegetation). Analysis of means (ANOM) (Ott, 1983) was then applied as a post-hoc test. In addition, Correspondence Analysis (Manly, 1994) was used to verify the possible relation between the landscape unit type and species composition (Hypothesis 1). Linear regressions (Zar, 2010) were used to detect the possible relationship between species richness and relative abundance of leaf-litter amphibians and the distance of the sampling units to the nearest water body and native vegetation using 14 sampling units located within *Eucalyptus* plantations of Três Lagoas Farm (Hypothesis 2).

The temporal variation in species richness and overall relative abundance was evaluated by t-test (Zar, 2010) and Mann–Whitney test (Siegel, 1956), respectively (Hypothesis 3). April through September was considered dry season whereas October through March was considered wet season. In addition, temporal variation along this study period is presented. At last, one-way ANOVA (Zar, 2010) were applied to check possible differences in snout-vent length (SVL) of dominant species among environments (Hypothesis 4). All analyzes were performed in the statistical software Minitab 16.

14.3 Results

Eighteen leaf-litter amphibian species were detected in the study area, totaling 1047 captured individuals. These species are distributed in the families Bufonidae, Craugastoridae, Cycloramphidae, Leiuperidae, Leptodactylidae and Microhylidae (Tab. 14.1). The total sampling effort was 4977 trap-night (*Eucalyptus* plantations = 3054 trap-night; native vegetation = 1030 trap-night; abandoned pasture = 893 trap-night) and the overall capture success was 21.05%. Our sampling effort was enough to detect from 83.3 to 93.5% of the estimated leaf-litter amphibian species richness of this study area (Fig. 14.1, Tab. 14.2).

No species detected in this study is considered threatened or endangered according to the IUCN Red List (IUCN, 2014) and the List of Endangered Fauna of the State of São Paulo (State Decree No. 60.133/2014). The leaf-litter amphibian assemblage in our study area is composed predominantly by *Physalaemus cuvieri* (58.6%), *Physalaemus nattereri* (13.1%) and *Physalaemus marmoratus* (13.0%). There was a significant difference in species richness and relative abundance of leaf-litter amphibians among environments (richness: $F_{2,29} = 7.20$; $p = 0.003$; abundance: $H_{2,29} = 11.89$; $p = 0.003$), with the lowest species richness and abundance surprisingly in the native vegetation (Fig. 14.2). In addition, there appears to be a consistent difference between native vegetation and anthropic environments (i.e. *Eucalyptus* plantations and abandoned pasture) in terms of amphibian diversity (i.e. species composition, richness and relative abundance – Fig. 14.3 and 14.4). We can thus reject the first null hypothesis. However, there was no clear relationship between species richness and relative abundance of leaf-litter amphibians both to the distance from nearest water bodies (richness: $F_{1, 13} = 0.71$; $p = 0.414$; $r^2 = 5.6$; relative abundance: $F_{1, 13} = 0.77$, $p = 0.397$; $r^2 = 6.1$) or the nearest native vegetation border (richness: $F_{1, 13} = 0.12$; $p = 0.733$; $r^2 = 1.0$; relative abundance: $F_{1,13} = 0.23$; $p = 0.644$; $r^2 = 1.8$; Fig. 14.5). Therefore, we cannot reject the second null hypothesis.

Tab. 14.1: Number of individuals of leaf-litter amphibians on a silvicultural landscape of Angatuba, state of São Paulo, Southeastern Brazil.

Family/Species	*Eucalyptus* Plantations	Native Vegetation	Abandoned Pasture	Total
Bufonidae				
Rhinella ornata	1	5	0	6
R. schneideri	3	0	0	3
Craugastoridae				
Haddadus binotatus	0	5	3	8
Cycloramphidae				
Odontophrynus americanus	7	1	0	8
Leiuperidae				
Physalaemus centralis	35	2	3	40
P. cuvieri	200	41	373	614
P. nattereri	80	4	54	138
P. marmoratus	122	0	15	137
P. olfersii	0	1	0	1
Leptodactylidae				
Leptodactylus cf. *furnarius*	0	1	0	1
L. fuscus	40	1	7	48
L. labyrinthicus	0	1	1	2
L. latrans	4	0	1	5
L. mystaceus	0	1	0	1
L. mystacinus	5	0	3	8
Microhylidae				
Chiasmocleis albopunctata	2	0	0	2
Elachistocleis cf. *bicolor*	2	0	2	4
E. cf. *ovalis*	9	0	12	21
Total	510	63	474	1047

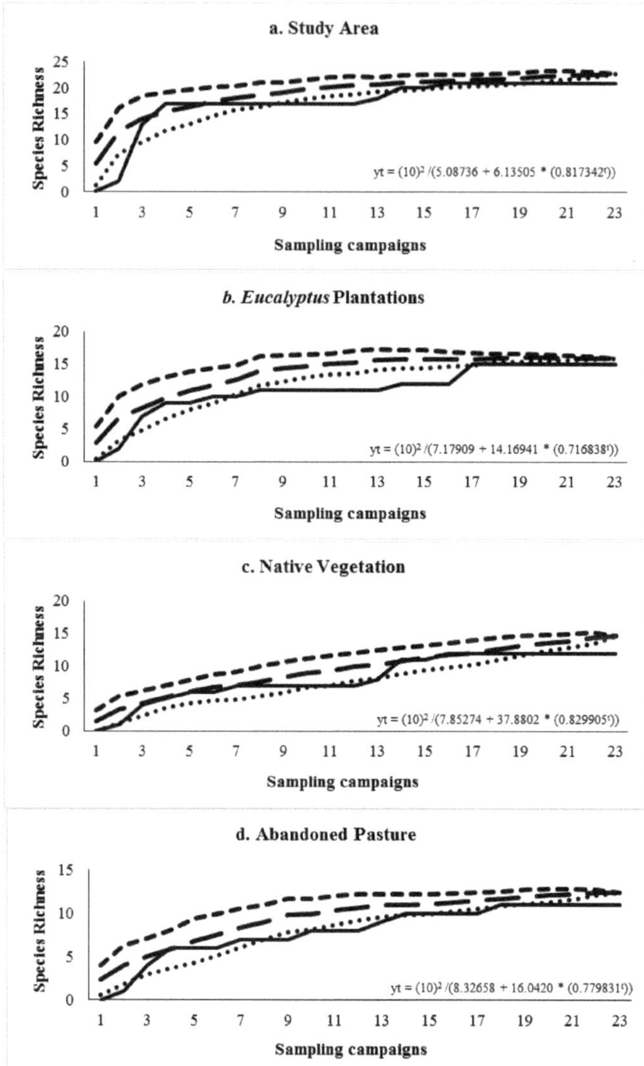

Fig. 14.1: Leaf-litter amphibians' species incidence curves, for the whole landscape and its environments.

Tab. 14.2: Estimated (r_{est}) and detected (r_{det}) species richness of leaf-litter amphibians on a silvicultural landscape of Angatuba, state of São Paulo, Southeastern Brazil.

Environment	r_{est}	r_{det}	$(r_{det}/r_{est}) \times 100$ (%)
Eucalyptus plantation	13.9	13	93.5
Abandoned pasture	12.7	11	86.6
Native vegetation	12	10	83.3
Total	19.7	18	91.3

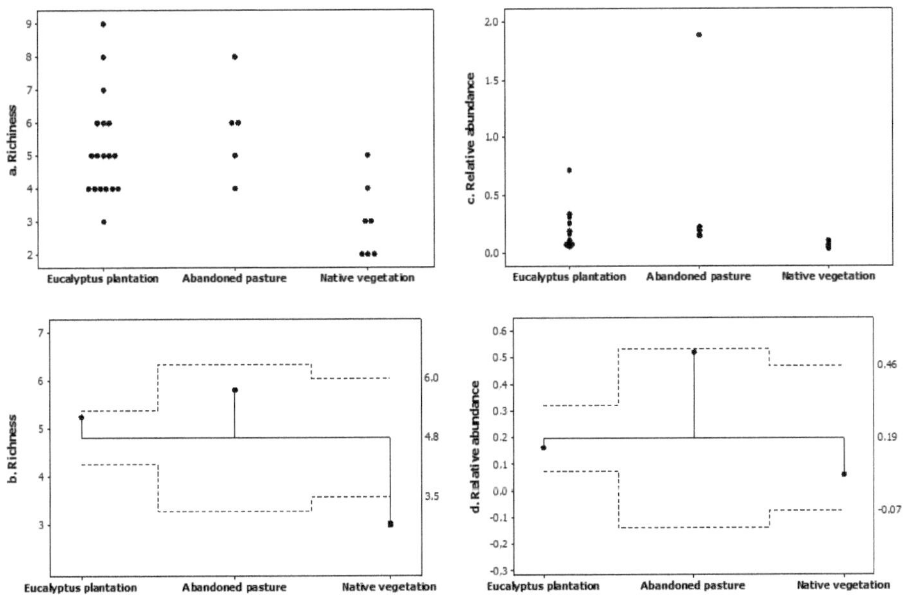

Fig. 14.2: Species richness and relative abundance of leaf-litter amphibians in a silvicultural landscape in Angatuba, state of São Paulo, Southeastern Brazil. In "A" and "C" black dots represent the collected data. In "B" and "D" mean and standard deviation are presented in the Analysis of Means.

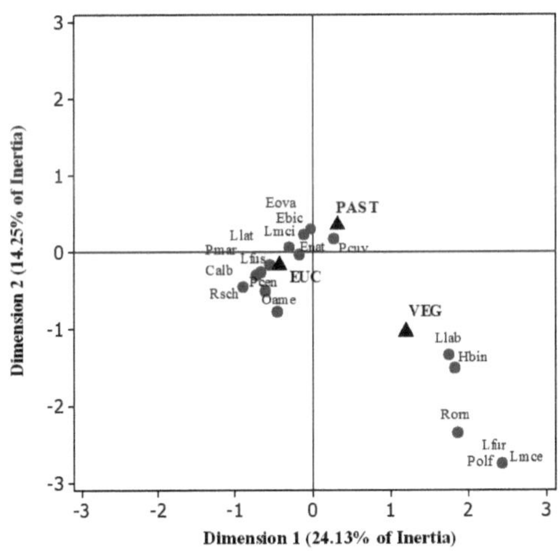

Fig. 14.3: Correspondence Analysis plot, relating the landscape unit types and species composition and relative abundance. EUC: *Eucalyptus* plantation, PAST: Abandoned pasture and VEG: Native vegetation. Calb - *Chiasmocleis albopunctata*, Ebic - *Elachistocleis cf. bicolor*, Eova - *E. cf. ovalis*, Hbin - *Haddadus binotatus*, Lfur - *Leptodactylus cf. furnarius*, Lfus - *Leptodactylus fuscus*, Llab - *L. labyrinthicus*, Llat - *L. latrans*, Lmci - *L. mystacinus*, Lmce - *L. mystaceus*, Oame - *Odontophrynus americanus*, Pcen - *Physalaemus centralis*, Pcuv - *P. cuvieri*, Pmar - *P. marmoratus*, Pnat - *Physalaemus nattereri*, Polf - *P. olfersii*, Rorn - *Rhinella ornata* and Rsch - *R. schnederi*.

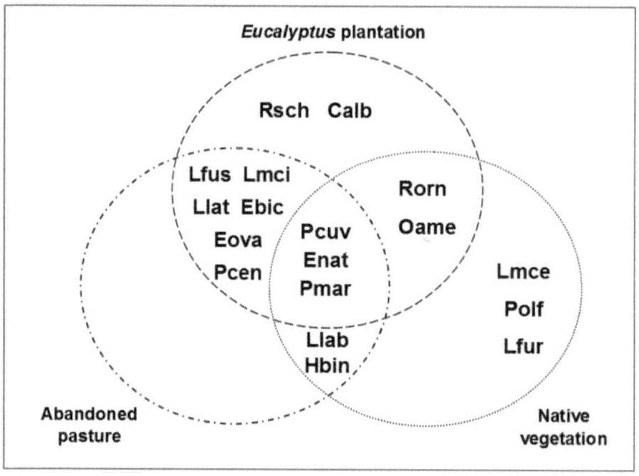

Fig. 14.4: Composition of leaf-litter amphibians by environment in a silvicultural landscape in Angatuba, state of São Paulo, Southeastern Brazil (Calb - *Chiasmocleis albopunctata*, Ebic - *Elachistocleis cf. bicolor*, Eova - *E. cf. ovalis*, Hbin - *Haddadus binotatus*, Lfur - *Leptodactylus cf. furnarius*, Lfus - *Leptodactylus fuscus*, Llab - *L. labyrinthicus*, Llat - *L. latrans*, Lmci - *L. mystacinus*, Lmce - *L. mystaceus*, Oame - *Odontophrynus americanus*, Pcen - *Physalaemus centralis*, Pcuv - *P. cuvieri*, Pmar - *P. marmoratus*, Pnat - *Physalaemus nattereri*, Polf - *P. olfersii*, Rorn - *Rhinella ornata* and Rsch - *R. schnederi*).

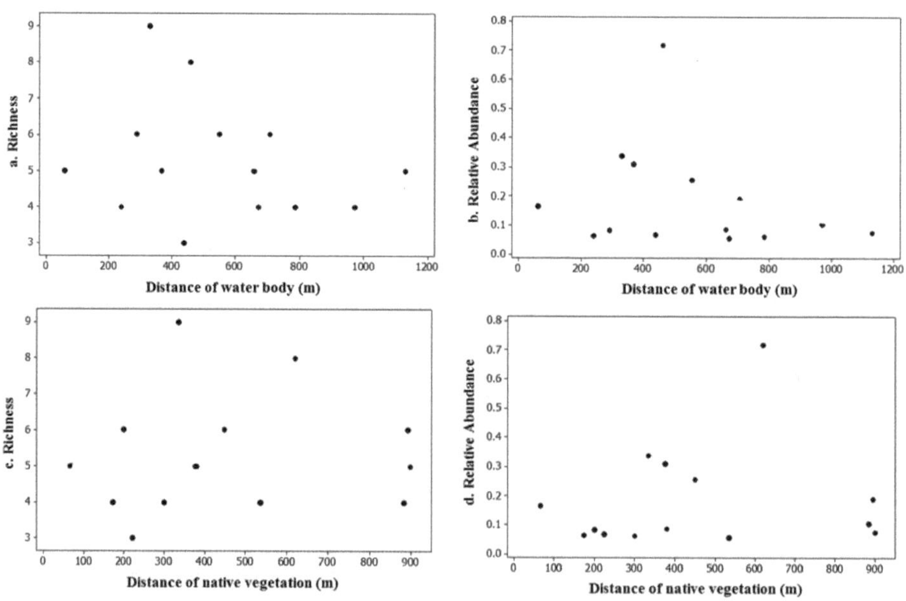

Fig. 14.5: Relationship between richness and relative abundance of litter amphibians and distance from nearest water body and native vegetation.

The species richness and overall relative abundance varied significantly between the dry and wet seasons sampled (richness: $t_{1,22} = 2.99$; $p = 0.007$; relative abundance: $W_{1,22} = 93.0$; $p = 0.001$), both higher in the wet season. In addition, there appears to be a consistent time variation along the study period in terms of leaf-litter amphibians in all landscape components (Fig. 14.6). We can, therefore, reject the third null hypothesis.

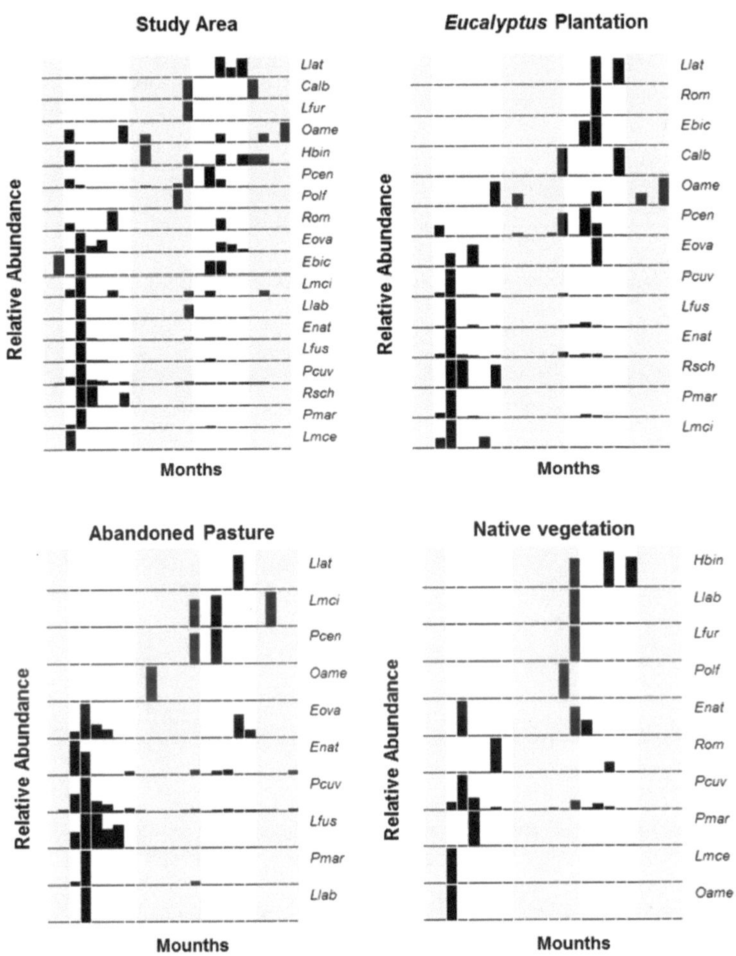

Fig. 14.6: Temporal variation in the relative abundance of the different litter amphibians species detected in the study area (Calb - *Chiasmocleis albopunctata*, Ebic - *Elachistocleis cf. bicolor*, Eova - *E. cf. ovalis*, Hbin - *Haddadus binotatus*, Lfur - *Leptodactylus cf. furnarius*, Lfus - *Leptodactylus fuscus*, Llab - *L. labyrinthicus*, Llat - *L. latrans*, Lmci - *L. mystacinus*, Lmce - *L. mystaceus*, Oame - *Odontophrynus americanus*, Pcen - *Physalaemus centralis*, Pcuv - *P. cuvieri*, Pmar - *P. marmoratus*, Pnat - *Physalaemus nattereri*, Polf - *P. olfersii*, Rorn - *Rhinella ornata* and Rsch - *R. schnederi*). The area shaded in gray represents dry season and the area without shadow represents wet season.

The dominant species, *P. marmoratus* and *P. nattereri*, showed no intraspecific differences in snout-vent length in relation to the environments (*P. marmoratus*: $F_{1,53}$ = 1.75; p = 0.191; *Eucalyptus* plantation = 3.6 ± 0.5 cm; abandoned pasture = 4.0 ± 0.4 cm; *P. nattereri*: $F_{2,104}$ = 1.01; p = 0.367; *Eucalyptus* plantation = 4.2 ± 0.5 cm; native vegetation = 4.2 ± 0.2 cm; abandoned pasture = 4.3 ± 0.6 cm). However, there was a significant difference for *P. cuvieri*, with snout-vent length slightly longer in *Eucalyptus* plantations, but shorter in smaller at abandoned pastures ($F_{2,235}$ = 8.85; p < 0.001; *Eucalyptus* plantation = 2.7 ± 0.6 cm; native vegetation = 2.7 ± 0.2 cm; abandoned pasture = 2.4 ± 0.6 cm) (Fig. 14.7). We can, therefore, partially reject the fourth null hypothesis.

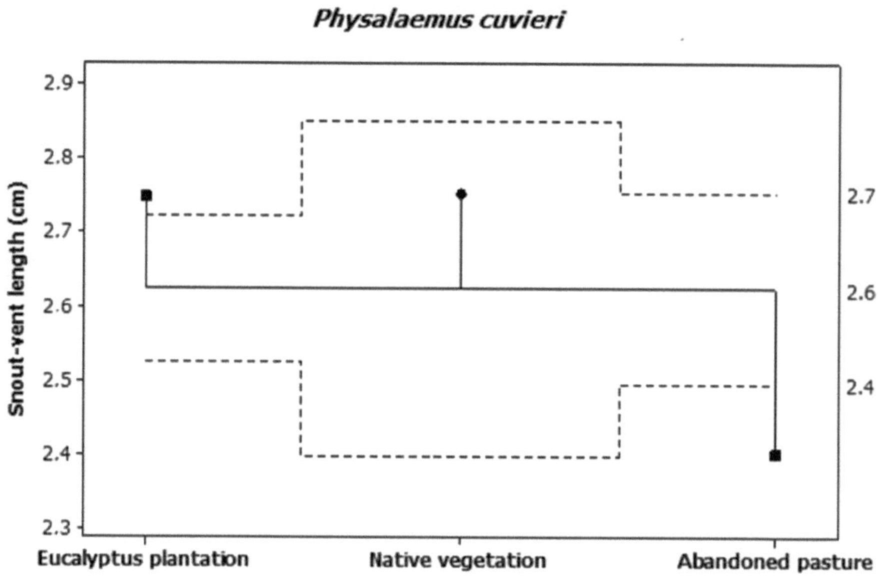

Fig. 14.7: Intraspecific variation of snout-vent length of leaf-litter amphibians in different components of a silvicultural landscape in Angatuba, state of São Paulo, Southeastern Brazil.

14.4 Discussion

Although we carried out monthly sampling campaigns during two years the number of species detected in each sampled environment (from 83.3 to 93.5%) suggest that more intense sampling effort (possibly longer than only 2 night captures per month) should be applied in silvicultural landscapes. However, for logistic reasons, in order to do so it might be necessary to restrict the number of sampling units per sampled environment. Future studies should prioritize sampling design and sampling methodology concerning leaf-litter amphibians in silvicultural (and agricultural) landscapes. Such concerns have been raised for mammals (Lyra-Jorge et al., 2014) and birds (Penteado et al., 2014) as traditional methods for vertebrate surveys have been generally developed for pristine environments from temperate regions (Sutherland, 2006).

The pattern of variation in species richness and relative abundance of leaf-litter amphibians among environments found in this study is possibly due to the expansion of the dominant species in anthropic areas at the beginning of the *Eucalyptus* cycle (first three years of implantation), possibly in response to the sudden increase in food supply and lack of predators. This hypothesis is corroborated when we reanalyze the data without the dominant species with consequently no difference in species richness and relative abundance among environments (richness: $F_{2,29} = 2.02$; $p = 0.153$; abundance: $H_{2,29} = 2.34$; $p = 0.311$). The conversion of pasture on *Eucalypts* plantations promotes the appearance of an arboreal extract, an increase on the litter amount, a decrease in soil compaction and a decrease in light incidence (Alves et al., 2006). Such factors can be beneficial to some species of amphibians. Future studies should investigate whether such patterns persist along the productive cycles of the *Eucalypus* plantations (i.e. during a few decades).

The species composition of leaf-litter amphibians appears to be different between the native vegetation and the anthropic environments, with no distinction in this study between *Eucalyptus* plantations and abandoned pastures. Such patterns could be due to the fact that the initial phase of *Eucalyptus* plantations has structural characteristics intermediate between open and forest environment. In such a circumstance, Cerrado species like *Physalaemus cuvieri*, *P. centralis*, *P. marmoratus* and *Physalaemus nattereri* can be benefitted (Bastos, 2007).

Surprisingly, native vegetation showed just few exclusive species in relation to the anthropic environments. This pattern may be the historical result of local land use and exploitation of natural resources (e.g. deforestation). The study area is located in a transition between Cerrado and Semideciduous forest that suffered cycles of deforestation and revegetation since 1870 (Lisboa, 2008). Since the early 1970's, land use has been dominated by pastures for livestock production which possibly resulted in local extinction of many forest species of leaf-litter amphibians. Different species may be distinctly affected by land use changes (Swihart et al., 2003; Rubbo & Kiezecker, 2005). Revegetation of conservation areas (i.e. APP and LR) might result in an increase in leaf-litter amphibian diversity. A network of long-term biodiversity monitoring sites should be established in agricultural/silvicultural landscapes in order to evaluate the impacts of changes on land use and agricultural management practices on biodiversity (Verdade et al., 2014a) including leaf-litter amphibians.

The distance from water bodies or native vegetation can be decisive for amphibian survival. However, no such pattern has been detected in this study, which can be partially explained by the presence of small fragments of native vegetation of irregular shape with relatively small species richness possibly due to border effect (i.e., increase in temperature and decrease in relative humidity). In addition, in this study we considered only permanent water bodies due to the logistical difficulties to monitor temporary water bodies in large areas long term. However, the distribution of temporary water bodies may have affected the present results, since most leptodactylids of the *Physalaemus* group use temporary ponds and puddles for reproduction (Brasileiro et al., 2005). Such temporary water bodies can increase the carrying capacity of *Eucalyptus* plantations for leaf-litter amphibian at least in the early phase of its first productive cycle.

The species richness and overall relative abundance varied between dry and wet seasons. The decrease in abundance during the dry season may be related to lower activity of amphibians during this period due to decrease in temperature, humidity conditions and food supply (Gibbs, 1998; Alford & Richards, 1999; Pinheiro et al., 2002; Vasconcelos & Rosa-Feres, 2005). Humidity and/or temperature have been considered the main determining factors of amphibian diversity (Aichinger, 1987; Maffei et al., 2011). Most amphibian species which inhabit regions without marked seasonality can reproduce throughout the year; however, Cerrado species tend to concentrate their reproductive activities during rainy season (Brasileiro et al., 2005) as they depend on temporary aquatic environments to breed (Duellman & Trueb, 1986). The temporal distribution of the species in this study seems to be consistent with this pattern.

Body size is related to age, gender, phylogeny, and environment (Calder, 1996; Morrison & Hero, 2003; Bidau et al., 2011). Habitat-related body-size variation can be an important indicator of a species' ability to adapt and survive to environmental change (Rosalino et al., 2013). As an example, the reduction in blacksnake head size has been associated with the consumption of smaller prey (Phillips & Shine, 2004). The relationship between snout-vent length and the environment found in this study for *P. cuvieri* may be due to a possible increase in environmental quality from the abandoned pasture to the native vegetation and the *Eucalyptus* plantation. However, this pattern can possibly be affected by silvicultural management practices like chemical weed control and mechanical harvest. Future studies should investigate the availability of food resources in the distinct environments of silvicultural landscapes along productive cycles of *Eucalyptus* plantations.

The results of this study suggest that silvicultural landscapes may have some conservation value, as long as natural water bodies are maintained along with conservation areas (i.e. APP/LR). In addition, the development of wildlife-friendly silvicultural management practices (i.e. the maintenance of sparse native trees inside *Eucalyptus* plantations and a relaxation in chemical weed control, as suggested by Athayde (2013) and Millan (2013), might possibly increase β-diversity in silvicultural landscapes (Verdade et al., 2014b). Such information should be incorporated in the certification process of *Eucalyptus* production (e.g. Forest Stewardship Council) in order to improve the conservation value of silvicultural landscapes.

Acknowledgements: This study was funded by the Biota Program/FAPESP (Proc. 2006/60954-4) and Ripasa S.A. Papel e Celulose. Paula C. Lopes and Thiago S. Marques had scholarships from FAPESP (Procs. 2007/01632-0 and Procs. 2013/11032-0, respectivelly). Luciano M. Verdade hold a CNPq Productivity Scholarship (309468/2011-6).

Bibliography

ABRAF (2012). Anuário estatístico da ABRAF 2012, ano base 2011. Retrieved from: http://www.abraflor.org.br/estatisticas/ABRAF12/ABRAF12-BR.pdf. Access on: 30 Jan. 2014.

Aichinger, M. (1987). Annual activity patterns of anurans in a seasonal neotropical environment. Oecologia, 71, 583-592.

Alford, R.A., & Richards, S.J. (1999). Global amphibian declines: a problem in applied ecology. Annual Review of Ecology and Systematics, 30, 133-165.

Athayde, E.A. (2013). Importância da heterogeneidade natural na conservação da diversidade de árvores em paisagem agrosilvicultural. Thesis in Biological Sciences. Universidade Estadual Paulista „Júlio de Mesquita Filho", Rio Claro, SP.

Alves, M.V.G., Koehler, H.S., & Mello Filho, B. (2006). Tendências e perspectivas para o setor florestal brasileiro. Retrieved from: http://www.iufro.org/ uploads/media/t1-alves-marcos-diag-for-br.pdf. Access on: 09 Mar. 2007.

Bastos, R.P. (2007). Anfíbios do Cerrado. In L. B. Nascimento & M. E. Oliveira (Eds.), Herpetologia no Brasil II (pp. 87-100). Belo Horizonte: Sociedade Brasileira de Herpetologia.

Bernarde, P.S., & Macedo, L.C. (2008). Impacto do desmatamento e formação de pastagens sobre a anurofauna de serapilheira em Rondônia. Iheringia, 98, 454-459.

Bidau, C.J., Martí, D.A., & Baldo, D. (2011). Inter- and Intraspecific Geographic Variation of Body Size in South American Redbelly Toads of the Genus *Melanophryniscus* Gallardo, 1961 (Anura: Bufonidae). Journal of Herpetology, 45, 66-74.

Block, M.B., Morrison, M.L., & Scott, P.E. (1998). Development and evaluation of habitat models for herpetofauna and small mammals. Forest Science, 44, 430-437.

Brasileiro, C.A., Sawaya, R.J., Kiefer, M.C., & Martins, M. (2005). Amphibians of an open Cerrado fragment in southeastern Brazil. Biota Neotropica, 5, 1-17.

Calder, W.A. (1996). Size, Function and Life History. Mineola, NY: Dover Publications Inc.

Campos, F.S., & Vaz-Silva, W. (2010). Distribuição espacial e temporal da anurofauna em diferentes ambientes no município de Hidrolândia, GO, Brasil Central. Neotropical Biology and Conservation, 5, 179-187.

Cechin, S.Z., & Martins, M. (2000). Eficiência de armadilhas de queda (pitfall traps) em amostragens de anfíbios e répteis no Brasil. Revista Brasileira de Zoologia, 17, 729-740.

Colwell, R.K. (2004). EstimateS: statistical estimation of species richness and shared species from samples. Retrieved from: http://viceroy.eeb.uconn.edu/estimates/EstimateSPages/ EstSUsersGuide/EstimateSUsersGuide.htm. Access on: 10 Jan. 2013.

Colwell, R.K., & Coddington, J.A. (1994). Estimating terrestrial biodiversity through extrapolation. Philosophical Transaction of the Royal Society of London, 345, 101-118.

Corn, P.S. (1994). Straight-line drift fences and pitfall traps. In W.R. Heyer, M.A. Donnely, R. W. Mcdiarmid, L.C. Hayek & M.S. Foster (Eds.), Measuring and monitoring biological diversity: standard methods for amphibians (p.109-117). Washington, DC: Smithsonian Institution.

Demaynadier, P.G., & Hunter Jr., M.L. (1998). Effects of silvicultural edges on the distribution and abundance of amphibians in Maine. Conservation Biology, 12, 340-352.

Diamond, J. (1997). Armas, germes e aço: os destinos das sociedades humanas. Rio de Janeiro, RJ: Record.

Dixo, M.B.O., & Verdade, V.K. (2006). Herpetofauna de serrapilheira da Reserva Florestal de Morro Grande, Cotia (SP). Biota Neotropica, 6, 1-20.

Duellman, W.E., & Trueb, L. (1994). Biology of amphibians. New York, NY: McGraw-Hill Publishing.

Eterovic, P.C., & Sazima, I. (2000). Structure of an anuram community in a montane meadow in southeastern Brazil: effects of seasonality, habitat and predation. Amphibia-Reptilia, 21, 439-461.

Feder, M.E., & Burggren, W.W. (1992). Environmental physiology of the amphibians. Chicago: The University of Chicago Press.

Gibbs, J.P. (1998). Amphibian movements in response to forest edges, roads, and streambeds in Southern New England. Journal of Wildlife Management, 62, 584-589.

Gray, M.J., Smith, L.M., & Brenes, R. (2004). Effects of agricultural cultivation on demographics of southern high plains amphibians. Conservation Biology, 18, 1368-1377.

I.F. (Instituto Florestal do Estado de São Paulo) (2009). Inventário florestal da vegetação nativa do estado de São Paulo. Retrieved from: http://www.iflorestal.sp.gov.br/imagindex/ mapain-ventario.pdf. Access on: 10 Jan. 2013.

IUCN (2014). The IUCN Red List of Threatened Species. Version 2014.1. Retrieved from: http://www.iucnredlist.org. Access on: 12 Jun. 2014.

Jim, J. (1980). Aspectos ecológicos dos anfíbios registrados na região de Botucatu, São Paulo (Amphibia, Anura). Thesis in Zoology, Universidade de São Paulo, São Paulo, SP.

Lisboa, M.A.M. (2008). A política dos coronéis e a difusão do ensino primário em Angatuba/SP (1870-1930). Thesis in Education, Universidade Estadual de Campinas, Campinas, SP.

Lyra-Jorge, M.C., Gheler-Costa, C., Piña, C.I., Rosalino, L.M., & Verdade, L.M. (2014). Wildlife Surveys in Agricultural Landscapes: Terrestrial Medium- to Large-Sized Mammals. In L.M. Verdade, M.C. Lyra-Jorge, & C.I. Piña (Eds.), Applied Ecology and Human Dimensions in Biological Conservation (p.133-147). EUA: Springer-Verlag Berlin Heidelberg.

Machado, L.M., Mathias, P.V.C., Oliveira Jr., J.M.B., Sousa, P.C.F., & Calvão, L.B. (2011). O impacto de uma monocultura de eucalipto sobre a comunidade de anfíbios (anuros) no município de Luziânia-GO, Brasil. Enciclopédia Biosfera, 7, 1-16.

Maffei, F., Ubaid, F.K., & Jim, J. (2011). Anurofauna em área de cerrado aberto no município de Borebi, estado de São Paulo, Sudeste do Brasil: uso do habitat, abundância e variação sazonal. Biota Neotropica, 11, 221-233.

Magnusson, W.E., Lima, A.P., Luizão, R., Luizão, F., Costa, F.R.C., Castilho, C.V., & Kinupp, V.F. (2005). RAPELD: a modification of the Gentry method for biodiversity surveys in long term ecological research sites. Biota Neotropica, 5, 1-6.

Manly, B.F.J. (1994). Multivariate statistical methods: a primer. London: Chapman & Hall Statistics Texts.

Martin, P.S., Gheler-Costa, C., Lopes, P.C., Rosalino, L.M., & Verdade, L.M. (2012). Terrestrial non-volant small mammals in agro-silvicultural landscapes of Southeastern Brazil. Forest Ecology and Management, 282, 185-195.

Millan, C.H. (2013). Influência de práticas de manejo e contexto da paisagem sobre a ocorrência de aves em plantio exótico de eucalipto. Master in Ecology, University of São Paulo, São Paulo, SP.

Morrison, C., & Hero, J.M. (2003). Geographic variation in life-history characteristics of amphibians: a review. Journal of Animal Ecology, 72, 270–279.

Noris, K. (2008). Agriculture and biodiversity conservation: opportunity knocks. Conservation Letters, 1, 2–11.

Ott, E.R. (1983). Analysis of means: a graphical procedure. Journal of Quality Technology, 15, 10-18.

Peltzer, P.M., Lajmanovich, R.C., & Beltzer, A.H. (2003). The effects of habitat fragmentation on amphibian species richness in the floodplain of the middle Parana River. Herpetological Journal, 13, 95-98.

Penteado, M., Silva, W.R., & Verdade, L.M. (2014). Point counts method for bird surveys in agroecosystems of the State of São Paulo, southeastern Brazil. In L.M. Verdade, M.C. Lyra-Jorge, & C.I. Piña (Eds.), Applied Ecology and Human Dimensions in Biological Conservation (p.149-158). EUA: Springer-Verlag Berlin Heidelberg.

Phillips, B.L., & Shine, R. (2004). Adapting to an invasive species: toxic cane toads induce morphological change in Australian snakes. PNAS, 101, 17150–17155.

Pinheiro, F., Diniz, I.R., Coelho, D., & Bandeira, M.P.S. (2002). Seasonal pattern of insect abundance in the Brazilian cerrado. Austral Ecology, 27, 132-136.

Pough, F.H., Andrews, R.M., Cadle, J.E., Crump, M.L., Savitzky, A.H., & Wells, K.D. (2004). Herpetology. New Jersey, NJ: Pearson Prentice Hall.

Renken, R.B., Wendy, K.G., Debra, K.F., Stephen, C.R., Timothy, J.M., Kevin, B.R., Bradley, R., & Wang, X. (2004). Effects of forest management on amphibians and reptiles in Missouri Ozark Forests. Conservation Biology, 18, 174-188.

Rocha, C.F.D., Bergallo, H.G., Sluys, M.V., & Alves, M.A.S. (2006). Biologia da conservação: Essências. São Carlos, SP: Rima Editora.

Rosalino, L.M., Martin, P.S., Gheler-Costa, C., Lopes, P.C., & Verdade, L.M. (2013). Allometric relations of neotropical small rodents (Sigmodontinae) in anthropogenic environments. Zoological Science, 30, 585-590.

Rubbo, M.J., & Kiezecker, J.M. (2005). Amphibian breeding distribution in an urbanized landscape. Conservation Biology, 19, 504-511.

Segalla, M.V., Caramaschi, U., Cruz, C.A.G., Garcia, P.C.A., Grant, T., Haddad, C.F.B., & Langone, J. (2012). Brazilian amphibians - List of species. Sociedade Brasileira de Herpetologia. Retrieved from: http://www.sbherpetologia.org.br. Access on: 30 Jan. 2014.

Siegel, S. (1956). Nonparametric statistics for the behavioral sciences. New York, NY: McGraw Hill Publishing.

Silvano, D.L., Colli, G.R., Dixo, M.B.O., Pimenta, B.V.S., & Wiederhecker, H.C. (2003). Anfíbios e Répteis. In D.M. Rambaldi & D.A.S. Oliveira (Eds.), Fragmentação de Ecossistemas: causas, efeitos sobre a biodiversidade e recomendações de políticas públicas (p.183-200). Brasília: Ministério do Meio Ambiente/Secretaria de Biodiversidade e Florestas.

Stebbins, R.C., & Cohen, N.W. (1995). A Natural History of Amphibians. New Jersey, NJ: Princeton University Press.

Sutherland, W.J. (2006). Ecological Census Techniques – A Handbook. Cambridge, UK: Cambridge University Press.

Swihart, R.K., Gehring, T.M., Kolozsvary, M.B., & Nupp, T.E. (2003). Responses of "resistant" vertebrates to habitat loss and fragmentation: the importance of niche breadth and range boundaries. Diversity and Distributions, 9, 1-18.

Valan, D. (2002). Effects of anthropogenic environmental changes on amphibian diversity in the rain forests of eastern Madagascar. Journal of Tropical Ecology, 18, 725-742.

Vasconcelos, T.S., & Rossa-Feres, D.C. (2005). Diversidade, distribuição espacial e temporal de anfíbios anuros (Amphibia, Anura) na região noroeste do Estado de São Paulo, Brasil. Biota Neotropica, 5, 1-14.

Verdade, L.M., Penteado, M., Gheler-Costa, C., Dotta, G., Rosalino, L.M., Pivello, V.R., & Lyra-Jorge, M. C. (2014a). The conservation value of agricultural landscapes. In L.M. Verdade, M.C. Lyra-Jorge & C.I. Piña (Eds.), Applied Ecology and Human Dimensions in Biological Conservation (p.91-103). EUA: Springer-Verlag Berlin Heidelberg.

Verdade, L. M., Lyra-Jorge, M.C., & Piña, C.I. (2014b). Redirections in conservation biology. In L.M. Verdade, M.C. Lyra-Jorge & C.I. Piña (Eds.), Applied Ecology and Human Dimensions in Biological Conservation (p.3-17). EUA: Springer-Verlag Berlin Heidelberg.

Vianna, L.G.G., Sato, A.M., Fernandes, M.C., & Netto, A.L.C. (2007). Fronteira de expansão dos plantios de eucalipto no geoecossistema do médio vale do rio Paraíba do Sul (SP/RJ). In Seminário de recursos hídricos da bacia hidrográfica do Paraíba do Sul: O Eucalipto E O Ciclo Hidrológico. Taubaté: IPABHI.

Victor, M.A.M., Cavalli, A.C., Guillaumon, J.R., & Serra-Filho, R. (2005). Cem anos de devastação revisitada 30 anos depois. Brasília: Ministério do Meio Ambiente/ Secretaria de Biodiversidade e Florestas.

Young, B.E., Lips, K.R., Reaser, J.K., et al. (2000). Population declines and priorities for amphibian conservation in Latin America. Conservation Biology, 15, 1213-1223.

Young, B.E., Stuart, S.N., Chanson, J.S., Cox, N.A., & Boucher, T.M. (2004). Joyas que estan desapareciendo: El estado de los anfibios en el Nuevo Mundo. Arlington, VA: NatureServe.

Zar, J.H. (1996). Biostatistical analysis. Upper Saddle River: Prentice Hall.

Zug, G.R., Vitt, L.J., & Caldwell, J.P. (2001). Herpetology: an introductory biology of amphibians and reptiles. San Diego, CA: Academic Press.

Marli Penteado, Carlos Yamashita, Thiago S. Marques, Luciano M. Verdade

15 Bird Diversity in Relation to Land Use in Agricultural Landscapes of Southeastern Brazil

Abstract: The Atlantic forest has been dramatically reduced in the state of São Paulo to 7% of its original area with less than 1% still considered pristine. Agricultural landscapes formed by agroecosystems and small remaining forest fragments are now widespread. The Passa-Cinco river basin can be considered as an agricultural landscape model for the Eastern-Central region of the State of São Paulo, in Southeastern Brazil, for presenting relatively well preserved native forest fragments alongside local crops (i.e., *Eucalyptus* and sugarcane plantations, and exotic pastures). Birds are usually considered effective indicators of environmental change. The main goal of this study was to evaluate spatial variation in bird diversity inhabiting agricultural landscapes in the State of São Paulo. We surveyed birds across 16 study sites from September 2003 to January 2005 and compared species richness and abundance by nested-ANOVA in relation to these environments. We detected 3,679 individuals belonging to 224 species (138 passerine and 86 non-passerine) in the whole study area. Native forest fragments presented the largest species richness (164) and exotic pastures had the largest abundance of individuals. Native forest fragments presented a distinct pattern compared to the other three landscape elements considering species composition and abundance in multidimensional scaling analysis. The results stress that native forest fragments should be considered crucial elements of agricultural landscapes.

15.1 Introduction

Land use changes due to anthropic actions have resulted in the replacement of 80% of the original vegetation of the State of Sao Paulo, in Southeastern Brazil (Câmara, 1990). The Atlantic Forest has been reduced to 7% (Mittermeier et al., 1999), with less than 1% considered pristine (SMA, 1996). This widespread destruction and fragmentation is the main cause of species loss and has driven conservation biologists to warn that a large number of bird species will rapidly become extinct, including a great number of rare and endemic taxa (Brooks & Balmford, 1996; Stotz et al., 1996; Mittermeier et al., 1999).

Considering the drastic reduction of natural areas in the State of São Paulo and the present landscape in a mosaic composed of natural and agricultural areas, the avifauna survey and the conservation programs must target not only to the pristine ecosystems, but also the secondary forest fragments (Willis, 1979; Aleixo & Vielliard, 1995; Pozza & Pires, 2003; Donatelli et al., 2004), altered environments (Aleixo,

1997; Willis, 2003) and mosaic habitats (Carbonari, 1999). However, the State of São Paulo has approximately 700 bird species, which represent approximately 45% of the species that occur in Brazil, with approximately one quarter still occurring in altered environments, like cattle ranches, plantations and even urban areas (Silva, 1998). For these reasons, altered environments should be considered for conservation programs (Zafra-Calvoa, 2010) and birds should be used as bio-indicators of environmental change (Verner, 1981). In addition, the environmental effects of agriculture intensification should be monitored (Flore et al., 2011) in order to maintain ecosystem processes in agricultural landscapes (Holffman & Greef, 2003; Jeanneret et al., 2003; Verhulst et al., 2004; Moreira et al., 2005; Verdade et al., 2014a).

An agricultural landscape generally presents a mosaic structure, composed of patches of various sizes and different uses (Forman, 1995) with the coexistence of species on a global, regional and local scale (Hurlbert & Jetz, 2010). The avifauna, density and distribution of a population in a mosaic depends on the movement capacity of the individuals (Wiens et al., 1993). The spatial structure of a population depends, therefore, on the integration between the spatial pattern of the landscape and the dispersion capacity of organisms (Bell et al., 1991; Fahrig & Merrian, 1994; Fahrig & Grez, 1996). In addition, it is also important to consider the influence of local spatial-temporal heterogeneity on the species diversity when planning the management of natural reserves in heterogeneous landscapes (Forman, 1995; Buger et al., 1995; Verdade et al., 2014a). In such context, river basins are appropriate units for biodiversity studies (Magnusson, 2001).

The main goal of this study was to evaluate spatial variation in bird diversity inhabiting agricultural landscapes of the State of São Paulo, in Southeastern Brazil. The following null-hypotheses have been tested: 1) There is no difference in species richness and relative abundance among the different components of agricultural landscapes (native vegetation, *Eucalyptus* plantations, pastures and sugarcane plantations); and, 2) There is no difference in species composition of birds among the agricultural landscape components.

15.2 Material and Methods

15.2.1 Study Area

The study was conducted in the Passa-Cinco river basin, located in the Eastern-Central region of the State of São Paulo (22°05' – 22°30'S; 47°30' – 47°50'W; Fig. 15.1), in Southeastern Brazil. It is the largest (527,576 ha) and the most preserved sub-basin of the Corumbatai river basin, which is one of the five most important tributaries of the Piracicaba River. The average temperature varies from 17 to 22°C (Salati, 1996) and the average annual rainfall is 1422.3 mm (Zavanti & Cano, 1993), with dry and cool weather between April and September and wet and warm weather between October and March.

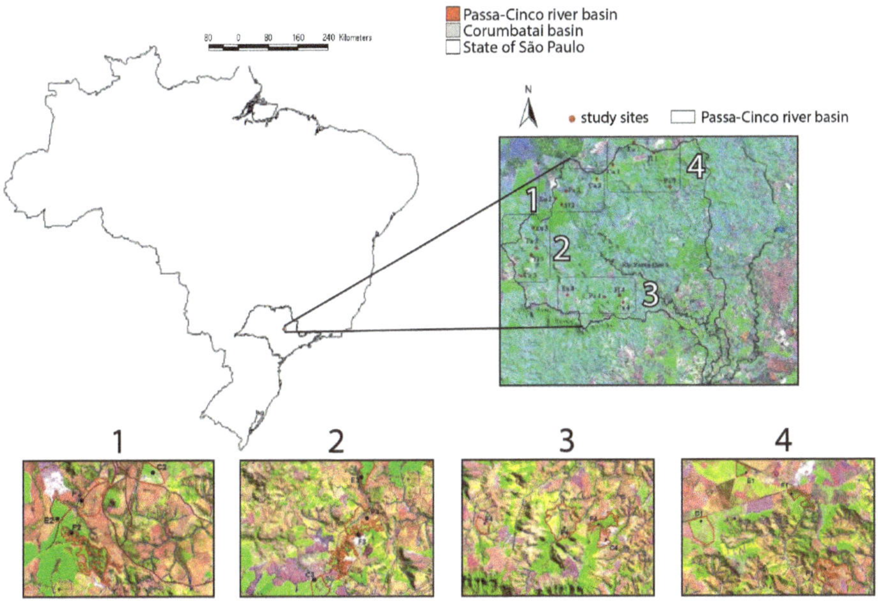

Fig. 15.1: Study areas in Passa Cinco River Basin located in the eastern-central region of the state of São Paulo in southeastern Brazil. The land use (Valente, 2001) is present for four micro-geographic regions sampled (F: native vegetation; E: *Eucalyptus* plantations; P: Pastures; C: Sugarcane plantations).

The local native vegetation is predominantly composed of semi-deciduous seasonal forest and Cerrado, with smaller areas of riparian and swamp forest (Koffler, 1993; Rodrigues, 1999). The local forest degradation began in the early 20th century, with the implantation of coffee and subsistence crops. Afterwards, coffee was replaced by pasture and, more recently, by sugarcane, fruit orchards and forestry (Garcia, 2005), resulting in the fragmentation of the forest structure, with agroecosystems as the landscape matrix (Valente, 2001). At the present, the majority of the remaining native forest fragments are smaller than 50 ha and present some degree of isolation and other anthropogenic alterations (Rodrigues, 1999). The native vegetation on the Passa-Cinco River Basin has been reduced to 16.4% of its original area (15.6% semi-deciduous seasonal forest and 0.74% cerrado) spread over an agricultural (*lato sensu*) matrix (51.8% pastures and 14% sugarcane), with 10.8% of the area composed of planted *Eucalyptus* plantations (Valente, 2001) (Fig. 15.1).

The semi-deciduous seasonal forest fragments constitute more structurally complex habitats with three to four extracts of vegetation, presenting the arboreal extract from 4 to 23 morph species and the shrub extract from 14 to 40 morph species. Bromeliads, lianas, bamboos and palms are present in 50 to 100% of the sites, contributing to their structural complexity. The mulch thickness varied from 5 to 12 cm. The studied fragments are composed of heterogeneous secondary vegetation with signs of anthropic pressure (e.g. exotic species of grass, secondary shrubby vegetation, and sediment deposition), even in their core areas. However, the native forest fragments present some connectivity with areas of higher slope and more preserved vegetation.

The areas of *Eucalyptus* plantations are mainly composed by *Eucalyptus grandis*, with plantations between 3 to 10 years old (*Eucalyptus grandis* 10 to 18 m high), with some understory vegetation (including 6 – 14 morph species of shrub and 5 -12 morph species of sparse native trees with 4 – 7 cm of mulch thickness) under periodic chemical control.

The agricultural management of sugarcane plantations includes intensive use of agrochemicals (i.e., fertilizers, insecticides and herbicides) and pre-harvest fire (Gheler-Costa et al., 2013). Local pastures, on the other hand, have low management intensity with mechanical weed control, but virtually no use of agrochemicals, resulting in Cerrado-like shrub vegetation (with 3 – 7 dominant species) and sparse trees (4 – 8 m high).

15.2.2 Sampling Methodology

15.2.2.1 Study Sites

The study sites comprised the most representative elements of the local agricultural landscapes as follows (semi-deciduous seasonal forest, *Eucalyptus* and sugarcane plantations, and pastures) with four replicates each, grouped around the four largest fragments of semi-deciduous seasonal forest, considering the least perimeter/area ratio and their average distance to the other environments, forming four distinct micro-geographic regions (Fig. 15.1, Appendix 1). Due to logistical or operational reasons (e.g. lack of landowners' permission to access), some of the *a priori* chosen sites needed to be replaced under the same criteria above.

The average area of the study sites was 156.6 ± 86.8 ha (47.7 to 313.8 ha). The perimeter/area ratio varied from 0.6 to 2.7. The average distance between one study site and the closest neighboring study site was 19.1 ±.12.8 m (1.0 to 50.5 m). All study sites have similar topographic conditions (620 to 980 m height) and anthropic pressure (presence of external disturbance, unpaved roads, trails, clearings, domestic animals, exotic animals and fences), with an average distance from urban areas of 4594 meters (sd = 2476.75, from 1450 to 9150 m), and 81.3% of the sites are close to a highway (Appendix 1).

15.2.2.2 Bird Surveys

We carried out eleven campaigns in each study site, six of them during the rainy season (October to March) and five during the dry season (April to September) in the period between September 2003 and January 2005. Species identification followed guidelines by Sick (1985) and the American Ornithologist Union (AOU, 1998). The quantitative research was carried out by Fixed Score Method (Blondel et al., 1970) where we considered all detections (seen and/or heard) with an unlimited distance, unless in patches of small size. In these cases, we considered only individuals detected inside the environment in question (Scott et al., 1981).

The trails to demarcate the sampling points were cleared in the nuclear areas of each habitat (with a minimum distance of 200 m from the border) and the number of points was defined according to the smallest area of the native forest fragment of each replicate. In this manner, we sampled five points in each site, located at least 200 m from each other. Sampling started 30 minutes after sunrise and lasted 10 minutes in each point. The initial point of each campaign and, thus, the sampling order were randomly chosen.

Birds flying over the study sites, but not perching or foraging, were not included in the records. A Sony recorder DAT TCD-D100™ and a directional microphone Sennheiser System K6-ME67™ were used to record all the vocalizations in each point. Zeiss Deltrinten™ binoculars (8 × 30) were used for visual contacts.

15.2.3 Analytical Methodology

The sampling effort sufficiency was evaluated for each landscape unit and for the whole study area by the species incidence curves (adapted from Colwell & Coddington, 1994) where the time-series sigmoid model bootstrapped asymptote was considered the estimated number of species. We used the non-parametric estimator Bootstrap calculated by EstimateS statistical software (Colwell, 2004).

The species richness was considered the number of species detected and the relative abundance index was considered the number of individuals detected per species. In addition, the Shannon diversity index was calculated for each agricultural landscape component (Magurran, 1988) as follows:

$$H' = -\sum_{i=1}^{S} P_i \ln P_i$$

where S is the species richness and Pi proportion of S made up of the ith species.

Prior to statistical analysis we tested data distribution and homoscedasticity, respectively by log-normal abundance test and Levene's tests. Nested analysis of variance (ANOVA) (Zar, 1999) was used to test whether bird species richness, relative abundance and Shannon diversity index differ among the micro-geographic regions and among the four distinct agricultural landscape components. Analysis of means (ANOM) (Ott, 1983) were then applied as a post-hoc test (Hypothesis 1).

The pattern of species composition and relative abundance across the environments was determined by non-metric multidimensional scaling (NMDS), using the reciprocal of the Morisita-Horn similarity index (Magurran, 1988). The hierarchical cluster analysis of the similarities between species and habitats were done based on species relative abundance. The clustering method used was Ward's and the similarity index was Pearson's r. The most evident clusters were highlighted and compared to the results of NMDS (Hypothesis 2). All analysis was carried out in Statistica 6.0 (StaSoft Inc., 2001) and Minitab 14 (Minitab Inc., 2003).

15.3 Results

A total of 224 species of birds have been recorded in the study area (native vegetation: 164, *Eucalyptus* plantations: 104, pasture: 112, sugarcane plantations: 52), totaling 3679 detected individuals (Appendix 2). These species are predominantly distributed in the Passerine order (n = 138) with 68 sub-oscine and 70 oscine species. Our sampling effort was enough to detect from 85.9 to 96.2% of the estimated bird species richness for different agricultural landscape components (Tab. 15.1; Fig. 15.2) and 91.4% for the whole study area (Fig. 15.3).

Tab. 15.1: Species richness detected (R_{det}) and estimated by non-parametric estimator bootstrap (R_{est}), number of detected individuals, number of exclusive species and Shannon index (H´) of agricultural landscapes of Southeastern Brazil.

	Native Vegetation	*Eucalyptus* Plantations	Pasture	Sugarcane Plantations	Total
R_{det}	164.0	104.0	112.0	52.0	224.0
R_{est}	185.0	121.0	119.0	54.0	245.0
$(R_{det}/R_{est}) \times 100$ (%)	88.6	85.9	94.1	96.2	91.4
N° of detected individuals	911.0	447.0	1122.0	626.0	3679.0
N° of exclusive species	57.0	6.0	14.0	2.0	-
Shannon index (H´)	4.5	4.2	4.1	2.9	-

There was no significant difference in species richness and relative abundance of birds among the four micro-geographic regions (richness: $F = 0.20$; $gl = 63$; $p = 0.890$; abundance: $F = 0.66$; $gl = 63$; $p = 0.588$). However, there was a significant difference among the agricultural landscape components (richness: $F = 10.36$; $gl = 63$; $p < 0.001$; abundance: $F = 4.13$; $gl = 63$; $p < 0.001$), with the highest species richness on the native vegetation and the highest abundance on the pasture (Fig. 15.4). The species ordination curve in decreasing order of abundance in each habitat shows distinct patterns, nevertheless a few abundant species and many rare species can be observed mainly in sugarcane and *Eucalyptus* plantations (Fig. 15.5). In addition, the Shannon diversity index varied among the agricultural landscape components ($F = 19.68$, $gl = 15$, $p < 0.001$), with the native forest and pastures as the most diverse and the sugarcane plantation as the least diverse environment. Therefore, we can reject the first null hypothesis.

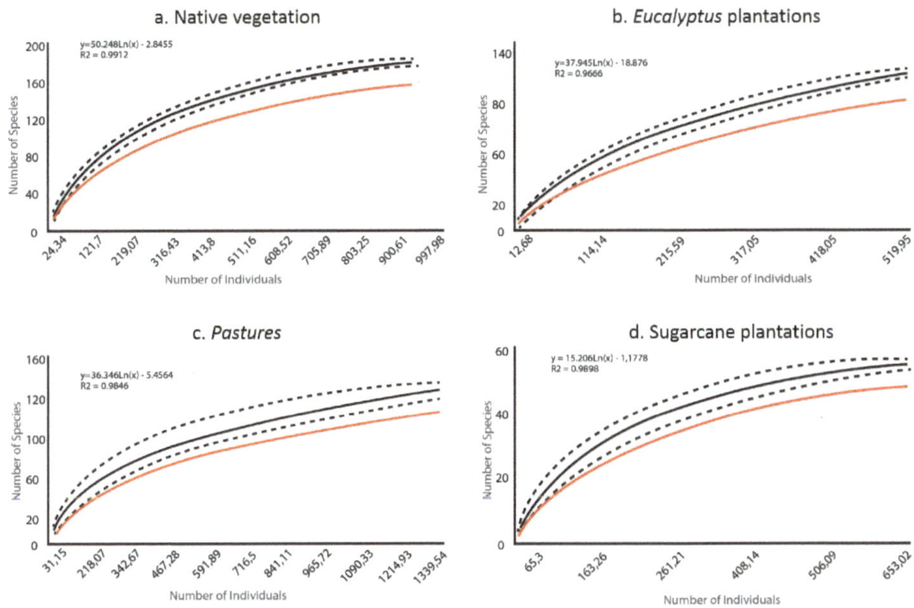

Fig. 15.2: Birds' species incidence curves for the agricultural landscape components (black line: species richness estimated, red line: species richness detected, dotted line: standard deviation).

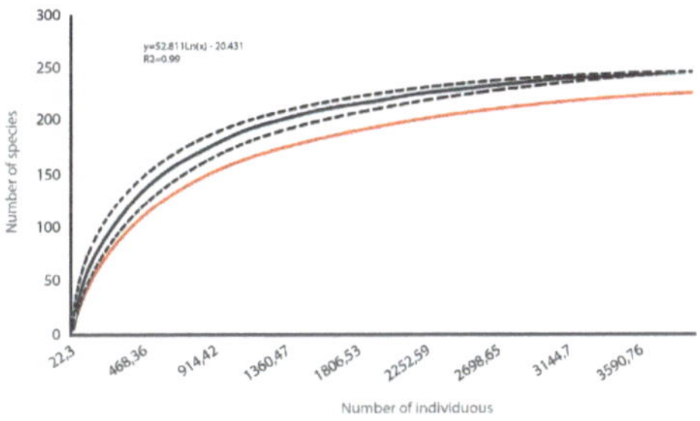

Fig. 15.3: Birds' species incidence curves for the whole agricultural landscape (blue line: species richness estimated, red line: species richness detected, black line: standard deviation).

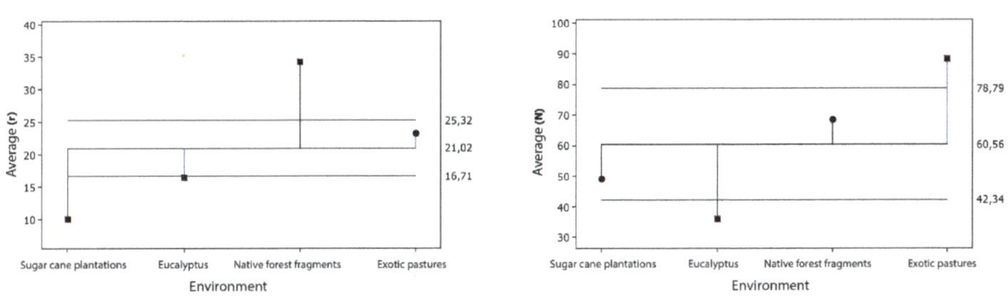

Fig. 15.4: Species richness and relative abundance of birds in a agricultural landscape of Southeastern Brazil.

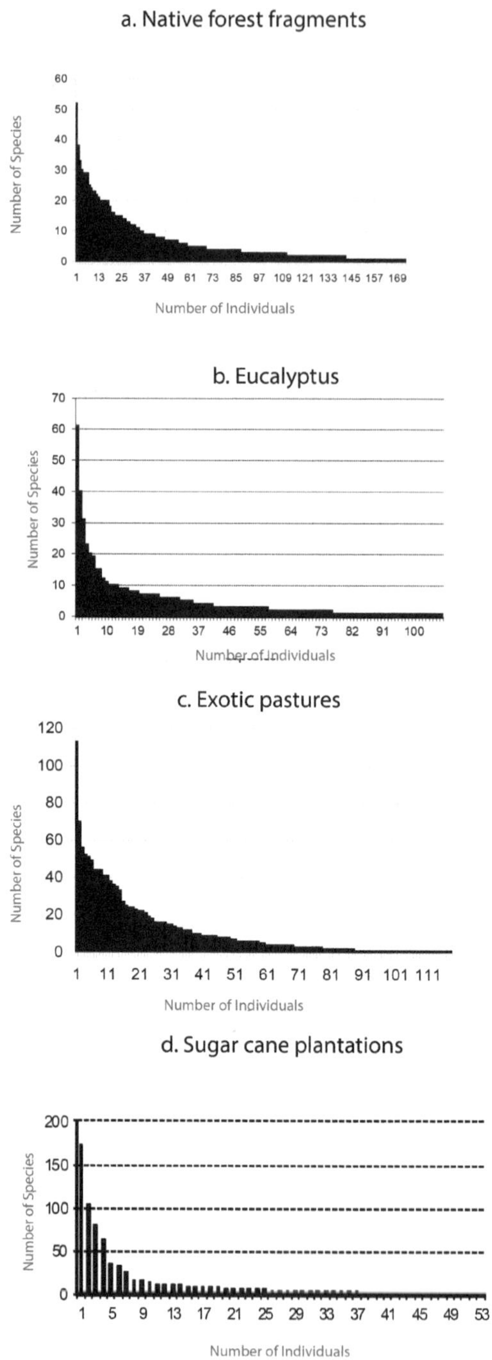

Fig. 15.5: Species ordination curve of the number of species registered by descending order of abundance.

The non-metric multidimensional scale analysis suggests a consistent difference between native vegetation and anthropic environments (i.e. *Eucalyptus* and sugarcane plantations, and pasture) in terms of bird diversity (i.e., species composition, richness and relative abundance) (Fig. 15.6, Fig. 15.7). Cluster analyses showed a similar pattern (Fig. 15.7). The anthropic environments present marked inter-relation between *Eucalyptus* plantations and pasture, with sugarcane plantations forming a distinct group (except for sugarcane plantation 1 - C1). The percentage of exclusive species per environment was as follows: 34.7% in native vegetation fragments, 10.7% in pastures, 9.6% in sugarcane plantations and 5.8 in *Eucalyptus* plantations. We can thus reject the second null hypothesis.

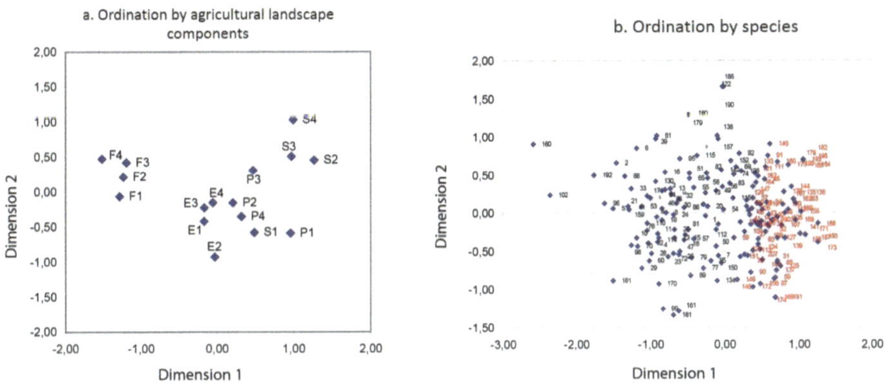

Fig. 15.6: Non-metric multidimensional scaling (NMDS) ordination by habitats and species, with illustration of the clusters obtained in the analysis of the clusters

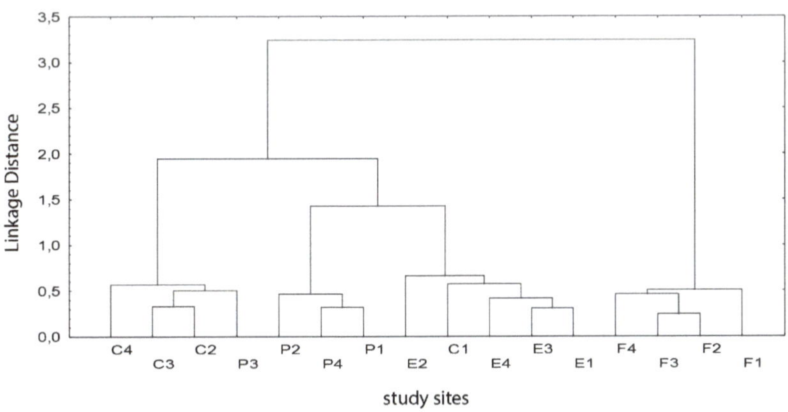

Fig. 15.7: Dendrogram of the analysis of the hierarchic clustering for similarities of richness, abundance and composition of bird species in the 16 sampled study sites, located on the Passa-Cinco River Basin, São Paulo, Brazil. Ward Clustering Method and Pearson r similarity index. Analysis of the hierarchic clustering for the agricultural habitats.

15.4 Discussion

The species incidence curves reached an asymptote for the whole study area, but not for the individual landscape elements. This pattern suggests that the overall sampling effort was enough to detect the local species pool. However, the species that remained undetected in specific environments appear to be those that have already been detected in the landscape as a whole. These species appear to be using the landscape as a whole, although heterogeneously. Three possible consequences may then arise:

a) Strictly specialist species may have already gone extinct, possibly resulting in the process of species homogenization (see Magnusson, 2006); or,
b) The remaining species are experiencing an anthropic selective pressure and consequently adapting to these new environments (see Rosalino et al., 2014); or,
c) The current landscape β-diversity is more relevant than each environment's α-diversity for biodiversity conservation efforts in these agricultural landscapes (see Verdade et al., 2014b).

Passa-Cinco river basin is characterized by a mosaic landscape, which maintains a diversity of resident and some migrant bird species able to use both the agroecosystems (i.e., the landscape matrix) and the remnant patches of native vegetation. This heterogeneous landscape maintains a considerable diversity of avifauna, when compared to other altered environments in the State of São Paulo (Tab. 15.2), including locally endangered species like *Amazona aestiva* and *Antilophia galeata* detected in fragments of native vegetation and *Cyanocompsa brissonii* and *Oryzoborus angolensis* detected in the pastures (Willis, 1979; Donatelli et al., 2004).

The similarity in terms of species richness and abundance among the micro-geographic regions of the present study suggests that the sampling unities are representative of the river basin. On the other hand, the bird diversity variation among the distinct environments of the local agricultural landscape appeared to have been captured by the present nested sampling design.

Not surprisingly, the native vegetation was observed to be the most complex environment. Its greatest species richness is predominantly related to its higher diversity of ecological niches that can be occupied by bird species due to its more complex vegetation structure. The remaining fragments of native vegetation play an important role in the maintenance of the local diversity of birds, being to promote the conservation of the biodiversity in the agricultural landscapes.

The pastures were shown to be the most diverse agroecosystem of this study, with the highest abundance of individual birds. This environment has arboreal/shrubby elements typical of the original vegetation that contribute to its structural complexity as a Cerrado-like habitat. For this reason, the following Cerrado bird species are relatively abundant in pastures: *Crypturellus parvirostris, Rynchotus rufescens, Nothura maculosa, Falco sparverius Cariama cristata, Melanerpes candidus, Lepdocolaptes angustirostris, Xolmis cinerea, Elaenia chiriquensis, Suiriri suiriri,*

Tab. 15.2: Previous birds' surveys in altered environments of the state of São Paulo, Southeastern Brazil.

Location	Environment	Area (ha)	Method	Species richness	Period	Source
Passa-cinco River basin	Atlantic forest fragments	78-313	Point survey	224/164	2002-2005	Present study
Anhembi	Atlantic forest fragments	4890	Collection	178	1957-1966	MZUSP
Anhembi and Campinas	Atlantic forest fragments	21-400	Transect	202,146,93	1975-1978	Willis (1979)
Anhembi	Atlantic forest fragments	4890	Collection	338	1975-1996	Magalhães (1999)
Lençóis Paulista	Atlantic forest fragments	600	Point survey	216/74	2001-2002	Donatelli et al. (2004)
Campinas	Atlantic forest fragments	251	Point survey	134/97	1992-1994	Aleixo & Vielliard (1995)
São Carlos	Atlantic forest fragments	75-100	Point survey	145/60, 176/72	2000-2001	Pozza & Pires (2003)
Rio Claro	Atlantic forest fragments	75	Transect	156/117	1986-1988	Carbonari (1990)
Rio Claro	Eucalipt plantation with understory vegetation	2314	Transect	255	1982-2001	Willis (2003)

Volatinia jacarina, Coryphospingus cuculatus and *Cyanocorax cristatellus*. In addition, the large number of individuals in this habitat is favored by the presence of species that live in intraspecific flocks and species which are associated to the presence of cattle (e.g. *Crotophaga ani, Guira guira, Colaptes campestris, Melanerpes candidus, Pseudoleistes guirahuro, Aratinga leucophthalmus, Vanellus chilensis e Bulbucus îbis*) and intraspecific gramivorous flocks (e.g. *Volatinia jacarina, Sporophila caerulescens* and *Sporophila lineola*).

The native vegetation and pastures in our study area seem to be more structured habitats, with a higher number of species with mean frequency of use (less abrupt fall in accumulation curves; Fig. 15.5). They can therefore show more steady species that make frequent use or present a resident population, characteristic of balanced communities (Magurran, 1988). However, sugarcane and *Eucalyptus* plantations show low homogeneity, that is, the distribution of abundance among the species is not balanced (Magurran, 1988) or it is less balanced than the abundance of communities sampled in the native vegetation and pastures.

Homogeneous plantations like *Eucalyptus* and sugarcane, as the landscape matrix, have less spatial heterogeneity, but more temporal heterogeneity than the native vegetation and pasture (see Driscoll et al., 2013 and Verdade et al., 2014b). However, *Eucalyptus* plantations have a longer productive cycle than sugarcane (harvested, respectively, every 6-7 years and annually). In addition, *Eucalyptus* plantations, in general, have a more strict certification process which results in more effective stimulation for the maintenance of conservation areas as part of the agricultural landscape, as well as for the use of wildlife-friendly management techniques, like the maintenance of sparse trees and understory vegetation in the matrix (Willis, 2003; Athayde, 2013; Millan, 2013).

The species composition of birds in this study is different between the native vegetation and the anthropic environments. The remaining native vegetation, which is structurally more complex, can have more specialist species, whereas the anthropic environments have a group of generalist bird species dependent on the agricultural landscape use (most recorded species), which benefit from the habitat heterogeneity and from the interaction of the agricultural matrix with less altered environments.

Other relevant factors to be considered in agricultural landscapes are movement and migration of species and individuals. Although this study has not been directed to detect specific variations related to migration, it is worth mentioning that the results may have been influenced by the addition of regional migrant bird populations that temporarily join the local population or substitute part of it, which might have migrated to other regions (Pacheco & Gonzaga, 1994). Besides this, circular migration of hummingbirds, psistacidae and columbidae can influence relative abundance of use dependence in the different habitats of the mosaic. This way the avifauna movement inside and among the mosaics, either due to climate change (regional migration) or to the availability of resources (local and regional migration) must be considered as relevant factors for the conservation of the agricultural landscapes.

The results of this study suggest that agricultural landscapes may have important value for bird conservation, as long as native vegetation fragments are maintained along with conservation areas (i.e., Areas of Permanent Protection – APP and Legal Reserve - LR). In order to increase bird diversity of the avifauna, habitats with greater structural diversity must be prioritized and management strategies must be implemented on a regional spatial scale (Signor & Pinho, 2011), aiming at the conservation of the environmental mosaic. Last but not least, long-term biodiversity monitoring programs should investigate the possible effects of spatial-temporal heterogeneity in the adaptive process of the bird species throughout the agricultural production cycle (Verdade et al., 2014a).

Acknowledgements: This study has been funded by FAPESP (Proc. No. 2001/13251-4). The University of São Paulo, IBAMA and RIPASA S.A. Papel e Cellulose helped with logistics. Prof. Pedro Gerhard and Guilherme Carbonari assisted with statistical analyses and maps, graphics and tables. Edson Davanzo, José Henrique Rocha, Roberto Nogueira, Luzia Machitte, Graziela Dotta, Carla Gheler and Fábio Comin assisted in field work. Thiago S. Marques has scholarships from FAPESP (Procs. 2013/11032-0).

Bibliography

Aleixo, A. (1997). Estrutura e organização de comunidades de aves em áreas de Mata Atlântica primitiva e explorada por corte seletivo. MSc. Dissertation, Universidade de Campinas, Campinas, Brasil.

Aleixo, A., & Vielliard, J.M.E. (1995). Composição e dinâmica da avifauna da mata de Santa Genebra, Campinas, São Paulo, Brasil. Revista Brasileira de Zoologia, 12, 493-511.

AOU (American Ornithologists´ Union). (1998). Check list of american birds. New York, 7th ed.

Athayde, E.A. (2013). Importância da heterogeneidade natural na conservação da diversidade de árvores em paisagem agrosilvicultural. Rio Claro, SP, Brazil: Universidade Estadual Paulista.

Bell, S.S., Mccoy, E.D., & Mushinsky, H.R. (1991). Habitat structure: the physical arrangement of objects in space. London: Chapman Hall.

Brooks, T., & Balmford, A. (1996). Atlantic forest extinctions. Nature, 380, 115-132.

Burger, J., Gochfeld, M., & Niles, L.T. (1995). Ecotourism and birds in coastal New Jersey: contrasting responses of birds, tourism and managers. Environmental Conservation, 22, 56-65.

Blondel, J., Ferry, C., & Prochot, B. (1970). La méthode des indices ponctuels d'abondance (I.P.A.) ou des releves d' avifaune par "stations d'écoute". Alauda, 38, 55-71.

Câmara, I.G. (1990). Plano de ação para a Mata Atlântica. São Paulo: Ipsis Gráfica e Editora.

Carbonari, M.P. (1990). Estudo da avifauna de vegetação natural e implantada no município de Rio Claro, SP. Rio Claro, Brasil: UNESP Rio Claro

Colwell, R.K. (2004). EstimateS: statistical estimation of species richness and shared species from samples. Retrieved from: http://viceroy.eeb.uconn.edu/estimates/EstimateSPages/ EstSUsersGuide/EstimateSUsersGuide.htm. Access on: 10 Jan. 2013.

Colwell, R.K., & Coddington, J.A. (1994). Estimating terrestrial biodiversity through extrapolation. Philosophical Transaction of the Royal Society of London, 345, 101-118.

Donatelli, R.J., Vieira da Costa, T.V., & Ferreira, C.D. (2004). Dinâmica da avifauna em fragmento de mata da Fazenda Rio Claro, Lençóis Paulista, São Paulo, Brasil. Revista Brasileira de Zoologia, 21, 97-114.

Driscoll, D.A., Banks, S.C., Barton, P.S., Lindenmayer, D.B., & Smith, A.L. (2013). Conceptual domain of the matrix in fragmented landscapes. Trends in Ecology and Evolution, 1729, 1-9.

Fahrig, L., & Grez, A.A. (1996). Population spatial structure, human-caused landscape changes and species survival. Revista Chilena de História Natural, 69, 5-13.

Fahrig, L., & Merrian, G. (1994). Conservation of fragmented populations. Conservation Biology, 8, 50-59.

Flore, A., Fischer, C., Avik, T., Bengtsson, J., et al. (2011). Agricultural intensification and biodiversity partitioning in European landscapes comparing plants, carabids, and birds. Ecological Applications, 21, 1772-1781.

Forman, R.T.T. (1995). Land mosaics, the ecology of landscapes regions. Cambridge: Cambridge University Press.

Garcia, L.B.R. (2005). Ocupação e desenvolvimento econômico da Bacia do Corumbataí – séculos XVIII a XX: Atlas ambiental da Bacia do rio Corumbataí. Retrieved from: www.rc.unesp.br/igce/ceapla. Access on: 12 Nov. 2005.

Gheler-Costa, C., Sabino-Santos, G., Amorim, L.S. Rosalino, L.M. Figueiredo, L.T.M., & Verdade, L.M. (2013). The effect of pre-harvest fire on the small mammal assemblage in sugarcane fields. Agriculture, Ecosystems and Environment, 171, 85-89.

Hoffmann, J., & Greef, J.M. (2003). Mosaic indicators-theoretical approach for the development of indicators for species diversity in agricultural landscapes. Agriculture, Ecosystems & Environment, 98, 387-394.

Hurlbert, A.H., & Jetz, W. (2010). More than "more individuals": the non-equivalence of area and energy in the scaling of species richness. American Naturalist, 176, 50-65.

Jeanneret, P.H., Schüpbach, B., & Luka, H. (2003). Quantifying the impact of landscape and habitat features on biodiversity in cultivated landscapes. Agriculture, Ecosystems & Environment, 98, 311-320.

Koffler, N.F. (1993). Uso das terras da bacia do rio Corumbataí em 1990. Geografia, 18, 135-150.

Magnusson, W.E. (2001). Catchments as basic units of management in conservation biology courses. Conservation Biology, 15, 1464-1465.

Magnusson, W.E. (2006). Homogeneização biótica. In C.F.D. Rocha, H.G. Bergallo, M. van Sluys & M.A.S. Alves (Eds.), Biologia da Conservação: Essências (p.211-229). São Carlos: RiMa Editora.

Magurran, A.E. (1988). Ecological Diversity and its Measurement. New Jersey: Princeton University Press.

Millan, C.H. (2013). Influência de práticas de manejo e contexto da paisagem sobre a ocorrência de aves em plantio exótico de eucalipto. São Paulo, Brazil: Universidade de São Paulo.

Mittermeier, R.A., Myers, N., Thomsen, J.B., Da Fonseca, G.A.B., & Olivieri, S. (1998). Biodiversity hotspots ans major tropical wilderness areas: approaches to setting conservation priorities. Conservation Biology, 12, 516-520.

Moreira, F., Beja, P., Morgado, R., Reino, L., Gordinho, L., Delgado, A., & Borralho, R. (2005). Effects of field management and landscape context on grassland wintering birds in Southern Portugal. Agriculture, Ecosystems and Environment, 109, 59-74.

Ott, E.R. (1983). Analysis of means: a graphical procedure. Journal of Quality Technology, 15, 10-18.

Pacheco, J.F., & Gonzaga, L.P. (1994). Tiranídeos do estado do Rio de Janeiro provenientes de regiões austrais da américa do sul. Nótulas Faunísticas, 63, 1-4.

Pozza, D.D., & Pires, J.S.R. (2003). Bird communities in two fragments of semideciduous forest in rural São Paulo State. Brazilian Journal Biology, 63, 307-319.

Rodrigues, R.R. (1999). A vegetação de Piracicaba e municípios de entorno. Circular técnica IPEF, 189, 1-18.

Rosalino, L.M., Lyra-Jorge, M.C., & Verdade, L.M. (2014). Adaptation and evolution in changing environments. In L.M. Verdade, M.C. Lyra-Jorge, & C.I. Piña (Eds.), Applied ecology and human dimensions on biological conservation (p.53-71). EUA: Springer-Verlag Berlin Heidelberg.

Salati, E. (1996). Análise ambiental sintética e qualidade de água do rio Corumbataí (SP) como subsídio para o planejamento regional integrado da bacia de drenagem do rio Corumbataí. São Carlos, Brazil: Universidade Federal de São Carlos.

SMA (Secretaria de Estado do Meio Ambiente) (1996). Mata Atlântica: ciência, conservação e políticas: workshop científico sobre a Mata Atlântica. São Paulo: Documentos Ambientais, Secretaria do Meio Ambiente do Estado de São Paulo.

Scott, J.M., & Ralph, C.J. (1981). Estimating birds: introduction. In C.J. Ralph, & J.M. Scott (Eds.), Estimating numbers of terrestrial birds: studies in avian biology (p.1-2). Kansas: Allen Press.

Sick, H. (1985). Ornitologia Brasileira, Uma Introdução. Brasília: Editora Universidade de Brasília.

Signor, C.A. & Pinho, J.B. (2011). Spatial diversity patterns of birds in a vegetation mosaic of the Pantanal, Mato Grosso, Brazil. Zoologia, 28, 725-738.

Silva, W.R. (1998). Bases para o diagnóstico e o monitoramento de aves no estado de São Paulo, In R.M.C. Camargo, C.A. Joly, C.E.M. Bicudo (Eds.), Biodiversidade do Estado de São Paulo, Brasil: síntese do conhecimento ao final do sáculo XX (p.39-50), São Paulo: FAPESP.

Stotz, D.E., Fitzpatrick, J.W., Parker, T.A., & Moskovits, D.K. (1996). Neotropical birds: ecology and conservation. Chicago: University of Chicago Press.

Valente, R.O.A. (2001). Análise da estrutura da paisagem na bacia do rio Corumbataí, SP. MSc. Dissertation, Escola superior de Agricultura "Luiz de Queiroz", Universidade de São Paulo, Piracicaba, Brazil.

Verdade, L.M., Lyra-Jorge, M.C., & Piña, C.I. (2014a). Redirections in conservation biology. In L.M. Verdade, M.C. Lyra-Jorge, & C.I. Piña (Eds.), Applied ecology and human dimensions on biological conservation (p.3-17). EUA: Springer-Verlag Berlin Heidelberg.

Verdade, L.M., Penteado, M., Gheler-Costa, C., Dotta, G., Rosalino, L.M., Pivello V.R., & Lyra-Jorge, M.C. (2014b). The conservation value of agricultural landscapes. In L.M. Verdade, M.C. Lyra-Jorge, & C.I. Piña (Eds.), Applied ecology and human dimensions on biological conservation (p.91-102). EUA: Springer-Verlag Berlin Heidelberg.

Verner, J. (1981). Measuring responses of avian communities to habitat manipulation. Studies in Avian Biology, 6, 543- 47.

Verhulst, J., Baldi, A., & Kleijn, D. (2004). Relationship between land-use intensity and species richness and abundance of birds in Hungary. Agriculture, Ecosystems and Environment, 104, 465-473.

Wiens, J.A., Stenseth, N.C., Van Horne, B., & Ims, R.A. (1993). Ecological mechanisms and landscape ecology. Oikos, 65, 369-380.

Willis, E.O. (1979). The composition of avian communities in remanescent woodlot in Southern Brazil. Papeis Avulsos de Zoologia, 33, 1-25.

Willis, E.O. (2003). Birds of a eucalyptus woodlot in interior São Paulo. Brazilian Journal of Biology, 63, 1-31.

Zafra-Calvoa, N., Cerro, R., Fuller, T., Lobo, J.M., Rodríguez, M.A., & Sarkar, S. (2010). Prioritizing areas for conservation and vegetation restoration in post-agricultural landscapes: A Biosphere Reserve plan for Bioko. Equatorial Guinea. Biological Conservation, 143, 787-794.

Zar, J.H. (1999). Biostatistical analysis. Prentice Hall: Upper Saddle River.

Zavanti, J.A., & Cano, H. (1993). Variações do ritmo pluvial na bacia do rio Corumbataí – SP. Boletim de Geografia Teorética, 32, 215-240.

Appendix 1

Descriptors of the study sites (local scale) and of the landscape metrics for the 1000 m plot (F: native vegetation, E: *Eucalyptus* plantations, P: pastures and C: sugarcane plantations).

	F1	F2	F3	F4	E1	E2	E3	E4	P1	P2	P3	P4	C1	C2	C3	C4
Area (ha)	178.7	217.7	246.2	96.6	47.7	93.6	159.5	155.2	256.7	210.1	168.7	313.8	180.4	123.7	130.4	152.4
Area/Perimeter	1.0	0.9	0.6	0.8	1.3	1.8	1.8	1.4	1.2	1.7	1.7	2.7	2.7	2.3	2.2	0.9
Central area (ha)	12.3	28.2	12.6	20.9	13.8	44.5	66.6	18.3	64.0	86.7	69.4	176.7	111.3	66.8	69.9	4.8
Topography (altitude in meters)	730 -840	680 -740	780 -910	620 -660	835 -840	700 -740	720 -760	940 -980	620 -730	680 -780	700 -840	640 -840	815 -840	760 -820	940	620 -745
First order river	1	0	1	0	1	2	1	1	4	0	2	2	1	1	4	1
Second order river	0	2	0	1	0	1	1	1	1	1	0	1	0	0	0	0
Third order river	0	1(3ª)	2 (4ª)	0	0	0	0	0	0	0	1(4ª)	1(3ª)	0	0	0	0
Total area (ha)	312.4	312.5	312.6	310.1	249.8	312.4	312.3	312.6	312.3	312.3	312.3	312.3	312.6	312.3	312.3	312.3

Appendix 2

List of birds' species in agricultural landscapes of Southeastern Brazil.
Codes adopted in the systematic list:
Occurrence: relates the habitat where species where recorded [F = Native forest, E = *Eucalyptus* Plantations, P = Pastures, C = Sugar Cane Plantations, O = others (lakes, swamps, plantation, etc.)].
Category: refers to the frequency of use the species present in the agricultural habitat and/or the information of migration of species (R = Resident, Ms = Summer Migrant, Mw = Winter Migrant, V = Vagrant).
Trophic Category: refers to the feeding habit predominantly observed during the survey and those collected in the bibliography (F = Frugivorous, I = Insectivore, C = Carnivorous, D = Detritivorous, E = Exudivorous, G = Granivorous, O = Omnivore).
Number of records: refers to all records of species during the quantitative survey, by study site.

Species	Occurrence	Records				Category	Feeding Habit
		F	E	P	C		
Tinamidae							
Crypturellus parvirostris (Temminck, 1815)	F, E, P, C	18	7	14	11	R	F, I
Crypturellus tataupa (Temminck, 1815)	F, E	4	1			R	F
Rhyncothus rufescens (Temminck, 1815)	P			4		R	G
Nothura maculosa (Temminck, 1815)	P, C			6		R	I, G
Podicipedidae							
Podilimbus podiceps (Linnaeus, 1758)	O					R	C
Phalacrocoracidae							
Phalacrocorax brasilianus (Gmelin, 1789)	O					R	C
Anhinga anhinga (Linnaeus, 1766)	O					V	I, C
Ardeidae	O						
Casmerodius albus (Linnaeus, 1758)	O					R	C
Egretta thula (Molina, 1782)	O					R	C
Butorides striatus (Linnaeus, 1758)	O					R	C
Bulbucus íbis (Linnaeus, 1758)	E, P		2	113		R	I
Syrigma sibilatrix (Temmick, 1824)	E, P, C		1	4	1	R	I
Pilherodius pileatus (Boddaert, 1783)	F, E	2	2			V	I
Nycticorax nycticorax (Linnaeus, 1758)	O					R	C
Threskiornithidae							
Mesembrinibis cyennensis (Gmelin, 1789)	O					V	O
Theristicus caudatus (Boddaert, 1783)	O					V	

Species	Occurrence	Records				Category	Feeding Habit
		F	E	P	C		
Anatidae							
Dendrocygna viduata (Linnaeus, 1766)	O					R	G, I
Amazonetta brasiliensis (Gmelin, 1789)	O					R	F, I, C
Cairina moscata (Linnaues, 1758)	O					V	C
Oxyura dominica (Linnaeus, 1766)	O					V	
Cathartidae							
Coragyps atratus (Bechstein, 1793)	E, P, C		7	10	6	R	D
Cathartes aura (Linnaeus, 1758)	F, P					V	D
Accipitridae							
Elanus leucurus (Vieillot, 1818)	C				1	V	C,I
Leptodon cayanensis (Latham, 1790)	F	1				V	C
Harpagus diodon (Temminck, 1923)	F					V	C
Ictinia plumbea (Gmelin, 1788)	C				1	Mv	I, C
Buteo albicaudatus (Vieillot, 1816)	F, E, P, O	1	1	1		R	I, C
Buteo magnirostris (Bertoni, 1901)	F, E, P, C, O	1	6	16	6	R	C, I
Buteo brachyurus (Vieillot, 1816)	F, P, C	2		1	1	R	C, I
Heterospizias meridionalis (Latham, 1790)	E, C		1		1	R	C, I
Falconidae							
Herpetoheres cachinnans (Linnaeus, 1758)	F, P	1		1		R	C
Milvago chimachima (Vieillot, 1816)	F, E, P, C	1	15	19	5	R	C, I
Caracara plancus (Miller, 1777)	E, P, C		10	16	17	R	D, C
Falco femoralis (Temminck, 1822)	E, C		3		4	R	I, C
Falco sparverius (Linnaeus, 1758)	P, C			6	4	R	I, C
Cracidae							
Penelope superciliaris (Spix, 1825)	F, P, C	21		3	3	R	F
Rallidae							
Aramides cajanea (Muller, 1776)	F	1				R	O (I, C, G)
Aramides saracura (Spix, 1825)	F, P	3		1		R	O (I, C, G)
Porzana albicollis (Vieillot, 1819)	F	2				V	
Gallinula chhloropus (Linnaeus, 1758)	O					V	I, F,G
Porphyrula matinica (Linnaeus, 1766)	O					V	

Species	Occurrence	Records				Category	Feeding Habit
		F	E	P	C		
Cariamidae							
Cariama cristata (Linnaeus, 1766)	E, P, C		20	44	11	R	I, C
Jacanidae							
Jacana jacana (Linnaeus, 1766)	O					V	I
Charadriidae							
Vanellus chilensis (Molina, 1782)	E, P, C		2	49	10	R	I
Scolopacidae							
Tringa flavipes (Gmelin, 1789)	O					RV	I
Gallinago undulata (Boddaert, 1783)	O					V	
Columbidae							
Columba picazuro (Temminck, 1813)	F, E, P, C	5	15	18	8	R	G
Columba cayennensis (Vielliot, 1818)	F, E, C	20	2		3	R	F, I,G
Zenaida auriculata (Des Murs, 1847)	F, E, P, C	1	2	2	14	V	G, F
Columbina talpacoti (Temminck, 1811)	F, E, P, C	2	5	9	14	R	G
Claravis pretiosa (Ferrari-Perez, 1886)	E, P, C		2	2	2	V	G
Scardafella squammata (Lesson, 1831)	F, E, P, C	1	3	4	5	R	G
Leptotila verreauxi (Giglioli & Salvadori, 1870)	F, E	23	3			R	F, G
Leptotila rufaxilla (Pelzeln, 1870)	F	20				V	F, G
Geotrygon montana (Linnaeus, 1758)	F	6				R	F, G
Psittacidae							
Aratinga leucophthalmus (Muller, 1776)	F, E, P, C	38	31	70	10	R	F
Forpus xanthopterygius (Spix, 1824)	F, E, P	4	19	24		R	F
Brotogeris chiriri (Vieillot, 1817)	F, P	9		7		R	
Brotogeris versicolurus (Vieillot, 1817)	P			12		V	F
Forpus xanthopterygius (Spix, 1824)	F, E, P, O	4	19	24		R	F
Pionus maximiliani (Miranda Ribeiro, 1920)	F, E, P	14	2	12		R	F
Amazona aestiva (Linnaeus, 1758)	F, P	9		2		R	F
Cuculidae							
Coccyzus americanus (Linnaeus, 1758)	F	1				Mi	
Piaya cayana (Gambel, 1849)	F	3				R	I
Crotophaga ani (Linnaeus, 1758)	E, P, C		3	9	9	R	I
Guira guira (Gmelin, 1788)	P			44		R	I
Tapera naevia (Vieillot, 1817)	F, E, O	3	2			R	I

Species	Occurrence	Records				Category	Feeding Habit
		F	E	P	C		
Tytonidae							
Tyto alba (Scopoli, 1769)	O					R	C
Strigidae							
Otus choliba (Vieillot, 1817)	F	3				R	I, C
Glaucidium brasilianum (Gmelin, 1788)	F	2				V	C, I b
Athene cunicularia (Temminck, 1822)	P			43		R	I
Rhinoptynx clamor (Vieillot, 1807)	F	1				R	I, C
Asio flammens (Vieillot, 1817)	F	1				V	I
Caprimulgidae							
Lurocalis semitorquatus (Temminck, 1822)	F	1				Mv	I
Chordeiles minor (Forster, 1771)	F					V	I
Nyctidromus albicollis (Guold, 1838)	F	3				R	I
Caprimulgus rufus (Burneister, 1856)	F	2				V	I
Hydropsalis brasiliana (Vieillot, 1817)	F	2				R	I
Apodidae							
Streptoprocne zonaris (Shaw, 1796)	F, O	20				R	I
Chaetura Andrei (Berlepsch & Hartert, 1902)	F, P	23		2		Mv	I
Trochilidae							
Phaetornis pretrei (Lesson & Delattre, 1839)	F, E	7	1			R, V	E, I
Eupetomena macroura (Gmelin, 1788)	F, E, P	5	2	1		R	E, I
Melanotrochilus fuscus (Vieillot, 1817)	E		1			V	E, I
Colibri serrirostris (Vieillot, 1816)	F, E	2	1			R	E, I
Anthracothorax nigricollis (Vieillot, 1817)	F, P	2		1		R	E, I
Chrysolampis mosquitus (Linnaeus, 1758)	F	2				V	
Chlorostilbon aureoventris (Bourcier & Mulsant, 1848)	F, E	3	3			R	E, I
Thalurania glaucopis (Gmelin, 1788)	F	4				R	
Hylocharis chrysura (Shaw, 1812)	E		1			V	
Amazilia lactea (Lesson, 1829)	F, E	3	3			R	E, I
Amazilia versicolor (Ruschi, 1959)	E		1			R	E, I
Aphantochroa cirrochloris (Vieillot, 1818)	F, E, C	3	2		2	R	E, I
Heliomaster squamosus (Temminck, 1823)	F	1				V	E, I
Calliphlox amethystina (Boddaert, 1783)	F	2				V	E, I

Species	Occurrence	Records				Category	Feeding Habit
		F	E	P	C		
Trogonidae							
Trogon surrucura (Vieillot, 1817)	F	8				V, R	O
Alcedinidae							
Ceryle torquata (Linnaeus, 1766)	O					R	C
Chloroceryle amazona (Latharn, 1790)	O					R	C, I
Chloroceryle americana (Gmelin, 1788)	O					R	C, I
Bucconidae							
Nystalus chacuru (Vieillot, 1816)	F, P	2		1		R	I
Malacoptila striata (Spix, 1824)	F	3				V	I
Ramphastidae							
Ramphastos toco (Cabanis, 1862)	F, E	4	2			R	O
Picidae							
Picumnus minutissimus (Snethlage, 1924)	F	13				R	I
Celeus flavescens (Gmelin, 1788)	F	2				V	I
Colaptes campestris (Vieillot, 1818)	P, C			56	5	R	I
Colaptes melanochloros (Malherbe, 1848)	F, P	3		3		R	I
Dryocopus lineatus (Linnaues, 1766)	F, P	2		9		R	I
Melanerpes candidus (Otto, 1796)	F, E, P	1	10	33		R	I
Veniliornis passerinus (Natterer & Malherbe, 1845)	F, E	2		1		R	I
Dendrocolapidae							
Sittasomus griseicapillus (Temminck, 1821)	F, E, P	7	6	1		R	I
Dendrocolaptes platyrostris (Spix, 1825)	F	4				V	I
Lepidocolaptes fuscus (Vieillot, 1818)	F, E	4	4			R	I
Lepdocolaptes angustirostris (Lichtenstein, 1822)	P			1		R	I
Furnariidae							
Furnarius rufus (Lichtenstein, 1823)	F, E, P	3	1	9		R	I
Synallaxis spixi (Sclater, 1856)	E, P, C		6	12	2	R	I
Synallaxis ruficapilla (Vieillot, 1819)	F, P	11		5		R	I
Synallaxis frontalis (Pelzeln, 1856)	F, E, P, C	14	7	3	3	R	I
Synallaxis albescens (Temminck, 1823)	F, E, P	3	1	6		R	I
Certhiaxis cinnamomea (Vieillot, 1817)	F	6				R	I
Cranioleuca vulpina (Pelzeln, 1856)	F, E, P	4	7	4		R	I
Automolus leucophthalmus (Lichtenstein, 1823)	F	4				R	I
Xenops rutilans (Temminck, 1821)	F	1				V	I

Species	Occurrence	Records				Category	Feeding Habit
		F	E	P	C		
Formicariidae							
Mackenziaena severa (Lichtenstein, 1823)	F, E	7	1			R	I
Taraba major (Vieillot, 1816)	F		3			V	I
Thamnophilus doliatus (Vieillot, 1816)	F, E, P, C	2	1	7	1	R	I
Thamnophilus caerulescens (Vieillot,1816)	F	12				R	I
Thamnophilus torquatus (Swainson, 1825)	F	3				V	I
Thamnophilus ruficapillus (Vieillot, 1816)	F	9				V	I
Thamnophilus punctatus (Hellmayr, 1924)	F	3				R	I
Dysithamnus mentalis (Temminck, 1823)	F	5				V	I
Formicivora rufa (Lafresnaye & d´Orbigny, 1837)	F	1				V	I
Drymophila ochropyga (Pelzeln, 1874)	F	5				V	I
Pyriglena leucoptera (Vieillot, 1818)	F	2				V	I
Conopophagidae							
Conopophaga lineata (Ménétri'es, 1885)	F	29				R	I
Rhynocryptidae							
Melanopareia torquata (Hellmayr, 1924)	F	2				V	I
Tyrannidae							
Xolmis cinerea (Vieillot, 1816)	P				15	R	I
Xolmis velata (Lichtenstein, 1823)	O					R	I
Colonia colonus (Vieillot, 1818)	E, P		12	4		Mv	I
Gubernetes yetapa (Vieillot, 1818)	E, P		1	4		Mv	I
Alectrurus tricolor (Vieillot, 1816)	O					V	I
Knipolegus cyanirostris (Vieillot, 1818)	F	1				Mi	I
Fluvicola nengeta (Linnaeus, 1766)	O					V	I
Arundinicola leucocephala (Linnaues, 1764)	O					V	I
Satrapa icterophrys (Vieillot, 1818)	F, E, P	1	4	2		R	I
Machetornis rixosus (Vieillot, 1819)	C				1	R	I
Tyrannus savana (Linnaeus, 1766)	F, E, P	2	3	9		Mv	I, F
Tyrannus melancholicus (Vieillot, 1819)	F, E, P	8	11	6		Mv, R	I, F
Tyrannus albogullaris (Burmeister, 1856)	E, P		1	1		V	I
Empidonomus variius (Vieillot, 1818)	F, E, P	6	4	6		Mv	I, F (O)
Megarhynchus pitangua (Linnaues, 1766)	F, E	15	2			R	I, F (O)
Myiodinastes maculatus (Vieillot, 1819)	F, E, P	7	2	2		Mv	I, F (O)
Myiozetetes similis (Spix, 1825)	F, E, P	6	7	1		R	I, F (O)

Species	Occurrence	Records				Category	Feeding Habit
		F	E	P	C		
Pitangus sulphuratus (Cabanis & Heine, 1859)	F, E, P	3	4	4		R	I, F (O)
Casiornis rufa (Vieillot, 1816)	F, E	2	4			V	I
Myiarchus ferox (Hellmayr, 1927)	F	3				R	I
Myiarchus swainsonii (Cabanis & Heine, 1859)	F, E, P	1	1	1		Mv	I
Empidonax euleri (Cabanis, 1868)	F, E	5	3			R	I
Cnemotriccus fuscatus (Lafresnaye & d´Orbigny, 1837)	F, P	2		1		R	I
Myiophobus fasciatus (Temminck, 1822)	F, E, P	2	1	1		R	I
Hirundinea ferruginea (Vieillot, 1819)	P			10		R	I
Tolmomyias sulphurescens (Spix, 1825)	F	5				R	I
Todirostrum poliocephalum (Wied, 1831)	F	4				R	I
Todirostrum cinereum (Linnaeus, 1766)	F, E, P	3	1	1		R	I
Myiornis auricularis (Vieillot, 1818)	F, P		2			R	I
Tyrannidae (continuação)							
Polystictus pectoralis (Vieillot, 1817)	F	1				V	I
Serpophaga subcristata (Vieillot, 1817)	F, E	3	3			R	I
Elaenia flavogaster (Thumberg, 1822)	E, P		5	3		R	I, F
Elaenia obscura (Zimmer, 1941)	F	2				V	I, F
Elaenia chiriquensis (Pelzeln, 1868)	F, P	1		1		V	I, F
Myiopagis caniceps (Swainson, 1835)	F, E	2	1			R	I
Myiopagis viridicata(Vieillot, 1817)	F					Mv	I
Suiriri suiriri (Vieillot, 1818)	E		3			V	I
Phaeomyias murina (Spix, 1825)	F, E	2	1			V	I
Camptostoma obsoletum (Temminck, 1824)	F, E, P	4	8	1		R	I, F (O)
Leptopogon amaurocephalus (Tschudi, 1846)	F	3				V	I
Cotingidae							
Pachyramphus polychopterus (Swainson, 1837)	F	8				Mv	F
Tityra cayana(Swainson, 1837)	F	4				V	F, I (O)
Pipridae							
Antilophia galeata (Lichtenstein, 1823)	F	4				R	F (O)
Chiroxiphia caudata (Shaw & Noddler, 1793)	F	17				R	F (O)
Manacus manacus (Linnaeus, 1766)	F	30				R	F (O)
Hirundinidae							
Tachycineta albiventer (Boddaert, 1783)	O					R	I

Species	Occurrence	Records				Category	Feeding Habit
		F	E	P	C		
Tachycineta leucorrhoa (Vieillot, 1817)	F, E, P, C	1	5	3	1	Mi, R	I
Progne tapera (Vieillot, 1817)	E, P, C		3	4	103	Mv, R	I
Progne subis (Linnaeus, 1758)	P, C			8	32	RMv	I
Progne chalybea (Vieillot, 1817)	F, E, P	3	4	3		Mv	I
Notiochelidon cyanoleuca (Vieillot, 1817)	P, C			51	172	Mv,R	I
Stelgidopteryx ruficollis (Vieillot, 1817)	E, P, C		23	27	2	R	I
Hirundo rustica (Linnaeus, 1758)	P			13		Mv	I
Trogloditidae							
Troglodyes musculus (Naumann, 1823)	F, E, P, C	10	9	36	6	R	I
Mimidae							
Mimus saturninus (Hellmayr, 1903)	P, C			25	9	R	I, F, G (O)
Donacobius atricapillus (Linnaeus, 1766)	O					R	I
Turdidae							
Turdus nigriceps (Cabanis, 1874)	F, E	4	1			V	F, I
Turdus rufiventris (Vieillot, 1818)	F, E	4	8			R	F, I
Turdus leucomelas (Vieillot, 1818)	F, E	22	1			R	F, I
Turdus amaurochalinus (Cabanis, 1851)	F, P	7		1		R	F, I
Turdus albicollis (Vieillot, 1818)	F, E, P	16	6	1		M	I
Emberizidae							
Saltator similis (Lafresnaye & d´Orbigny, 1837)	F, E, P	5	1	1		V	G, I (O)
Saltator atricollis (Vieillot, 1817)	P			1		V	G, I
Pitylus fuliginosus (Daudin, 1800)	F	1				V	G, I (O)
Passerina brissonii (Lichtenstein, 1823)	E, P		2	2		V	G, I (O)
Volatinia jacarina (Linnaeus, 1766)	E, P, C		9	52	63	R	G, I
Tiaris fuliginosa (Wied, 1831)	O					V	G
Sporophila lineola (Linnaeus, 1758)	P			15		R	G, I
Sporophila caerulescens (Vieilliot, 1817)	E, P, C		3	41	6	R, M	G, I
Sporophila leucoptera (Vieillot, 1817)	P			8		V	G
Sporophila bouvreuil (Statius Muller, 1776)	E, P, C		1	13	3	V	G
Oryzoborus angolensis (Linnaues, 1766)	P			6		V	G
Sicalis flaveola (Gmelin, 1766)	P, C			16	11	R	G,I
Coryphospingus cucullatus (Swainson, 1825)	P			1		R	G
Arremon flavirostris (Des Murs, 1856)	F, P	15		1		R	I, F

Species	**Occurrence**	**Records**				**Category**	**Feeding Habit**
		F	**E**	**P**	**C**		
Ammodramus humeralis (Bosc, 1792)	E, P, C		6	41	26	R	G, I
Zonotrichia capensis (Swainson, 1837)	E, P, C		61	44	81	R	G, I
Emberizoides herbicola (Vieillot, 1817)	P			3		R	I, F
Schistochlamys ruficapillus (Vieillot, 1817)	F	3				V	I, F
Schistochlamys melanopis (Latham, 1790)	F	2				V	I, F
Thraupidae							
Euphonia chlorotica (d´Orbigny & Lafresnaye, 1837)	F, E, P	7	1	2		R	I, F (O)
Euphonia violacea (Bertoni, 1901)	F	1				V	I, F (O)
Pipraeidea melanonota (Vieillot, 1819)	F	9				Mi, R	I, F (O)
Tangara seledon (Statius Muller, 1776)	F	4				V	I, F
Tangara cayana (Vieillot, 1819)	F, E	5	2			R	I, F (O)
Thraupis sayaca (Linnaues, 1766)	F, E, P, C	9				R	F, I (O)
Thraupis palmarum (Wied, 1823)	F, E	8	1			V	I, F (O)
Ramphocelus carbo (Hellmayr, 1920)	F	15				R	F, I (O)
Piranga flava (Spix, 1825)	P			5		V	F, I
Habia rubica (Vieillot, 1817)	F, E, P	12	9	3		R	I, F (O)
Tachyphonus coronatus (Vieillot, 1822)	F, P	13		2		R	I, F (O)
Trichothraupis melanops (Vieillot, 1818)	F	2				V	F, I (O)
Nemosia pileata (Chubb, 1910)	F, E, P	6	1	1		R	I, F (O)
Hemithraupis ruficapilla(Vieillot, 1818)	F, E	2	3			V	F, I
Thlypopsis sordida (d´Orbigny & Lafrenaye, 1837)	F, E, P	20	1	1		R	I, F (O)
Neothraupis fasciata (Lichtenstein, 1823)	F	9				R	F, I
Conirostrum speciosum (Temminck, 1824)	F, E	5	2			R	I, F
Dacnis cayana (Chubb, 1910)	F, E, P	7	1	1		R	I, F (O)
Coeribidae							
Coereba flaveola (Cabanis, 1851)	F, C	29			3	R	I, E
Parulidae							
Parula pitiayumi (Vieillot, 1817)	F	4				R	I
Geothlypis aequinoctialis (Vieillot, 1807)	F, E, P, C	11	3	21	8	R	I
Basileuterus flaveolus (Baird, 1865)	F, P	52		1		R	I
Basileuterus culicivorus (Lichtenstein, 1830)	F, E, P	72	2	3		R	I
Basileuterus leucoblepharus (Vieillot, 1817)	F	7				V	I

Species	Occurrence	Records				Category	Feeding Habit
		F	E	P	C		
Vireonidae							
Cyclarhis gujanensis (Baird, 1866)	F, E, P, C	26	1	2	2	R	I
Vireo olivaceus (Linnaues, 1766)	F, E	33	3			Mv	I, F (O)
Hylophilus poicilotis (Temminck, 1822)	F, E	4	1			V	I
Icteridae							
Molothrus bonariensis (Gmelin, 1789)	F, E, P, C	5	1	23	13	R	G, I, F (O)
Gnorimopsar chopi (Vieillot, 1819)	F, E, P	2	5	22		R	G, F, C (O)
Agelaius ruficapillus (Vieillot, 1819)	C				7	R	G, I (O)
Pseudoleistes guirahuro (Vieillot, 1819)	P, C			38	36	R	I, G, F (O)
Frigillidae							
Carduelis magellanica (Vieillot, 1805)	F	2		10		V	G, I
Corvidae							
Cyanocorax cristatellus (Temminck, 1823)	F, E, P, C	24	40	24	11	R	O
Cyanocorax chrysops (Vieillot, 1818)	E, P		6	8		V	O
Estrildidae							
Estrilda astrild (Linnaeus, 1768)	P, C			32	4	R	G

Maria Cecília Lunardelli, Maria Carolina Lyra-Jorge,
Carla Gheler-Costa

16 Howler Monkeys in an Urban Brazilian Forest: Ecology and Daily Feeding Patterns

16.1 Introduction

The primates of the genus *Alouatta* are popularly known as howler monkeys and have wide geographical distribution, occurring from Mexico to northern Argentina, from sea level to altitudes around 2,300 m (Hill, 1953). According to some authors, the genus *Alouatta* belongs to the family Atelidae (Hirsch *et al*, 1991; Ford & Davis, 1992; Rosenberger & Strier, 1992; Horovitz, 1999), the classification adopted in this study. The number of species of *Alouatta* is very controversial and several taxonomic arrangements have been proposed, but the most accepted (Hill, 1953) contains 6 species: *A. villosa* (= *A. pigra*), is found in Mexico and Central America; *A. palliata*, is found in Central and South America; *A. seniculus*, is found in the Amazon forest, more precisely in the west and north of the Amazon River; *A. belzebul*, distributed only in Brazil, originally from the Western Amazon to the Northeastern Atlantic forest; *A. caraya*, is found in the Central highlands of Brazil to the Paraguay basin and *A. guariba* (= *A. fusca*), in Southeastern and Southern Brazil to Northern Argentina. The genus *Alouatta* is one of the most studied groups of Brazilian primates and has shown to be one of the most adaptable to disturbance and fragmentation of habitats (Bicca - Marques, 2003; Chiarello, 2003).

The species *A. fusca* was found originally along the Atlantic forest, semi-deciduous forest, Araucaria forest and subtropical forest of Southeastern and Southern Brazil (Galletti & Chiarello, 1994). Agricultural activities and uncontrolled growth of cities, especially in Southeastern and Southern Brazil, promoted for many years deforestation of natural areas turning them into fragments. These are usually associated with sharp reliefs or the watershed areas (Leitão-Filho, 1995). In the remaining forests, flora and fauna may be terminated or develop alternative mechanisms of adaptation to the new environment. The altered vegetation can present floristic changes (Chiarello, 1993; Leitão-Filho, 1995) and influence the activity pattern and diet of animals, especially herbivores. In São Paulo, in Southeastern Brazil, howler monkeys are found in the Serra do Mar State Park, in the Cantareira State Park, Serra do Japi in Serra da Mantiqueira (Silva, 1981; Marinho-Filho, 1992; Galletti & Chiarello, 1994), Estação Ecológica do Bananal (Brandão, 1999), in Fazenda Intervales (Steinmetz, 2000), in which areas are almost continuous; and rural and urban forest fragments such as Barreiro Rico (Assumption, 1983), Reserva de Santa Genebra (Chiarello, 1993; Galletti & Chiarello, 1994) and Mata do Ribeirão Cachoeira, both in Campinas (Gaspar, 1997), Fazenda Rio Claro, in Lençóis Paulista (Martins, 1997).

An herbivory diet and the prevalence of leaves characterize the genus *Alouatta* as the most folivorous among neotropical primates, with fruits and flowers being seasonal resources in the diet. The folivory probably contributes to survival in small forest fragments, where the presence of flowers and fruits can be fickle (Rylands & Keuroghlian, 1988; Bicca-Marques & Calegaro-Marques, 1994). Due to the high consumption of leaves in the diet, the metabolism is slow and way of life sedentary, with rest being the main daily activity (Milton, 1981). The howler monkeys are also considered important seed dispersers because, due to the slow passage of food through the digestive tract, the seeds are deposited away from the parent plants (Estrada & Coates-Estrada, 1984). Most studies on *A. fusca* aim to analyze the pattern of activity and diet, quantifying food items and identifying the plant species consumed by direct observations (Mendes, 1989; Prates et al., 1990; Chiarello, 1993; Gaspar, 1997; Marques, 1996; Martins, 1997; Perez, 1998; Steinmetz, 2000) or indirectly, through analysis of feces; however, there are few in-depth studies on the feeding ecology of howler monkeys. In this work, in addition to quantifying and identifying the food items present in the diet, an investigation was carried out on the chemical composition of dietary plants, quantifying the phenolic compounds of the items consumed and not consumed by a group of *A. fusca*, thus allowing for a greater understanding of these animals and the feeding selectivity of the chemical aspect of primate-plant interaction. Activity patterns were studied by determining the percentage of time that the animals allocated for daily activities and results were compared with other studies with *A. fusca* and other species of the same genus. The study area, Serra da Cantareira, which is characterized by the high density of howlers, with 80 individuals per square kilometer (Silva, 1981). However, this is being increasingly threatened by the advance of real estate subdivisions, due to its proximity to major urban centers such as the City of São Paulo, one of the largest cities in Latin America.

16.2 Feeding Ecology

16.2.1 The Feeding Selectivity and the Study of Diet

Dietary studies are important for understanding how primates use the environment and select food. The feeding selectivity may depend on several factors, such as the "risk" of demand or feed intake, the proportion of each item in the diet and food abundance. When food is plentiful, the animals are more selective; however, when there are food shortages, they tend to be generalists (Schoener, 1971). Herbivores can be influenced by the nutritional characteristics of plants because plants generally are rich in some nutrients and deficient in others. Because of these variations, the diet of herbivores tends to be diverse and generalist. In addition to the nutritional content of plants, food selectivity can be influenced by toxic or inhibitory substances, such as tannins (Freeland & Janzen, 1974; Kinzey, 1997).

The genus *Alouatta* differs in foraging strategy when compared with others atelides. While *Ateles*, *Brachyteles* and *Lagothrix* maximize energy gain, howler monkeys have developed a strategy to save energy and spend much of the day resting (Strier, 1992). Howler monkeys, despite having many plant species in the diet, tend to use a few very intensely, selecting the worn parts. This behavior can be explained as a way to balance daily nutrition and metabolic requirements (Smith, 1977; Milton, 1979; Nagy & Milton, 1979; Milton et al., 1980; Milton, 1981). In forests where there is a seasonality of food resources, the fruiting of some species usually coincides with rain, therefore the consumption of fruit is higher during the wet season and items such as buds and flowers are eaten in the dry season (Milton, 1981; Braza et al.,1983; Bonvicino, 1989; Mendes, 1989; Alves & Guix, 1992). Even if there is eventually an increase in the consumption of seasonal resources, the diet of the howler remains folivorious. Studies show that these animals regularly spend 40 to 90% of feeding activity eating shoots (Milton et al.,1980; Bonvicino, 1989; Mendes, 1989; Julliot & Sabatier, 1993; Galetti et al., 1994; Stoner, 1996), and particularly young leaves, which usually have a high protein content (Milton, 1981; Silva, 1981; Braza et al.,1983; Bonvicino, 1989; Mendes, 1989).

16.2.2 The Ecological Implications of an Herbivorous Diet with Emphasis on Primate Herbivores

Among the herbivores, the primates herbivore represents a very small group, and of the 391 genera of mammals, only 42 are arboreal herbivores. According to a study conducted in a hectare of tropical forest, only 2% of the plant biomass is represented by leaves, approximately 28% consists of the root system and the other 70% are represented by branches and trunks. Thus, about 98% of the plant biomass of a tropical forest is unavailable to herbivores (Queiroz, 1995). The digestive tract of mammals is unable to digest cellulose, the major structural constituent of the cell wall. However, herbivores exhibit morphological adaptations such as enlargement of the digestive chambers and the presence of fermenting micro flora (Chivers & Hladik, 1980). Fermenting microorganisms are important for making plant nutrients available to animals (Waterman, 1984), as they act in the degradation of cellulose and hemicellulose producing energy by volatile fatty acids (Milton & McBee, 1983). It is estimated that the howler obtains about 26 to 36% of daily energy requirement by using such substances (Milton et al., 1980). Herbivore primates can have two types of adaptations (Parra, 1978): 1) gastric fermenters such as primates of the genera *Presbytis*, *Colobus* and *Nasalis*, which have extended stomachs divided into chambers; 2) the post- gastric fermenters such as primates *Lepilemur, Propithecus, Avahi, Indri, Alouatta, Gorilla*, which have an extended cecum and large colon. Despite being post gastric fermenters, howlers have a large stomach, without chambers. The pH of the cecum and colon is around 6.8 and the stomach around 4.5, which makes fermentation difficult (McBee & Milton, 1983).

In addition to anatomical adaptations, the rate of food passage through the digestive tract is slow, allowing a more efficient extraction of nutrients and fermentation processes (Crissay et al., 1990). Some behaviors serve to conserve energy and perhaps increase the energy spent on foraging return, such as: standard daily inactivity, the use of energy- rich foods when available (fruit), an efficient location of food, body postures that conserve or dissipate heat and division of labor in parental care between males and females. Another important dietary aspect of herbivores that has been previously mentioned is that the compounds present in plants can work against herbivory (Whittaker & Fenny, 1971; Rhoades & Cates, 1976).

The plant-herbivore interaction, without doubt, was the most chemically studied type of ecological relationship, giving rise to a new discipline: chemical ecology. However, these same compounds can also act as mediators of other ecological interactions such as pollination, seed dispersal, allelopathy and the relationships between plants and their pathogens (Waterman & Mole, 1994), without necessarily acting as "defense" substances. Primary compounds are essential to cell metabolism, while secondary compounds are important to the physiological role for survival. However, a compound may be essential for some species and secondary to others. Another important consideration is the association of secondary compounds with toxicity, since not all secondary substances are toxic to herbivores. The fact that certain substances do not harm some herbivores does not mean that they have no defense function for plants (Glander, 1982).

When secondary compounds act against herbivory, they do so in two ways: causing toxicity or reducing digestibility. Toxins are substances that exhibit low molecular weights and are present in small quantities interfering with intracellular physiological processes. Substances that reduce digestion are different from toxins because they are formed by molecules of varying sizes which do not cross cell membranes, acting only in the digestive tract (Freeland & Janzen, 1974; Rhoades & Cates, 1976). Some compounds have a negative impact on herbivores, especially insects, and include alkaloids, rotenoids, terpenes, saponins, cyanogenic glycosides, tannins, phenolic resins, among others. Many of these compounds also act as "protection" for plants against vertebrate herbivores, for example (Glander, 1982):

- terpenes and phenolic resins present in adventitious roots of *Betula papyrifera*, *Populus tremuloides*, *P. balsamifera* and *Alnus crispa*, repel *Lepus americanus*;
- cyanogenics glicosides hold rodents and rabbits that feed respectively *Lotus comiculatus* and *Trifolium repens*.

Herbivores may have the ability to detoxify these substances through specific metabolism and excretion, and can also, as in some insects, "kidnap" secondary plant compounds and use these in self-defense. They can also present two types of feeding strategies, acting as specialists or as generalists. Specialists feed on one or a few species, such as koalas (*Phascolarctos cinereus*), who eat only eucalyptus and thus need only to deal with a few compounds. The generalists feed on many species of plants; however, they need to deal with various types of compounds. As it is impossible to maintain many detox systems, generalists tend to reduce the intake of chemicals, selecting items or plant species that have few compounds or ingesting small amounts of various types of plants (Glander, 1982).

Some authors consider the compounds as derivatives of secondary substances obtained from the primary metabolism (Mann, 1994; Waterman & Mole, 1994). Tannins are complex substances that are very widespread in the plant kingdom, occurring in almost all plant families. Tannins are located in the leaves, immature fruit, stems or cortex. Chemically, the tannins are polyphenols with high molecular weight characterized primarily by astringent and by precipitating proteins *in vitro*, forming insoluble complexes and resistant to proteolytic enzymes. There are two distinct groups of tannins: condensed tannins and hydrolysable tannins. Condensed tannins are found in plant cellulose walls, acting as fungicides and bactericides (Waterman & Mole, 1994). They may also act as inhibitors of larval growth for certain species of insects, such as condensed tannins found in buds from *Croton* sp cause a growth reduction of 84% in larvae of *Heliothis virescens* (Waterman, 1984). Hydrolyzable tannins serve to inactivate the digestive enzymes of herbivores, interfering with the digestive processes of insects, birds, mammal ruminants and non-ruminants.

Diet composition and food selectivity are important themes in ecological studies of primates. However, to understand how they survive and adapt, the primatologists needs to further emphasize certain aspects of primate interaction with plants. It is a complex problem which requires the determination of the chemical composition of plants, nutrients or secondary metabolites, the consumed items and items not consumed by primates.

There are many data on insect-plant interactions documenting the effects of chemicals on the feeding behavior of these animals. However, one cannot infer that primates present the same effects as insects. Thus, to understand the feeding behavior of primates, it is necessary to carry out specific studies. Many studies have been conducted in order to assess the impact of some plant chemical compounds on the feeding selectivity of primates, especially for the Old World (Hladik, 1978; McKey et al., 1978; Gartlan et al., 1980; Oates et al., 1980; Davies et al., 1988; Mowry et al., 1996). For primates of the genus *Alouatta* some studies on the diet and feeding behavior were conducted, and the species *Alouatta palliata* was the most studied from the perspective of chemical ecology (Glander, 1978; Nagy & Milton, 1979; Milton, 1980). The species of howler monkeys found in Brazil lack detailed studies on feeding ecology. In this study, an investigation was conducted on the chemical composition of plants, quantifying the phenolic compounds of the items consumed and not consumed by a group of *A. fusca*, thus allowing a greater understanding of these animals and feeding selectivity of the chemical aspect of primate plant interaction.

16.3 Daily Pattern of Activity

Studying activity patterns is very important because it provides estimates of the percentage of time allocated to different activities carried out by an animal. The time for each activity may vary depending on the habitat, the seasonal and climatic factors such as precipitation and temperature (Milton, 1980). The howler monkeys are considered low activity animals because their daily activities are reduced (Bonvicino, 1989; Mendes, 1989; Chiarello, 1993). Being predominantly herbivores, in general, they are slower than animals with diverse eating habits, since the digestion of leaves is time-consuming and most often involves bacterial fermentation to degrade cellulose (Bauchop, 1978; McNab, 1978). The feeding activity is usually associated with the movement of animals through the living area and almost exclusively associated with food (Milton, 1980).

Social interactions such as playing and grooming are neither frequent nor long lasting. Vocalizations are very common, especially the characteristic aggressive roar that is used as a mechanism to maintain the distance among the groups. They are frequent at dawn and usually precede the day's activities. Adult males vocalize more, but the female and male sub-adults can also participate. The groups can vary from 3 to 10 individuals, typically having a greater number of females than adult males (Mendes, 1989).

16.4 Methods

16.4.1 Study Area

The Cantareira State Park, located in the hills of the same name, is locatedin São Paulo (23°22'S/46°26'W) and also covers the areas of the municipalities of Caieiras, Guarulhos and Mairiporã, and is an area of approximately 7,900 ha (Fig. 16.1). The name was given to Cantareira between the sixteenth and seventeenth centuries, due to the presence of many springs. Problems with the water supply for the city of São Paulo and the substitution of forests by coffee plantations led the state government to expropriate areas of the Serra da Cantareira in 1890. In 1965, due to its role as watershed protection, the region's forests were considered for permanent protection. The Cantareira State Park was created in 1972 and currently is being threatened by development, mainly in the municipalities of Caieiras and Mairiporã. According to the phytogeographical classification (Huek, 1972), the vegetation is a Tropical Rain Forest of Mountainous slopes. The floristic studies (Tabarelli, 1994), revealed a ecotonal character due to the presence of species of the Mountainous Tropical Rain Forest, characteristic of slope regions, combined with elements of Forest Semicaducifolia. Common plant families were: *Myrtaceae, Lauraceae, Rubiaceae, Leguminosae* and *Euphorbiaceae*, which feature high richness in both Atlantic slope forests and in the

Cloud Forests of São Paulo. Besides *Alouatta fusca*, other primate species such as capuchin monkeys (*Cebus apella nigritus*), titi monkeys (*Callicebus personatus*) and an introduced a kind of marmoset (*Callithrix jacchus*), can be found in the Cantareira State Park.

Fig. 16.1: The Cantareira State Park (From: http://www.tuhic.com/estacao/descricao,adriano).

16.4.2 Study Group

The study group was composed of two adult males, one sub-adult male, three adult females, one juvenile and two infants. The animals were observed with the aid of a binocular and the data was recorded in the field contained the start time of data collection and categorized the subjects according to defined age-sex classes and the exact time of each record.

16.4.3 Data Collection: Scan Sampling

To study the pattern of activity and to quantify the food resources of the diet, the method of instant scan sampling was used (Altmann, 1974). This method consists of the instantaneous recording of activity performed by each individual in the group at intervals of predetermined time, so the displayed animals are recognized and their activities recorded every 10 minutes, each "scan" lasted 3 minutes (Chiarello, 1993; Mendes, 1989). This technique quantifies the time that the entire group allocates to the different behavioral categories and is well employed in studies with howler monkeys because individuals in a group tend to coordinate their activities and time spent on individual observations is minimal (Milton, 1980). The animals were followed for one year and monthly data collections were made. Data were analyzed bimonthly because the dry and rainy seasons are not defined (Koppen, 1948). The categories of activities sampled in this study were classified as resting (R), moving (T), feeding (F), social interactions (play, vocalization and grooming) (SI). The relative frequency of observation was calculated for a particular activity:

$$\text{Frequency of observation} = \frac{\text{Number of records for the activity}}{\text{Total number of registrations}}$$

Each activity is represented by the relative frequency or a percentage of the total time sampled. During the feeding time the items consumed were: YL (young leaves), ML (mature leaves), FR (fruits) and FL (flowers, buds and inflorescences) and for each item the relative frequency or percentage of the consumption was calculated. To investigate possible differences between marking periods, both in relation to the pattern of activities regarding the frequencies of consumption of food items, the Kruskal-Wallis nonparametric analysis of variance and the Tukey Test was applied *a posteriori* to locate where the marking periods occurred differences (Zar, 1999).

16.4.4 Chemical Analysis

In order to quantify the amount of total phenols and tannins, plant samples were separated into two categories: eaten and uneaten items. The 17 most consumed species were selected and the young leaves, flowers and fruits were analyzed. Uneaten items consisted of mature leaves from 5 species, and the fruit of a species that the howlers avoided throughout the fruiting period. Plant samples were dried at 60°C for 72 hours (Waterman & Mole, 1994). They were crushed with an electric macerator and stored in a dry and dark place in plastic bags. For the extraction and quantification of total phenols (Folin – Dennis method), we followed the methodology described by Waterman & Mole (1994).

For the extraction and quantitative analysis of tannins we used the volumetric method for dosing potassium permanganate, which is a simple method that allows quantification of tannins, both hydrolysable and condensate tannins. The Kolmogorov-Smirnov test was used to verify the normality of the quantitative data of total phenols for both categories of samples mentioned above. To detect possible differences in the amounts of total phenols between consumed and unconsumed items, the student t test for independent samples (Zar, 1999) was applied. Quantitative tannin data were not statistically analyzed because for each species sampled we obtained only one extract and therefore an insufficient sample size for statistical analysis.

16.5 Results

16.5.1 Daily Activity Pattern

In this study 13,774 records ("scans") for observed activities were obtained. The principal activity was resting (60.6%), followed by feeding (21.4%), moving (13.8%) and social interactions (play, vocalizations, grooming) which were uncommon (4.3%). Even when activities were grouped into two-month periods, the time allocation patterns for each activity remained the same, with resting being predominant, followed by feeding, moving and social interactions (Tab.16.1).

Tab. 16.1: Percentages (%) of time allocated to activities by the study group in two month periods at the Cantareira State Park, SP, Brazil.

Two month period	Resting	Feeding	Moving	Social interactions
1	64.0	19.6	11.5	4.7
2	62.4	16.5	15.4	5.6
3	61.3	17.5	17.1	4.2
4	61.5	23.6	11.8	3.1
5	52.4	27.5	15.3	4.7
6	61.6	23.4	11.6	3.3

Comparing the two-months periods, there were no differences in the time allotted for resting ($H = 8.81$, $gl = 5$, $p = 0.12$). The greatest time allocation for feeding occurred in the fifth period, and the least in the second. Differences were observed for the time allotted between marking periods ($H = 12.8$, $gl = 5$, $p = 0.02$), specifically between the second and fifth month periods (Tukey, $p = 0.02$). The variations between the frequencies of observations for moving did not differ between marking periods ($H = 7.69$, $gl = 5$, $p = 0.17$). Social interactions were reduced in general, with a slight increase in time allocation in the second period, however, the variations over the marking periods were not significant ($H = 5.62$, $gl = 5$, $p = 0.34$).

16.5.2 Feeding Ecology

We have obtained 2,932 food records (scans) for the annual diet of the howler group. Young leaves (50.3%) were predominant, flowers (29.3%) and fruits (15.4%) were, respectively, the second and third most consumed items. Mature leaves (5.1%) were less consumed. Stems and tree barks have not been greatly explored throughout the year, so the consumption of these items has not been analyzed. The percentages of food items consumed in two-month periods are shown in Table 16.2.

Tab. 16.2: Percentage (%) of food items consumed by the study group in two month periods at the Cantareira State Park, SP, Brazil.

Two month period	Young leaves	Mature leaves	Fruits	Flowers
1	76.8	2.3	5.7	15.1
2	51.6	0.5	47.8	0
3	50.1	4.7	45.2	0
4	34.5	9.2	4.5	51.9
5	23.4	4.6	10.0	61.8
6	69.0	7.5	0.3	23.3

The young leaves were preferentially consumed in the first three two-month periods and in the last two-month period. However, among periods total consumption of young leaves did not differ ($H = 10.42$; $gl = 5$; $p = 0.06$). Mature leaves were the least consumed item throughout the study period. For this food item, differences were observed in the frequencies of observations among periods ($H = 11.29$, $gl = 5$, $p = 0.04$), but the Tukey test did not identify the differences. The highest consumption of fruits occurred in the second and third periods, with the latter showing the largest variation of observation frequencies. Among periods, consumption of fruits showed significant differences ($H = 19.15$; $gl = 5$; $p < 0.001$) (Tab. 16.3).

Tab. 16.3: Tukey test results for the differences between two month periods measuring consumption of fruits by the study group in the Cantareira State Park, SP, Brazil.

Comparison between each two month period	Value "p" (Tukey test, p < 0.05)
2 - 1	0.01
2 - 4	0.02
2 - 6	0.01

The greatest consumption of flowers occurred in the fifth and fourth periods, with the latter showing the largest variation in observation frequencies. Consumption of flowers was also different among periods ($H = 28.5$; $gl = 5$; $p < 0.001$) except for the last period of study (Tab. 16.4).

Tab. 16.4: Tukey test results for the differences between two month periods measuring consumption of flowers by the study group in the Cantareira State Park, SP, Brazil.

Comparison to each two month period	Value "p" (Tukey test, p < 0.05)
2 - 4	0.001
2 - 5	0.000
1 - 5	0.007
3 - 4	0.001
3 - 5	0.000

The study of howler diet recorded 41 plant species of 23 families (Tab. 16.5) and the most representative families were: *Leguminosae* (12%), *Sapindaceae* (19%), *Euphorbiaceae* (10%) and *Moraceae* (9%); and together account for half of total food records. Despite the young leaves being the total most consumed resource, the flowers of *Jacaranda puberula, Luehea candicans* and *Cupania oblongifolia* presented in absolute terms, are the largest food records. When consumed species were observed every period, it was found that in the first two months, young leaves of *L. candicans, Guapira opposita, Machaerium nictitans* were more consumed while *Sorocea bonplandii, Alchornea glandulosa* and *Allophylus edulis* were less consumed. In this first period, the flowers of *Jacaranda puberula* were the second most consumed item while the fruits were little consumed. Among the mature leaves, *Ficus insipida* were the most consumed. In the second period the young leaves mainly consumed belonged to the species *Mollinedia schottiana, Platymiscium floribundum* and *F. insipid*, and the fruits of *Myrcia* sp. and *A. edulis*.

Tab. 16.5: Plant species and the number of records of items consumed by the study group in Cantareira State Park, SP, Brazil.

Family and species	**Food item consumed (records)**
Araceae	
Anthurium sp.	YL (21)
Balsaminaceae	
Impatiens balsamina L.	YL (36)
Bignoniaceae	
Jacaranda puberula Cham.	FL (221)
Cactaceae	
Rhipsalis sp.	Stark (35)
Dilleniaceae	
Davilla rugosa Poir	YL (11)
Euphorbiaceae	
Alchomea glandulosa Poepp.	YL (85)
Alchornea triplinervia (Spreng.) Müll. Arg.	FR (43)
Croton üoribundus Spreng	YL (130)
Sapium glandulatum (Vell.) Pax	YL (13)
Flacourtiaceae	
Casearia obliqua Sprengel	YL (7)
Casearia sylvestris Sw.	YL (9)

Family and species	Food item consumed (records)
Leguminosae	
Dalbergia frutecens (Veil.) Britton	YL (4)
Inga marginata Willd.	YL (22), FL (27)
Machaerium nictitans (Vell.) Benth	YL (140)
Machaeríum aculeatum Raddi	YL (80)
Piptadenia gonoacanyha (Mart.) Macbr	ML (12)
Platymiscium florihundum Vogel	YL (55)
Senna bicapsularis (Benth.) Irw & Bam	YL (7)
Malpighiaceae	
Heteropteris sp.	ML (18)
Melastomataceae	
Miconia cabucu Hoehne	YL (7)
Meliaceae	
Guarea macrophylla (Vell.) Penn	YL (70), ML (45)
Monimiaceae	
Mollinedia schottiana (Spreng.) Perkins	YL (76)
Moraceae	
Ficus insipida Willd	YL (119),ML (56)
Cecropia sp.	YL (21), FR (13)
Sorocea bonplandii (Baill.) W. Burger	YL (50)
Nyctaginaceae	
Guapira opposita (Vell.) Reitz	YL (102)
Olacaceae	
Heisteria silvianni Schwacke	YL (3)
Proteaceae	
Roupala brasiliensis Klotzsch	YL (7)
Rosaceae	
Prunus scllowii Koehne	ML (10)
Rutaceae	
Zanthoxyllum sp.	ML (8)
Sapindadaceae	

Family and species	Food item consumed (records)
Allophylus edulis (St.Hil) Radlk	YL (35), FR (121)
Cupania oblongifolia Mart.	FL(250), FR (79)
Cupania emarginata Cambess	YL (14)
Matayba guianensis Aubl.	YL (16)
Matayba juglandifolia (Camb.) Radlk	YL (13)
Serjania grandiflora Cambess	YL (12)
Solanaceae	
Cestrum schlechtendalli D. Don	FR (113)
Tiliaceae	
Luehea candicans Mart. & Zucc.	YL (184), FL (360)
Urticaceae	
Urera nitída (Veil.) Brack	YL (16)
Verbenaceae	
Aegiphyla sellowiana Cham.	FL (7)
Lantana sp.	FL (13)

Young leaves consumed in the third period were of *F.insipida, M. nictitans* and *Croton floribundus*, while fruits were of *Cupania oblongifolia* and *Cestrum schlechtendalli*. In this period the howler monkeys explored another resource, stalks of Rhipsalis sp. In the fourth period the consumption of *L. candicans* flowers was higher than that of young leaves of the species *C. floribundus* and *S. bonplandii*. The fruits of *C. oblongifolia* and *C. schlechtendalli* were less eaten, probably because they started becoming scarce. In contrast, mature leaves of *Guarea macrophylla* and *F.insipida* were more consumed. In the fifth period, the consumption of flowers of *L. candicans* increased, becoming even greater than the consumption of the young leaves of the same species. The consumption of *Alchornea triplinervia* fruits ranked third in the diet during this period. In the last two months, the consumption of young leaves increased, mainly *G. opposita, Inga marginata, G. macrophylla*. Young leaves of *Anthurium* sp and *Impatiens balsamina* were consumed mainly during the final hours of the afternoon. Individuals of *C. oblongfolia* started flowering and the flowers took second place in the diet, and the mature leaves of *F. insipida* were consumed again.

The chemical analysis results of total phenols and tannins for the items consumed and not consumed are related in Tables 16.6 and 16.7, respectively. To quantify total phenols, we used 78 samples of the 17 species consumed (SD = 0.09, min = 0.55, mean = 0.78, max = 0.93) and 27 samples of not consumed items of 6 species (mean = 0.82,

SD = 0.08, min = 0.67, max. = 0.94). To quantify tannins, a single sample of each species was used.

Tab. 16.6: Total phenol and tannin content in consumed items of plant species by the study group in Cantareira State Park, SP, Brazil.

Species	Consumed items	Total phenols (% dry weight)	Total tanins (% dry weight)
Machaerium nictitans	YL	0.78	2.3
Ficus insipida	YL	0.86	13.0
Sorocca bonplandii	YL	0.58	2.1
Croton üoribundus	YL	0.83	3.8
Alchomea glandulosa	YL	0.85	16.8
Urerea nitida	YL	0.83	2.1
Impatiens balsanodna	YL	0.70	0.4
Anthurium sp	YL	0.69	1.7
Luehea candicans	YL	0.73	1.3
Luehea candicans	FL	0.90	14.3
Jacaranda puberula	FL	0.94	
Guapira opposita	YL	0.84	1.3
Rhipsalis sp.	starks	0.80	0.4
Mollinedia schottiana	YL	0.82	2.5
Cupania oblongifolia	FL	0.79	50.4
Cupania oblongifolia	FR	0.84	6.7
Alchomea triplinervia	FR	0.87	-
Allophylus edulis	FR	0.83	-
Cestrum sp.	FR	0.91	8.0

Tab. 16.7: Total phenol and tannin content in not consumed items of plant species by study group in Cantareira State Park, SP, Brazil.

Species	Not consumed items	Total phenols (% dry weight)	Total tanins (% dry weight)
Machaerium nictitans	ML	0.80	-
Sorocea bonplandii	ML	0.78	4.6
Croton floribundus	ML	0.83	5.0
Alchornea glandulosa	ML	0.85	37.8
Luehea candicans	ML	0.93	13.0
Solanum sp.	FR	0.73	1.3

The phenol amount of consumed items and not consumed items did not differ significantly ($t = -1.97$, $gl = 103$, $p = 0.05$). Quantitative tannin data showed a greater variation compared to total phenols. The analyzes showed that for some species, the amount of tannins (% dry weight) in mature leaves was not consumed in comparatively larger amounts than the consumed new leaves (Tab. 16.6 and 16.7). For example, for the *S. bonplandii* species, young leaves presented 2.1% of tannins and mature leaves 4.6%; young leaves of *C. floribundus* presented 3.8% and mature leaves 5.0%; young leaves of *A. glandulosa*, presented 16.8% and mature leaves 37.8%; young leaves of *L. candicans* presented 1.3% and mature leaves 13.0% - the flowers of this species were widely consumed and presented 14.3% of tannins.

Although some species show quantitative differences in their young and mature leaves, interspecies comparisons for levels of these compounds in items consumed were similar to those obtained for items that are not consumed, for example, the young leaves of *F. insipida* presented the same amount of tannins than not consumed mature leaves of *L. candicans* (13.0%). The fruits of *Solanum* sp. that were not consumed by howlers had a tannin content (1.3%) identical to the young leaves of *L. candicans* and *Guapira opposita*, which is small compared to the amount of tannins from other widely consumed items, such as young leaves of *F. insipida* (13.0%) and *A. glandulosa* (16.8%). The amount of tannins found in the flowers of *Cupania oblongifolia* (50.4 %) was very high, and a likely explanation for this is that the test used may have detected flavonoids which are usually abundant compounds in flowers and chemically very similar with tannins.

16.6 Discussion

In general, the pattern of group activities were similar to those obtained in other studies on the same species and with other species of the same genus, *A. palliata, A. seniculus, A. caraya, A. belzebul* (Tab. 16.8). The principal activity was resting, and when comparing the results obtained, it can be observed that the pattern of temporal allocation for resting is consistent for different populations in different areas. Feeding activity was the second greatest activity of the study group, and comparing this result with others obtained for the same species, we note that in this study the time allocation for feeding activity was higher. Among the studies of *A. fusca*, Martins (1997) observed that moving (13.4%) was higher than feeding (11.6%). Generally, group time allocation for moving was less (13.8%) and similar to results obtained by Chiarello (1993) (13%), Marques (1997) (13.3%) and Martins (1997) (13.4%). The time allocation for moving in this study has been surpassed in others studies, 18.2% (Oliveira & Ades, 1993) and 16.3% (Gaspar, 1997). Mendes (1989) obtained 11.0% for this activity.

Tab. 16.8: Activity patterns for several species of the genus *Alouatta* (the social interaction category was unrelated; R = resting; F = feeding; M = moving).

Species	Activities (% time allocation)			References
	R	F	M	
A. fusca	60.0	2.4	13.8	this study
A. fusca	71.8	17.3	11.0	Mendes, 1989
A. fusca	63.9	18.0	13.0	Chiarello, 1993
A. fusca	59.0	19.0	15.0	Perez, 1998
A. fusca	59.5	18.9	18.2	Oliveira & Ades, 1993
A. fusca	57.6	13.9	13.3	Marques, 1996
A. fusca	73.8	11.6	13.4	Martins, 1997
A. fusca	63.7	16.2	16.3	Gaspar, 1997
A. fusca (= A. clamitans)	45.2	28.0	21.7	Oliveira, 2011
A. fusca	62.94	17.05	15.51	Steinmetz, 2000
A. caraya	61.6	15.6	17.6	Bicca-Marques, 1991
A. caraya	77.4	15.2	2.4	Zunino, 1986
A. belzebul	58.7	20.0	18.2	Pinto, 2002
A. belzebul	59.1	10.0	19.1	Bonvicino,1989
A. seniculus	78.5	12.7	5.6	Gaulin & Gaulin, 1986
A. seniculus	61.7	21.8	16.6	Braza et al., 1981
A. palliata	63.7	24.72	9.7	Azkarate & Arroyo- Rodriguez, 2007
A. palliata	65.5	16.2	13.4	Milton, 1980

According to Milton (1980), animals move exclusively for feeding and coincide with environments that offer higher fruit resources. Fruits represent more energy resources when compared to leaves and they are generally more heterogeneously grouped (Mendes, 1989). In the Atlantic Forest and the mesophilic forest, fruit availability is seasonal (Develey & Peres, 2000). Thus, seasonal fruit should be available in the study area, which may explain the higher percentage of moving. Social interactions showed the lowest time allocation (4.3 %), characterizing the howlers as less socially active primates (Jones, 1979). with Vocalizations were very common and usually occurred when neighboring groups approached or when they were exploring the same food resource. In the absence of intergroup encounters, howler monkeys vocalize mainly in the morning and at the end of the day.

Playing was very common among the younger animals, occurring most often when adults were at rest. The participation of young individuals in play activity should be related to learning for more efficient locomotion and socialization among members of the group. It was observed that grooming involved mainly adult females and the dominant male as the receiver. According to Jones (1979), this activity is a form of appeasement and the most important in adulthood to maintain group cohesion.

The dietary study was very similar to other studies, showing that folivory is characteristic for the genus *Alouatta* and the inclusion of seasonal resources in the diet such as fruits and flowers. Howler monkeys have a diverse diet, but with a tendency to concentrate on a few species, in addition to selecting the consumed plant parts, corroborating the results obtained with other species of the genus (Nagy & Milton, 1979; Milton et al., 1980). Consumption of leaves by the group was predominant throughout the year; the same was found in other studies (Mendes, 1989; Chiarello, 1993; Oliveira & Ades, 1993; Marques, 1996; Gaspar, 1997; Martins, 1997). Flowers and fruits in this study took second (29.3%) and third position (15.4%) in the diet. However, the results obtained by Mendes (1989) and Marques (1997) show that the consumption of fruit was higher than that of flowers. Gaspar (1997) showed the consumption of flowers and fruits was practically the same, being 9.1% and 11.7% respectively.

Generally, young leaves are rich in proteins and water compared with the mature leaves while fresh fruits are rich in energy, such as carbohydrates and lipids (Waterman *et al*, 1980). Primates are selective in diet, spending more energy searching seasonal food resources according to their availability in the forest (Alves & Guix, 1992; Bicca-Marques & Calegaro-Marques, 1994). However, in times of scarcity of energy-rich foods such as fruit, howler monkeys have other strategies to save energy: a regular pattern of daily inactivity and body postures to conserve or dissipate heat.

Studies show that feeding was predominant in the morning and late in the day (Bonvicino, 1989; Mendes, 1989; Chiarello, 1993; Oliveira & Ades, 1993; Bicca- Marques & Calegaro -Marques, 1994; Marques, 1996; Gaspar, 1997; Martins, 1997). *A. seniculus* (Gaulin & Gaulin, 1982), used to eat fruits in the morning, probably because fruits provide energy and are easily digested, and they quickly restore energy demands after an overnight rest period. Leaves are rich in proteins, but digest more slowly, and are thus more consumed before the resting period.

The results presented in this study and in others, confirm that the howler monkeys are less active because diet is basically folivore and leaves are items that have low amounts of nonstructural carbohydrate (Milton, 1981) and therefore do not provide sufficient energy to immediate daily needs. While others atelides – *Ateles, Brachyteles* and *Lagothrix* - are considered predominantly frugivorous and present a strategy to maximize energy investment, the howler saves or conserves energy through inactivity (Strier, 1992). Howler monkeys are post-gastric fermenters presenting cecum and extended intestinal colon regions of the digestive tract where bacterial fermentation of structural carbohydrates occurs, providing part of the energy (26-36%) for metabolism (Milton, 1981). The morphological adaptations of howler monkeys do not exhibit the same characteristics and magnitude of specializations presented by colobineos, for example, the stomach of which is divided into chambers. Probably for this reason, the howler avoids the intake of mature leaves, which are generally rich in fiber (structural carbohydrates) and select food sources that provide energy balance while minimizing the cost of searching for food (Milton, 1979).

The phenolic compounds especially the tannins are considered an important plant defense against herbivores since they are capable of forming complexes that precipitate proteins into insoluble and indigestible proteins. (Rhoades & Cates, 1976). Many studies with *Colobus* sp. and *Presbytis* sp. (Oates et al., 1980; McKey et al., 1981; Davies et al., 1988) showed that, generally, these primates select young leaves by water and protein content. However, *C. badius* eventually ingested mature leaves with a lot of tannin. One possible reason for this was that these compounds are not always harmful so they can help in the detoxification of alkaloids by precipitating them. The influence of tannins on the food selectivity of primates does not seem very consistent. These substances should influence food choice because they have a strong astringency in the taste of food. But it was observed that they are unable to hold the choice of a food item, since the effect depends on the type and concentration of tannins (Waterman, 1984). However, McKey et al. (1981) observed *C. satanas* consuming considerable amounts of *Rauvolfia vomitoria* leaves containing toxic alkaloids. These toxins do not seem to present obstacles for food. Hladik (1978) noted that the species *P. entellus* primates fed on fruits of *Strychnos potatorum* that are also rich in alkaloids and suggested that bacterial flora present in the stomach of this species has the ability to detoxify these compounds. According to Oates (1978), detoxification occurs through the absorption of some toxins and probably geophagy as this is often observed in some primates. Chemical analysis showed that the most important factor in feeding selectivity of *Alouatta palliata* (Glander 1982) was the protein content of plants consumed, however, the animals chose less astringent plants, with low concentrations of tannins. Romio (2012), in a study with *A. fusca*, showed statistically that the quantities of essential oils present in consumed and unconsumed items were not different, despite an apparent preference for vegetables that contained less amounts of these substances. The essential oils can be toxic (also depending on the amount ingested) and have influences on digestion while possessing medicinal properties like protection against parasites, thus reducing the negative effects caused by the ingestion of other secondary compounds and antimicrobial action.

The results of the chemical analyses of this study showed that there is a tendency of *Alouatta fusca* in selecting items with low levels of total phenolic, but the results showed no significant differences in total phenolic content between food and non-food items. On the other hand, there was a substantial variation in the amount of tannins between food (0.4-50.4%) and non-food items (1.3-37.8%). Results obtained in this study indicate that levels of tannins seem to have little or no influence in the feeding selectivity of this population of howler monkeys. Secondary compounds are not the only factors involved in the selectivity of howler monkeys. The digestibility of plants and the nutrient content should also be considered. There are many factors that interact to increase or decrease the "value" of each item in the diet of primates. While certain items contain compounds that decrease digestion efficiency, these items can be rich in protein and energy. Folivory is associated with relatively small allocation areas for individual arboreal primates, since they have small areas of use (home range) relative to body size compared to frugivorous primates and omnivores (Milton & May, 1976). Many studies indicate that some of the adaptations of howler habitat fragmentation can be caused by increased population density and consequently a reduction in the home range. For the howler monkeys, the home range may vary from 4 to 20 ha (Milton & May, 1976; Milton, 1981). In studies with *A. fusca*, in Southeastern Brazil, similar sizes were obtained as showed in Table 16.9.

Tab. 16.9: Group size and home range of howler monkeys, *Alouatta fusca*, in Southeastern Brazil.

Location	Group size (number of individuals)	Home range (hectare)	Reference
Santa Genebra, Campinas, SP	6	4.1	Chiarello, 1993
Mata Ribeirão da Cachoeira, Campinas, SP	8	8.5	Gaspar, 1997
Lençóis Paulista, SP	3	12.5	Martins, 1997
Caratinga, MG	7	7.9	Mendes, 1989
Cantareira State Park, SP	8	4.5	this study

In a study of howler monkeys on an island at Los Tuxtlas, Mexico, Rodriguez-Luna et al. (2003), observed an increase in the time allocated for the consumption of leaves and lianas, after an increase in population density in this area. High densities of howler monkeys are more common in fragmented areas than continuous areas and, after isolation, forest fragments may serve as a refuge for primates. The main threat to the conservation of primates is the loss, alteration and fragmentation of habitat in which they live (Cowlishaw & Dunbar, 2000). Forest fragmentation is associated with a significant reduction in the size and quality of habitats (Marsh, 2003; Arroyo-Rodriguez & Mandujano, 2006).

However, the persistence of primates in fragmented areas depends on the ability to adapt their diets and their activity patterns to new modified habitats (Bicca-Marques, 2003). The success of primates remaining in fragmented habitats is related to the ability to adapt feeding to the dietary availability of species in the environment (Silver & Marsh, 2003); to increase the amount of leaves ingested in the diet (Rodriguez-Luna et al. (2003); to frequently consume exotic and secondary vegetation species, such as lianas and vines, in disturbed habitats (Chiarello, 2003; Rodriguez-Luna et al., 2003; Azkarate & Arroyo-Rodríguez, 2007); to have small home ranges (Neves & Rylands, 1991; Estrada & Coates-Estrada, 1996); and to minimize energy expenditure by adjusting the default activities (Milton, 1980; Strier, 1992; Silver & Marsh, 2003). Forest fragments play an important role in the conservation of howler monkeys since these primates do not reproduce easily in captivity, thus the zoological gardens should deploy small forested areas with plant species that could serve to nurture the endangered howler species, and hence subjects for management programs. Plant species in the diet of howler monkeys could serve as corridors between isolated forest fragments (Bicca-Marques & Calegaro-Marques, 1994).

Bibliography

Altmann, J. (1974). Observational study of behavior: Sampling Methods. Behavior, 49, 227- 267.

Alves, I.M.S.C., & Guix, J.C. (1992). Feeding habits of Alouatta caraya in a semi natural area (SE Brazil). Mammalia, 56 (3), 469-472.

Arroyo-Rodrıguez, V. & Mandujano, S. (2006). Forest fragmentation modifies habitat quality for *Alouatta palliata*. International Journal of Primatology, 27, 1079–1096.

Assumpção, C.T. (1983). Na Ecological Study of the Primates on Southeastern Brazil,with a Reppraisal of *Cebus apella* races. Edinburgh: University of Edinburgh.

Azkarate, J.C., & Arroyo-Rodríguez,V. (2007). Diet and activity pattern of howler monkeys (*Alouatta palliata*) in Los Tuxtlas, México: Effects of habitat fragmentation and implications for conservation. American Journal of Primatology, 69, 1013–1029.

Bauchop, T. (1978). Digestion of leaves in vertebrate arboreal folivores. In G.G. Montgomery (Ed.), The ecology of arboreal folivores. Washington, D.C.: Smithsonian Institution Press.

Bicca-Marques, J.C. (1991). Ecologia e comportamento de um grupo de bugios-pretos Alouatta caraya (Primates, Cebidae) em Alegrete, RS, Brasil. Brasil: University of Brasilia.

Bicca-Marques, J. C. (2003). How do howler monkeys cope with habitat fragmentation? In L.K. Marsh (Ed.), Primates in Fragments: ecology and conservation (p.283–303). New York: Kluwer Academic Plenum Publishers.

Bicca-Marques, J.C., & Calegaro-Marques, C. (1994). Activity budget and diet of *Alouatta caraya*: an age-sex analysis. Folia Primatologica, 63, 216-220.

Bonvicino, C.R. (1989). Ecologia e comportamento de *Alouatta belzebul* (Primates: Cebidae) na Mata Atlântica. Revista Nordestina de Biologia, 6 (2), 149-179.

Brandão, L.D. (1999). Distribuição altitudinal e ambiente preferencial de Callithrix aurita Humboldt, 1812 (Primates: Callitrichidae) na Estação Ecológica de Bananal, Serra da Bocaina, São Paulo. Livro de Resumos do IX Congresso Brasileiro de Primatologia, Santa Teresa, ES, Brasil.

Braza,F., Alvarez, F., et al.(1981). Behaviour of the red howler monkey (*Alouatta seniculus*) in the Llanos of Venezuela. Primates, 22, 459-473.

Braza,F., Alvarez, F., et al. (1983). Feeding habitats of the red howler monkeys (*Alouatta seniculus*) in the Llanos of Venezuela. Mammalia, 47 (2), 205-214.

Chiarello, A.G., (1993). Activity pattern of the brown howler monkey *Alouatta fusca*, Geoffroy 1812, in a forest fragment of southeastern Brazil. Primates, 34 (3), 289-293.

Chiarello, A.G. (2003). Primates of the Brazilian Atlantic Forest: the influence of forest fragmentation on survival. In L.K. Marsh (Ed.), Primates in Fragments: ecology and conservation (p.99-121). New York: Kluwer Academic Plenum Publishers.

Chivers, D.J., & Hladik, C.M. (1980). Morphology of the gastrointestinal tract in Primates: Comparisons with other mammals in relation to diet. Journal of Morphology, 166, 337-386.

Cowlishaw, G., & Dunbar, R.(2000). Primate conservation biology. Chicago: The University of Chicago Press.

Crissay, S.D., & Oftedal, O.T., et al. (1990). Gastro-intestinal tract capacity, food passage rates and the possible role of fiber in diets fed to captive Red Howler Monkeys (*Alouatta seniculus*) in Venezuela. Proceedings American Association of Zoo Veterinarians, 81-86.

Davies, A.G., & Bennett, E., et al. (1988). Food selection by two South-east Asian colobine monkeys (*Presbytis rubicunda* and *Presbytis melalophos*) in relation to plant chemistry. Biological Journal of the Linnean Society, 34, 33-56.

Develey, P.F., & Peres, C.A. (2000). Resource seasonally and the structure of mixed species bird flocks in a coastal Atlantic forest of southeastern Brazil. Journal of Tropical Ecology, 16, 33-53.

Estrada, A. (1984). Resource Use by Howler Monkeys (Alouattapalliata) in the Rain Forest of Los Tuxtlas, Veracruz, México. International Journal of Primatology, 5 (2), 105- 131.

Estrada, A., & Coates-Estrada, R.(1984). Fruit eating and seed dispersal by howling monkeys (*Alouatta palliata*) in the tropical rain forest of Los Tuxlas, Mexico. American Journal of Primatology, 6, 77-91.

Estrada, A., & Coates-Estrada, R.(1996). Tropical rainforest fragmentation and wild populations of primates at Los Tuxtlas, Mexico. International Journal of Primatology, 17, 759-783.

Ford, S.M., & Davis, L.C.S. (1992). Systematics and Body Size: Implications for feeding adaptations in New World Monkeys. American Journal of Physiology Antropology, 88, 415-468.

Freland, W.J., & Janzen, D. (1974). Strategies in herbivory by mammals: The role of plant secondary compounds. The American Naturalist, 108 (961), 269-289.

Gartlan, J.S., & McKey, D.B., et al (1980). A comparative study of the phytochemistry of two African forests. Biological Systematic Ecology, 8, 401-422.

Galleti, M., & Chiarello, A.G. (1994). Conservation of the brown howler monkey in southeastern Brazil. Oryx, 28, 37-42.

Galetti, M., & Pedroni, F., et al. (1994). Diet of the howler monkey *Alouatta fusca* in a forest fragment in southeastern Brazil. Mammalia, 58 (1), 111-118.

Gaspar, D.A. (1997). Ecologia e comportamento do bugio ruivo, *Alouatta fusca* (Geoffroy,1812, Primates: Cebidae), em um fragmento de Mata de Campinas. Campinas: Universidade Estadual de Campinas.

Gaulin, S.J.C, & Gaulin, C.K. (1982). Behavioral ecology of *Alouatta seniculus* in Andean cloud forest. International Journal of Primatology, 3, 1-32.

Glander, K.E. (1982). The impac t of secondary compounds on Primates feeding behavior. Yearbook of Physical Antropology, 25, 1-18.

Hladik, C.M., (1978). Adaptive Strategies of Primates in Relation to Leaf-eating. In G.G. Montgomery (Ed.), The Ecology of Arboreal Folivores. Washington, D.C.: Smithsonian Institute Press.

Hill, W.O.C. (1953). Primates: comparative anatomy and taxonomy. University Press.

Hirsch, A., Landau, E.C., Tedeschi, A.C.M., et al. (1991). Estudo comparativo das espécies do gênero *Alouatta* Lácèpede, 1799 (Platyrrhini, Atelidae) e sua distribuição geográfica na América do Sul. In A.B. Rylands & A.T. Bernardes (Eds.), A Primatologia no Brasil (p.239-263). Belo Horizonte: Fundação Biodiversitas e Sociedade Brasileira de Primatologia.

Huek, V. (1972). Florestas da América do Sul. Brasilia: Universidade de Brasília.

Jones, C.B. (1979). Grooming in the mantled howler monkey, *Alouatta palliata* Gray. Primates, 20 (2), 289-292.

Julliot, C., & Sabatier, D. (1993). Diet of Red Howler Monkey (*Alouatta seniculus*) in French Guiana. International Journal of Primatology, 14 (4), 527-550.

Kinzey, W.G. (1997). New World Primates: Ecology, Evolution and Behavior. New York: Aldine Transactions.

Köppen, W. (1948). Climatologia- com um estudio de los climas de la tierra. México, Fondo de cultura económica.

Leitão- Filho, H.F. (1995). A vegetação da Reserva de Santa Genebra. In H.F. Leitão-Filho & L.P.C. Morellato (Eds.), Ecologia e Preservação de uma Floresta Tropical urbana: Reserva de Santa Genebra.Campinas, SP: Editora da Universidade Estadual de Campinas.

Mann, J. (1994). Chemical Aspects of Bhiosynthesis. Oxford: Oxford University Press.

Marques, A.A.B. (1996). A espécie *Alouatta fusca clamitans* (Cabrera, 1940) (Primates,Cebidae) na Estação Ecológica de Aracuri, RS:Variações sazonais de forrageamento. Porto Alegre: Pontifícia Universidade Católica do Rio Grande do Sul.

Marsh, L.K. (2003).The nature of fragamentation. In L.K. Marsh (Ed.), Primates in fragments: ecology and conservation (p.1-10). New York: Kluwer Academics/Plenum Publishers.

Martins, C.S. (1997). Uso de habitat pelo bugio, *Alouatta fusca clamitans*, em um fragmento florestal em Lençóis Paulista- S.P. Campinas: Universidade Estadual de Campinas.

Marinho-Filho, J. (1992). Os mamíferos da Serra do Japi. História Natural da Serra do Japi: ecologia e preservação de uma área florestal no sudeste do Brasil. Campinas: Editora da Universidade Estadual de Campinas, Campinas.

McNab, B.K. (1978). Energetics of arboreal folivores: physiological problems and ecological consequences of feeding on a ubiquitous food supply. In G.G. Montgomery (Ed.), The ecology of arboreal folivores. Washington, D.C.: Smithsonian Institution Press.

McKey, D., & Waterman, P.G., et al. (1978). Phenolic content of vegetation in two African rain forests: ecological implications. Science, 202, 63-66.

McKey, D.B., & Gartlan, J.S., et al. (1981). Food selection by black colobus monkeys (*Colobus satanas*) in relation to plant chemistry. Biological Journal of the Linnean Society,16, 115-146.

Mendes, S.L. (1989). Estudo Ecológico de *Alouatta fusca* (Primates, Cebidae) na Estação Ecológica de Caratinga, M.G. Revista Nordestina de Biologia, 6 (2), 71-104.

Milton, K. (1979). Factors influencing leaf choice by howler monkeys: a test of some hypotheses of food selection by generalist herbivores. The American Naturalist, (114), 362-378.

Milton, K. (1980). The foraging strategy of Howler Monkeys. A study in primates economics. New York: Columbia University Press.

Milton, K. (1981). Food choice and digestive strategies of two sympatric primate species. The American Naturalist, 117, 496-505.

Milton, K. (1984). The role of Food-Processing factors in Primate Food Choice. In P.S. Rodman & J.G.H. Cant (Eds.), Adaptations for Foraging in Non-human Primates (p.249-279). New York: Columbia University Press.

Milton, K., & McBee, R.M. (1983). Rates of fermentative digestion in the Howler Monkey, *Alouatta palliata* (Primates: Cebidae). Comparative Biochemical Physiology, 74 (1), 29-31.

Milton, K., & Soest, P.J., et al. (1980). Digestive efficiencies of wild Howler Monkeys. Physiology Zoological, 53 (4), 402-409.

Mowry, C.B. & Decker, B.S., et al. (1996). The role of phytochemistry in Dietary choices of Tana River Red Colobus Monkeys (*Procolobus badius rufomitratus*). International Journal of Primatology, 17 (l), 63-83.

Miranda, J.M.D., & Passos, F.C. (2004). Feeding habits of the *Alouatta guariba* (Humboldt) (Primates, Atelidae) on a Araucaria Pine Forest, Paraná, Brazil. Revista Brasileira de Zoologia, 21 (4).

Nagy, K.A., & Milton, K. (1979). Energy metabolism and food consumption by wild (*Alouatta palliata*). Ecology, 60 (3), 475-480.

Neves, A.M.S., & Rylands, A.B. (1991). Diet of a group of howler monkeys, Alouatta seniculus, in an isolated forest patch in Central Amazonia. In A.B. Rylands & A.T. Bernardes (Eds.), A primatologia no Brasil – 3 (p.263-274). Belo Horizonte: Fundação Biodiversitas para a Conservação da Diversidade Biológica.

Oates, J.F. (1978). Water-plant and soil consumption by guereza monkeys (*Colobus guereza*): A relationship with minerals and toxins in the diet. Biotropica, 10, 241-253.

Oates, J., & Waterman, P.G., et al. (1980). Food selection by the South Indian Leaf-Monkey, *Presbytis johnii*, in relation to leaf chemistry. Oecologia, 45, 45-56.

Oliveira, A.B.M. (2011). Comportamento de bugios (*Alouatta clamitans*, Cabrera,1940) na Ilha Grande, RJ. Rio de Janeiro: Universidade do Estado do Rio de Janeiro.

Oliveira, D.A., & Ades, C. (1993). Aspectos do comportamento do bugio *Alouatta fusca* (Primates, Cebidae) no Parque Estadual da Cantareira (São Paulo). Revista do Instituto Florestal, 5 (2), 163-174.

Parra, R. (1978). Comparison of foregut and hindgut fermentation in herbivores. In G.G. Montgomery (Ed.). The ecology of arboreal folivores. Washington, D.C.: Smithsonian Institution Press.

Perez, D.M. (1998). Estudo ecológico do bugio-ruivo em uma floresta de Araucária do sul do Brasil (*Alouatta fusca*, Ihering, 1914 – Primates, Atelidae). São Paulo: Universidade de São Paulo.

Pinto, L.P. (2002). Dieta, padrão de atividades e área de vida de *Alouatta belzebul discolor* (Primates, Atelides) no sul da Amazônia. Campinas: Universidade Estadual de Campinas.

Prates, J.C., & Gayer, S.M.P., et al., (1990). Feeding habits of the Brown Howler Monkey *Alouatta fusca clamitans* (Cabrera,1940) (Cebidae, Alouattinae) in the Itapuã State Park: A preliminary report. Acta Biologica Leopoldensia, 12 (1), 175-188.

Queiroz, H.L. (1995). Preguiças e guaribas - Os mamíferos folivoros arboricolas do Mamirauá. Tefé, AM: Sociedade Civil Mamirauá.

Robbers, J.E., & Speedie, M.K. et al. (1997). Farmacognosia e Farmacobiotecnologia. São Paulo: Premier.

Rodriguez-Luna, E., Dominguez-Dominguez, L.E., Morales, M.J., et al., (2003). Foraging strategy changes in *Alouatta palliata mexicana* troop released on an island. In L.K. Marsh (Ed.), Primates in fragments: ecology and conservation (p.229-250). New York: Kluwer Academics/Plenum Publishers.

Rosenberger, A.L., & Strier, K.B. (1992). Adaptative radiation of the Ateline primates. Journal of Human Evolution, 18, 717-750.

Rhoades, D.F., & Cates, R.G. (1976). Toward a general theory of plant antiherbivore chemistry. Recent Advances in Phytochemistry, l0, 168-213. New York: Plenum Press.

Romio, C. (2012). Influência da Composição Química de Óleos Essenciais de Folhas na Seleção da Dieta de Bugios-Ruivos (*Alouatta guariba clamitans*). Revista da Graduação, 5 (2), http://revistaseletronicas.pucrs.br. Accessed on 04/05/2014.

Schoener, T.W. (1971). Theory of Feeding Strategies. Annual Review on Ecology, Evolution and Systematics, 2, 369-404.

Silva, E.C. (1981). A Preliminary Survey of Brown Howler Monkeys (*Alouatta fusca*) at the Cantareira Reserve (São Paulo, Brazil). Revista Brasileira de Biologia, 41 (4), 897-909.

Silver, S.C., & Marsh, L.K. (2003). Dietary flexibility behavioral plasticity, and survival in fragments: lessons from translocated howlers. In L.K. Marsh (Ed.), Primates in fragments: ecology and conservation (p.251-265). New York: Kluwer Academics/Plenum Publishers.

Smith, C.C. (1977). Feeding behaviour and social organization in the howling monkeys. In T.H. Clutton-Brock (Ed.), Primates ecology: studies of feeding and ranging behaviour inlemurs, monkeys and apes (p.97-126). London: Academic Press.

Steinmetz, S. (2000). Ecologia e comportamento do bugio (*Alouatta fusca clamitans*, Atelidae-Primates) no Parque Estadual Intervales - SP. São Paulo: Universidade de São Paulo.

Stoner, K.E. (1996). Habitat Selection and Seasonal Patterns Activity and Foraging of Mantled Howling Monkeys (*Alouatta palliata*) in Northeastern Costa Rica. International Journal of Primatology, 17 (1), 1-29.

Strier, K.B. (1992). Atelinae adaptations : behavioral and ecological constraints. American Journal of Physical Anthropology, 88, 515-524.

Tabarelli, M. (1994). Caracterização florística e estrutural de um trecho de floresta secundária e avaliação do papel das clareiras naturais na manutenção da riqueza e diversidade de espécies arbóreo-arbustiva e na dinâmica funcional da floresta. São Paulo: Universidade de São Paulo.

Tognon, F.R (. 2007). Comportamento alimentar de um grupo de bugios-pretos (*Alouatta caraya*, Humboldt,1812-Primates, Atelidae) em um fragmento de mata no Município de Jardinópolis-SP. Faculdade de Filosofia, Ciências e Letras de Ribeirão Preto-USP, Ribeirão Preto.

Waterman, P.G. (1984). Food acquisition and processing as function of plant chemistry. In D.J. Chivers, B.A. Wood & A. Bilsborough (Eds.), Food Acquisition and processing in primates (p.177-211). London: Blackwell Scientific Publications.

Waterman, P.G., & Mbi, C.N., et al. (1980). African rainforest vegetation and rumen microbes: phenolic compounds and nutrients as correlates of digestibility. Oecologia, 47, 22-33.

Waterman, P.G., & Mole, S. (1994). Analysis of Phenolic Plant Metabolites. In J.H. Lawton & G.E. Likens (Eds.) Methods in Ecology (p.66-103). Oxford, UK: Blackwell Scientific Publications.

Whittaker, P.G., & Fenny, P.P. (1971). Allelochemics: Chemical interations between species. Science, 171 (3973), 757-770.

Zar, J.H. (1999). Biostatistical Analysis. New Jersey: Prentice Hall Inc.

Zunino, G.E. (1986). Alguns aspectos de la ecologia y etologia del mono anullador negro (*Alouatta caraya*) em habitats fragmentados. Buenos Aires: Universidad de Buenos Aires.

Thiago P. de C. e Timo, Júlio César Costa, Carla Gheler-Costa, Maria Carolina Lyra-Jorge, Mariana R. Urbano, Luciano M. Verdade

17 Wildlife and Landscape Geometry in Silvicultural Landscapes

Abstract: Recent improvements in personal computers processing speed and capacity have led to the development of novel technics on landscape analysis and to improvements in statistical procedures which encompass intensive computational routines, some of those suggested long ago, but practically impossible to be implemented until present. In addition, satellite imagery has showed exponential increase in resolution, diversity of products and their availability making it feasible for a large number of researchers and students to address issues of conservation and management at wider spatial scales. These changes have shed light on questions as crucial as the dichotomy of pattern and process in ecology, and risen questions such as the appropriate scale in which patterns should be accessed to adequately illustrate the relationship among landscape patterns and biotic (taxonomic and functional) groups as part of the growing effort to develop knowledge that may form a solid basis for conservation and management planning and practice. Some of the more prominent and largely used landscape analysis technics are based on landscape metrics, which, in turn, encompass landscape geometry metrics and indices. In this chapter, we present an overview of landscape pattern evaluation and the influence of those accessible metrics on the estimation of wildlife patterns of abundance and distribution. We make use of an example where landscape geometry patterns are used as background for the distribution and abundance of mammals (large to medium size, ground dwelling) in Southwestern São Paulo State, Brazil, to address issues of theory and practice in the field of wildlife conservation research. Finally, we point to uncertainties in the concepts of habitat as human perceived patterns and the relevance of such metrics on the delimitation of management practices and public policies for the maintenance of wildlife diversity and functionality.

17.1 Geometry, Ecology and Landscape as Study Frame

17.1.1 Brief History of the Abstraction of Space or Bending the Cartesian Plane

Mankind abstractions on the characteristics of features and phenomena inherent to the space where we dwell are as ancient as the capability of rational thought itself. Greek geometricians inherited concepts and tools from Babylonians and Egyptians, who developed calculations for the description of space based on needs of land partitioning and taxation (Mlodinow, 2010). Early before them, a piece of bone with notches carved on its sides may have been a kind of ancient calculator used by those who lived at the margins of Lake Edward in what we know now as the Democratic Republic of Congo. This artifact, ca. 8,000 years old, found by archaeologists (De Heinzelin, 1962), may well represent the efforts of primitive human beings on quantifying and abstracting what is achieved by direct perception and transforming it into a deeper kind of cognition, the model.

From the *harpedonopta*, the rope stretchers who measured terrains to Egyptian architects, to nowadays topographers and further, astronomers and physicists, a long road was travelled, and not only in the forward direction. One of the most important steps ahead was given by Aristotle, as he perceived Earth's surface curvature by watching vessels vanishing on the horizon (Mlodinow, 2010; Guthrie, 2012). Although his perception was as early as 24 centuries ago, descriptions of space and shape were restricted to what is known as Euclidean, for most of western civilization history.

Thales of Miletus was the first to prove geometric theorems, in a first effort towards its systematization, around 2,500 B.C. (Guthrie, 1962, 2012). His work was the basis for Pythagoras and his followers, who took mathematics to a spiritual level, but contributed with the homonymous theorem (Guthrie, 1967). Later, (around 300 B.C.), Euclid compiled Ancient-Greek knowledge in geometry in the work known worldwide and exhaustively studied *Elements* (Heath, 1909; Guthrie, 1969). His was the first approach to the nature of two dimensional spaces with no account to the physical world, a premise for the abstraction in space description (Mlodinow, 2010).

René Descartes despised the Ancient-Greek's system of proofs and by consequence, translated geometry into algebra (Molland, 1995). The importance of the above cited abstraction is maximized by Descartes, as his algebraic descriptions of space and shape result, in first instance, in the plane representation where we place our x,y pairs. The underlying usefulness of geometry's algebraic description is the ability to relate features to each other, and in this sense, to compare different lengths, areas and shapes (Mlodinow, 2010), with obvious advantages for cartography and, more recently, landscape ecology.

It was Johan Carl Friedrich Gauss who finally broke with the "carved in stone" statements of traditional geometry. By recognizing that the sum of the inner angles of a triangle cannot result in 180° he established a new paradigm in science, the so called *hyperbolic geometry* (Dunnington, 2004; Mlodinow, 2010). It may sound strange and even unsettling that a triangle doesn't have the expected 180° in it, but, this is consistent with the reality perceived by Aristotle during his contemplative reflections on a harbor. The Euclidean postulate here discussed only holds for very small spaces, or large absolute scales (in this sense, the representation unity divided by the unity of interest).

The basis is settled and, then, Jules Henri Poincaré takes Gauss's abstractions to the next level of cognition, he models them (Greenberg, 1993). Poincaré's model of hyperbolic space and the work of Bernhard Riemann, a prominent Gauss student, on differential geometry contributed significantly to a new cornerstone of science,

Einstein's general theory of relativity (Mlodinow, 2010). Albert Eintein's contribution to physics and geometry incited a boiling effervescence to exact sciences and brought uncertainty, non-linearity and multiple dimensions to the front pages, thus creating the proper environment for the development and spread of discussion on criticality and self-affinity, basic concepts of fractals and Chaos Theory.

Two mathematicians, contemporary to Poincaré, worked on geometric abstractions that deeply influenced science in the most unexpected fields. Georg F. L. Phillip Cantor and Niels Fabian Helge von Koch envisaged complex structures with consistent basic properties in any scale they were observed (Gleick, 2011; Falconer, 2013). The structures are the Cantor Set and the Koch Curve. These figures present self-affinity, their structures self-replicates as they are extended (Voss, 1989; Mandelbrot, 1999, 2002).

The works of Cantor and Koch had a profound influence on two 20th century scientists in particular, Edward Lorenz and Benoit Mandelbrot, a meteorologist and a mathematician (Gleick, 2011). Lorenz developed graphical sets resultant from his attractors, pendulums attracted by magnets used to predict the behavior of weather conditions and coined the term "Butterfly Effect", based on the appearance of his graphical sets. Mandelbrot studied as many fields as is vast his contribution to post-modern science. He analyzed the variations in cotton prices, investigated game theory, applied Cantor Sets to noise in IBM data transmission and finally wrote "The Fractal Geometry of Nature" (1982) (Gleick, 2011).

Fractals are self-similar or self-affine structures that replicate themselves across every scale or nearly self-replicate in different scales (Mandelbrot, 1982). For this very particular property, fractals are widely used, and actually were developed, in the study of chaotic and noisy phenomena and fine scale complex structures (Mandelbrot, 1976, 1982, 1985, 1999, 2002). Criticality applies to situations of hierarchical non-linear relations of cause and effect in which slight alterations in one level may represent subtle large structural or organizational modifications of the next level, or not, as the interactions amongst components of a complex system are considered to result in non-predictable patterns (Bak & Bak, 1996).

Self-similarity and criticality are closely related and are both of great interest to natural sciences, such as ecology. Patterns of phenotypic characteristics, genetic frequencies, population distributions, community compositions and ecosystem functions are considered to be all levels of the same staircase (Holling, 1986, 1992; Holling & Allen, 2002). The influence of fractal geometry and Chaos Theory in many fields of science is wide and their applications range from geology (e.g. Kruhl, 2013) to astronomy (e.g. Persinger & Koren, 2013) and population (e.g. Nilsen et al., 2013), community and ecosystem ecology (e.g. Marsh & Ewers, 2013) as a few examples. Fractal Geometry and Chaos Theory impacts in Landscape Ecology are profound, and range from the development of satellite imagery instruments engineering and vehicle launching, to the study of environmental monitoring images and landscape geometric indices and metrics, as we will discuss in next sections.

17.1.2 Landscape in Ecology

17.1.2.1 Perceptions of Space and Cultural Influence

Geometry is the basic field of knowledge in the description of the physical world. As we are visual animals, every abstraction concerning to our surroundings starts with our perception of shape, perspective, length, and roughness as many other combinations of basic geometric features that we use to characterize the environment. To perform the exercise of translating it to others and making the information available through time and portable on space, perception has been modeled in different media. Useful examples for such a task are maps. Maps are as ancient as geometry itself and the practice of cartography has a long history, having been studied by Ptolemy in his book *Geographia* (Ptolemy, 1525; Marshall, 1972) written in the second century of Christian Age.

Partitioning of space, crop planning, war and trade are among the main promoters of map elaborations (Friendly, 2008). As humankind expanded their influential webs from small group settlements to cities, countries and continents, maps became more and more important to plan administration, routes and to present regional aspects of land management interests to decision makers (Sack, 1986). Along with cartography, positioning systems and instruments were developed. As an example, the combination of the sextant and compass aided in the global scale expansion of western cultural influence.

Landscape is defined by Alexander Von Humboldt as "the total character of a region of the Earth" (Zonneveld, 1995). Von Humboldt described landscapes of the New World with a heavy tone of the ideal aesthetics that dominated European cultures of his historical *momentum* (17th century), and the descriptive model's basic scheme that he adopted remained the same for the next 100 years (Antrop, 2005).

In the beginning of the 20th century, Eastern and Central European geographers approached landscape with more holistic concepts (Turner et al., 2001). Carl Troll, inspired by the advent of aerial photography and the ecosystem concept (Tansley, 1935), introduced the term landscape ecology and declared that it was "the marriage between biology and geography", to which he developed a methodological basis (Zonneveld, 1995; Antrop, 2005). Sukachef, Troll's contemporary, paralleled him but called it biogeocenology (Antrop, 2005).

The framework of landscape ecology, from the early 1940's to the end of 1950's, combined an integrated view of the human surroundings (Turner et al., 2001) with explicit accounting for sensory perception (Granö, 2003) and aesthetics. With the adoption of more deductive and quantitative research programs by geographers, landscape was rarely revisited (Antrop, 2005). Interest in landscape ecology was gradually revived in Central Europe during the 1980's and spread to North America by the same time (Turner et al., 2001).

Although satellite imagery was not a novelty, the access to their data and visualization during the early 1980's was expensive (powerful personal computers were not yet popular), time consuming (data processing and exchange needed high volumes of media) and computer intensive (there were no software for friendly interfaces widely available). The Cold War and Space Rush contributed substantially to the development of landscape ecology.

This time, there was a lag between the media and the capacity of cognition that was not a limiting factor in Troll's years. As the Cold War lost its breath, satellite imagery technology only available to military, along with cheaper, smaller and more efficient microchips, disclosed a spread of new environmental monitoring remote sensing products with finer resolution. The theoretical framework of landscape ecology by that given moment orbited the properties of middle scale resolution landscape features.

The application of comparative analyses to the geometric characteristics of landscape features evident at middle resolutions was solved long before by Gauss, and the curvature of the planet was no longer a problem in cartography, as projections could be interconverted through automated algorithms in geographical information systems, representing features from irregular surfaces in Euclidean plane graphical sets (Coppock & Rhind, 1991; Friendly & Denis, 2008; Friendly, 2008). But still, fine resolution satellite imagery made the investigation of questions, already accessed by theoretic ecology and macro ecology, such as self-affinity and self-criticality in ecosystem functions and biodiversity patterns, more prone to researchers, once their evaluation demanded hierarchical approaches and local higher quality remote sensing data that were at hand.

And there was the problem. Most theory and analytical framework relied on Euclidean measures and linear models for hypothesis testing and data comparisons, such as species-area relationships (Connor & Mccoy, 1979). The conceptual framework that included patches, matrix, corridors, permeability, edge, isolation and connectivity as the basis for fragmentation paradigm (Turner, 1989; Turner et al., 2001; Fahrig, 2003) had to invoke fractal geometry and chaos theory in order to balance the equation with ecology's theoretical advances. Multidimensional indexes of shape, such as the fractal dimension index (O'Neill et al., 1988), came to the attention of landscape ecologists and better captured finer resolution discrepancies between spatial features inside and across different extension scales. Their interpretation was not completely understood and, until presently, it is not readily absorbed by management practitioners or decision makers.

The conceptual lag that arises at this point is placed between theory and biodiversity conservation practice. Most countries, multinational corporations, commodities production and trade regulation agencies and international environmental treaties have their environmental protection premises supported by more conventional and simple aspects of a fragmentation paradigm already well-established (Primdahl & Swaffield, 2010). The widespread application of conservation unities design is an example of a political/socio-economical limited use of landscape ecology framework that is being reviewed in the light of novel findings fostered by the possibility of spatial complexity measures.

17.1.2.2 Concepts, Measures and Paradigms as Basis for the Decision Making Process

Fragmentation paradigm was influenced by concepts such as the Island Biogeography (Mcarthur & Wilson, 1967) and the Dynamic of Metapopulations (Hanski & Gilpin, 1991), which provided this landscape ecology subject with premises of relationships between area and diversity, species richness and genetic variation respectively. Habitat quality and qualitative interactions in fragmentation paradigm were supplied by discussions on habitat heterogeneity (Bazzaz, 1975) and complementation (Dunning et al., 1992).

Fragmentation has two basic meanings; it can refer to the spatial pattern or to the spatial-temporal process (Fahrig, 2003). Landscape Ecology Metrics and Indices obtained through measures of length, area and combination of both that result in shape indices, along with indices of land-use and land-cover diversity, are state-space measures of landscape features that represent snapshots of spatial characteristics based on fragmentation pattern, rather than provide immediate insights on the fragmentation process (Trzcinski et al., 1999; Lee et al., 2002; Fahrig, 2003).

Given the distinction between pattern and process, inferences on fragmentation processes are limited by the scarcity of biodiversity data at wide ranges in temporal scale. Samplings of large spatial extents, that represent different patterns and offer insights on intensity of habitat removal or loss, are used as surrogates. The exchange of time for space poses a challenge to the conduction of research and interpretation of results (Jelinski & Wu, 1996). Aiming to answer questions of management pertained by landscape design needs, that fulfill legal environmental regulations and private environmental certification agendas, decision makers rely on studies conducted at fine resolution, but small extent or, large extent with coarse resolution (Li & Wu, 2004; Li et al., 2005).

Therefore, most evaluations of the impact of land-cover and land-use change on patterns of biodiversity resulting from private enterprises are deficient in the regard of basic landscape ecology paradigm premises (Atauri & De Lucio, 2001). In addition, when large areas are sampled, qualitative differences of habitat cover composition and species intrinsic or community variations may pose bias to the overall evaluation results (Nielsen et al., 2007). Biases of such nature must be explicitly taken into account and discussed in order to avoid misleading predictions and decisions of management based on spurious model results.

The State of São Paulo is the economically and industrially most developed state of Brazil and holds only 13.94% of its total surface occupied by reminiscent native vegetation (Secretaria do Meio Ambiente, 2005). Legal regulation applied to such areas is the New Brazilian Forest Code (Novo Código Florestal - Lei Nº 12.651, 25th May 2012), a federal law that dictates, among other resolutions, proportions of native vegetation that must be protected by landowners. The great majority of the reminiscent native vegetation in the State is located in private areas and is submitted to CETESB (Companhia de Tecnologia de Saneamento Ambiental – Secretaria do Meio Ambiente/SP) for supervisory monitoring.

The New Brazilian Forest Code, unlike other environmental regulatory laws worldwide, had its elaboration guided by traditional well established and popularly accepted ecological theoretical background. It was considered by the scientific community as a relaxation of the demands from the former code, that dates back to 1965, with a few improvements concerning to declaration of the private area under protection by the landowners in an on-line database, that is still not yet up to date (Pardini et al., 2010). Moreover, it demands that smaller areas of overall protected native vegetation as riparian areas are summed to legal reserves, but the calculation of the riparian area to be protected was changed from the higher flood level to the lower shoreline. Advances in designs that should have aided in the preference for choosing to protect contiguous areas of neighboring properties completely disregarded as were more complex considerations concerning to recent advances of ecological theoretical and applied fields of research were not even considered.

The importance of riparian forests for wildlife conservation is exhaustively pointed as crucial (e.g. Lomolino & Perault, 2001; Mac Nally et al., 2001; Lindenmayer & Hobbs, 2004; Macdonald et al., 2006; Palmer & Bennett, 2006; Hawes et al., 2008; Keuroghlian & Eaton, 2008; Luck & Korodaj, 2008; Galliez et al., 2009; Rosalino et al., 2009; Berges et al., 2010; Smith-Ramirez et al., 2010), as much as the amount of native vegetation at regional scale (e.g. Edenius & Sjoberg, 1997; Chiarello, 1999; Jacquemyn et al., 2003; Numa et al., 2005; Beisiegel, 2006; Jiménez García et al., 2006). Ecological processes subjected to cross-scale self-critical effects are of dramatic influence on nutrient and energy flows that take place on rivers and streams, physical structures with fractal behavior by themselves (Holling, 1992). In addition, the combination of riparian habitats and upland native vegetation are complementary, henceforth provide wildlife with different resources (Dunning et al., 1992).

In the State of São Paulo, dramatic shifts in land-use and land-cover that distanced landscape from what was familiar to pre-Columbian inhabitants started around 500 years ago, but became much more intense and reached wider extensions on the edge of the 19th and beginning of the 20th century. The territory occupation by non-original human populations demanded the extirpation of indigenous groups that resemble Old West invasion in North America, and has railroads as driving forces of expansion in common to that region.

17.2 Wildlife and Planted Forests in the State of São Paulo, Brazil

17.2.1 Wildlife in South America and the State of São Paulo

17.2.1.1 Mammals and Human Colonization

Mammals are a monophiletic clade that derived from late permian cynodonta (Kemp, 2005). Present known wider groups (in particular Therians: Marsupials and Placentals) radiated after the late Mesozoic Great extinction Event, reaching its maximum diversity in late Cenozoic and achieving recent levels of known species numbers after Pleistocene/Holocene interface (Kemp, 2005). In South America, mammal evolution was subjected to events of isolation, dispersal and faunal interchange (mainly with North America) (Goin et al., 2012) in a history that can be traced to 250 Mya. In this context, Goin et al. (2012) identify at least five major events in which the primary driving forces were tectonics, climatic fluctuations and human arrival (at the Pleistocene dawn). The full length of this time span is beyond of the scope of our work, but some selected examples may help us explain a few patterns observed in mammal species present distribution in the light of continental faunal evolution.

Procyonids arrived from North America at ca. 7 - 6 Mya, during late Miocene (Eizirik, 2012; Goin et al., 2012), much before the Great American Biotic Interchange (GABI) (that began at ca. 4.5 Mya with the most recent connection of South and North Americas through Panamanian Isthmus). Other Carnivora came to the South later, but yet, before GABI, and radiated in many forms that included omnivorous species. Mustelids and Canids arrived early during GABI, and are still less recent in South America than other Carnivora Clades. *Leopardus* species diverged in South America after an ancestral northern taxon invaded neotropics post GABI, ca. 2.8 Mya (Eizirik, 2012). Canidae immigrated to South America in two separated taxonomic groups during the GABI, one included relatives of domestic dog and wolf, the other originated endemic South American species which included *Cerdocyon thous* (Eizirik, 2012). Anteaters are first found in the fossil record from 17 Mya and the *Myrmecophaga* genus, 6.8 – 4 Mya (Flynn et al., 2012). Dasypodidae, a family that includes the *Dasypus* genus, is present in a palaeofauna that ranges from 13.5 to 11.8 Mya (Flynn et al., 2012). Xenarthra, a group to which anteaters and armadillos belong, dates back to 102 Mya (Scally et al., 2001) in the Eocene of South America (Vizcaino et al., 2012). Tapirs and deer arrived in the continent at around 1.8 Mya, during GABI (Goin et al., 2012).

The South American Megafaunal Extinction (SAME) occurred around 10,000 years BP (Patterson & Costa, 2012) and the taxa above cited co-occurred with many elements of late Pleistocene hypercarnivores and other taxonomic groups larger in size than extant terrestrial mammal species. The first confirmed evidence of human presence in South America dates its most early settlement layer on ca. 10,000 years BP (Roosevelt et al., 1996). Although Johnson (2002) points out the arrival of humans as a contributing factor to SAME, Roosevelt et al. (1996) present evidence for a much wider spectrum of hunting, fishing and gathering habits of Paleoindians from the Amazon Basin. In addition, no evidence of Megafaunal remains or proper weapons for hunting such animals were found in the sediments studied by them.

According to Gouveia et al. (2002) it is possible that charcoal presence in soils under forest vegetation in São Paulo State that were dated in a range from middle Holocene to present, were associated to human occupancy in the last 6000 years. The analysis of the floristic composition of localities in Botucatu and Anhembí, São Paulo

State, through anthracology, and the identification of charcoal particles, indicate that, at early Holocene, both localities were dominated by forested vegetation and, moisture conditions increased to levels very similar to present day patterns of precipitation since at least 3000 years BP (Gouveia et al., 2002; Scheel-Ybert et al., 2003). Although, there is controversy on the productions systems and food-webs of human populations occupying uplands of South-Eastern Brazil in early and middle Holocene, Roosevelt et al. (1996) point to a high diversity of foraging strategies that ranged from fishing to horticulture in several combinations and different sequences of local and regional substitution.

Studies show the presence of Itararé-Taquara ceramic tradition in São Paulo State highlands until at least the fourth century of the recent age (de Mello Araujo, 2007) and Tupi-Guarany traditions were present since 1800 ya (Noelli, 2008). Lentz (2000) argues that Pre-Columbian Indians impacted landscapes mainly by interaction with domesticated plants, extending the range of organisms of economic interest, disseminating parasites and ruderal species, controlling competing species and reducing plant and animal prey populations. Moreover, indigenous South American groups practice diverse and complex modalities of silviculture (Peters, 2000). The influence of these human cultures on vegetation and landscape may have shaped animal communities (Dyer, 2007), in addition to hunting impact. In 1500 AD the Portuguese arrived on Brazilian Shores (de Abreu, 1999) and a process of decimation and enslavement of Amerindians pushed them to peripheral areas causing the disruption of traditional foraging systems of the surviving groups (Roosevelt et al., 1996).

17.2.1.2 From Slave Hunters to Railroads and Industry

Less than a century after first contact with South American Amerindians, European settlers established sugarcane plantations on coastal regions (Buarque de Holanda, 1994) and relied on the Amerindians as both slave labor and slave labor providers, as colonists used for their own sake the fluid system of ritualized wars of Tupi-Guarany (Fausto, 2008). Aiming to capture more slaves, as slaves had a fast turnover rate in the tropical developed Neo-European modality of monoculture, incursions were organized to hunt extant Amerindian populations in São Paulo State forests of the Central Plateau and beyond (Buarque de Holanda, 1990). Those who did not succumb to foreign diseases where evangelized by Jesuit priests (Dean, 1997) and, eventually, captured by colonists in attacks on Jesuit Missions (Davidoff, 1986).

From the 17th to 18th century, gold and diamond mines, in regions currently comprehended by Minas Gerais State and others in Northeastern Brazil, established trade routes for food, cattle and beasts of mount and burden (Buarque de Holanda, 1990). While beasts were raised in the Southern Regions of Brazil, livestock were bred in the São Paulo State countryside and Bahia and other Northeastern states (Buarque de Holanda, 1990).

More intense fires to open pastures and the introduction of African aggressive grasses contributed to the clearing of wide forest areas that were on the surroundings of emergent Neo-European settlements (Schmidt, 1959; Parsons, 1972; Buarque de Holanda, 1990; Dean, 1997). Saint-Hilaire (1972) visited the region in the first decade of the 19th century, and he described a landscape with several secondary forest fragments intermingled by African grasses dominated pastures, native savannas, and scrublands with riparian forests and, large areas of semi-deciduous and deciduous forests on the ridges of the Paulista Plateu.

The trade of livestock and beasts brought enough financial income to acquire African slaves by São Paulo Neo-European inhabitants granting expansion of sugarcane monoculture that later was completely substituted by coffee (Secretaria do Meio Ambiente, 2007). Coffee plantations spread to the Southeast, Southwest and West of São Paulo State (Brioschi, 1999). By the early 20th century there were no regions without settlers and the population increased from 170 thousand in the 17th century to more than one million with the immigration of salaried European laborers, as African and Amerindian slavery was abolished (Dean, 1997; Brioschi, 1999).

Expansion of coffee plantations was fostered by the rapid and massive transport capability of railroads, their construction and functioning demanded enormous quantities of wood supplied by native vegetation (Dean, 1997). In addition, the allegiance between estate owners, opportunist illegal settlers and the railroad company to pay for troops of Native American killers devastated with extreme brutality the remnant Amerindian populations (Ghirardello, 2002). Already reduced by extensive coffee plantations and firewood extraction, large traits of native vegetation were flooded for the operation of hydroelectric dams that supplied a growing industry and a nascent population that emigrated from other regions of Brazil, looking for job opportunities in the emergent market (Secretaria do Meio Ambiente, 2007).

17.2.2 Eucalypt Planted Forests and Wildlife

The first two decades of the 20th century saw nascent demand for wood to supply railroads. Predicting the rarity of the native commodity, one company from São Paulo State invested in the development of planted forests of exotic tree species. Eucalypt was the tree of choice and forestry engineer Edmundo Navarro de Andrade imputed enthusiasm on the task of disseminating its cultivation (Sampaio, 1949, 1957). Despite of Navarro's efforts, eucalypt was not enough to alleviate pressure on native vegetation in the face of demand not only for firewood, but also for timber for construction and furniture to supply the growing human population (Dean, 1997).

The coffee market in Brazil passed through two consecutive crises, and the second brought the end to this economic cycle, resulting in the crash of markets worldwide (Monbeig, 1984). However, more than 150 years after it fostered the conditions for coffee monoculture cycle, sugarcane spread again through the São Paulo State countryside, interrupting a brief period of diversification and becoming the dominant feature of the landscape in vast areas of the state (Mendonça, 2004). More recently, agricultural production of São Paulo State assumed a more defined regional pattern. International market demand, governmental agricultural programs and politics resulted in the increase of areas planted with sugarcane, wheat, soybeans and planted forests (Mendonça, 2004).

The expansion of the new sugarcane cycle was reinforced by the modernization of Brazilian agriculture, that took place from the 1960's and was characterized by the use of biological and chemical inputs and, by the intense use of machinery. Entrepreneurs closer to industrial funds became administrators of this new modality of agriculture and transformed landowner profiles through the creation of commodity trade cooperatives, trans and multi-national markets and increasing credit use (Tartaglia, 1988; Tilman et al., 2002; Butler et al., 2007).

Eucalypt plantations currently occupy over one million hectares in São Paulo, the richest and most developed state in Brazil. Their main commodities are the pulp and paper (ABRAF, 2013). Together with livestock production, the ethanol agroindustry primarily based on sugarcane, and the pulp and paper silvicultural industry primarily based on Eucalyptus, are the main agroindustrial sectors of São Paulo State. However, unlike the other two agroindustrial sectors, the pulp and paper industry follows Brazilian Environmental Law relatively well due to market pressure posed by certification organizations such as the Forest Stewardship Council (FSC) (Verdade et al., 2012).

FSC in Brazil has strict standards that concern to the observation of national, regional and local environmental legal regulation, and to the evaluation of environmental impact (FSC Principles and Criteria for Forest Stewardship). However, standardized worldwide accepted patterns may pose a threat to local biodiversity patterns, as traditional practices are lost and some general standards may not apply to fine-scale characteristics of particular remnant biota (Primdahl & Swaffield, 2010).

17.3 Mammals in Eucalypt Plantations of Southwestern São Paulo State

17.3.1 Study Area

This study was carried out at Southwestern São Paulo State and was divided in three sub-areas with different landscape patterns: i) matrix of native vegetation with low amount of eucalypt plantations (A); ii) intermediary quantities of both vegetation types (B); small amount of native vegetation and higher quantity of eucalypt plantations (C) (Fig. 17.1).

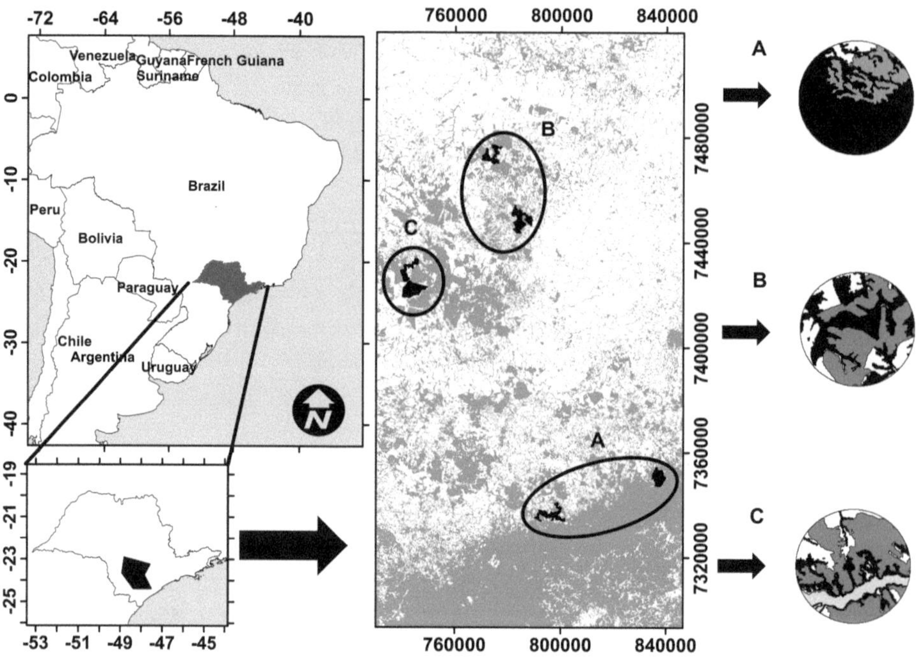

Fig. 17.1: Study area. State of São Paulo, Brazil. Region of the study area inside State of São Paulo limits. Locations of sampled Eucalyptus farms and forest cover (both native and planted). A – matrix of native vegetation with small eucalypt area; B – intermediate areas of eucalypt and native vegetation; C – eucalypt matrix with small area of native vegetation. Examples are extracts of 3000 m radius from center of mammal data collecting trails. In A, B and C: Black – Native Vegetation, Gray – Eucalypt and Light Gray – Water. State of São Paulo and Study Region are presented in GCS SIRGAS 2000, forest cover and 3000 m radius extracts are presented in SIRGAS 2000, UTM, zone 22S.

The area A has native vegetation mainly composed of moist broadleaf forest and is continuous to the larger remnants of local Atlantic Forest (Parque Estadual Carlos Botelho). Area B has native remnants of semi-deciduous forest and tropical savanna and scrubland ("Cerrado"). Area C presents the same vegetation types as area B, and is located at the margins of a large dam (Jurumirim Hydroelectric Dam).

17.3.2 Landscape Assessment

Classes of land-use and land-cover were obtained from Landsat TM 5, orbit 220, scenes 076 and 077; from 19/06/2006 and 20/06/2007. Images of the same date were mosaicked and standardized, geometric correction was performed with basis on land-cover maps of the sampled eucalypt farms and had Total RMS error of 12,1667 m.

Images were classified by the Maximum Likelihood algorithm (Jensen, 1996) using the bands 3 (0.63 – 0.69 μm), 4 (0.76 – 0.90 μm) and 5 (1.55 – 1.75 μm). Training samples were taken for seven classes, three of native vegetation, one eucalypt, two for open pastures and crops and one for exposed or paved soil.

After classification, we performed a weighted overlay of the two images 19/06/2006 and 20/06/2007 with equal weights (50%) for both. The resulting raster image was transformed in a vector file and misclassifications were corrected by photo interpretation, all classes of land-use and land-cover, except for native vegetation and eucalypt were joined and discarded from analyses.

Buffers of 500, 1000, 1500, 2000, 2500 and 3000 m radius were extracted around every central point estimated from the trace of footprint collecting trails. Landscape metrics and indices for both classes (native vegetation – NV and eucalypt – EU) were calculated for each of 16 points in areas obtained from above cited buffers.

Metrics estimated for each class on each buffer were: class perimeter (PE); class area (A); percentage of the class corresponding to the total area (ZA); perimeter-area ratio (PA); mean patch fractal dimension (FD) and area weighted mean fractal dimension (AWFD), which is adjusted to shape size (Mcgarigal & Marks, 1995; Rempel et al., 2012).

17.3.3 Analytical Design

Three events of data collecting took place between December 2006 and March 2007. On each event, sand plots were visited for three consecutive days, weather permitting, and camera traps remained triggered for approximately 15 days. A total of 15 sand plots were placed by point, five in eucalypt stands, five in native vegetation and five on the edge of the two classes. Two camera-traps were placed by point, one in the eucalypt stand and other in the native vegetation. Sand plots were placed in three parallel lines 25 m from each other.

The total number of species was calculated for each point by summing the number of species detected for the total time span of the sample collection. Total numbers of occurrences were the sums of detections on different days of each event. Detection histories were obtained by register of presence and absence on each event. Numbers of occurrences were also grouped by functional group: herbivorous, omnivorous, carnivorous and insectivorous (adapted from Eisenberg, 1981; Fonseca et al., 1996; Robinson & Redford, 1986a,b; Dotta, 2005; Dotta & Verdade, 2007).

To estimate a total number of species for the whole taxocenosis we used a hierarchical Bayesian approach as described by Dorazio et al. (2006) and Royle et al. (2007), with zero-inflation of 28 detection histories in order to achieve the sum of medium to large terrestrial mammal species known from São Paulo State that our sampling methods would be able to detected (de Vivo et al., 2011). General Linear Models with identity function of the family Poisson were used when mammal data had variance and means with close values, when these values were discrepant, Negative-Binomial models were used. Dependent variables in the models were Total Number of Species (TSpp), Total Number of Occurrences (Occ), and Number of Occurrences of Herbivores (Herb) and Omnivores (Omn). Independent variables were the landscape metrics for each class on each buffer radius. Estimation was performed by robust algorithm. Models were selected through comparison of Delta AICc (Anderson et al., 1998; Burnham & Anderson, 2004).

Hierarchical procedures for individual species were performed by Maximum Likelihood Estimation of probability of occurrence, based on detection history with account to missing data and logistic link function (Mackenzie et al., 2006). Only species with more than four detections were tested by this approach.

17.3.4 Results

Sixteen taxa of mammals (14 species and two genera with indistinct species) from 12 families in six orders (Wilson & Reeder, 2005) were detected during the three data collecting events. Total species number estimated from hierarchical Bayesian procedure that had the higher frequency in 1000 iterations was 18 species (Fig. 17.2). Species detected in each of the three sampling areas (A, B and C) are listed in Table 17.1.

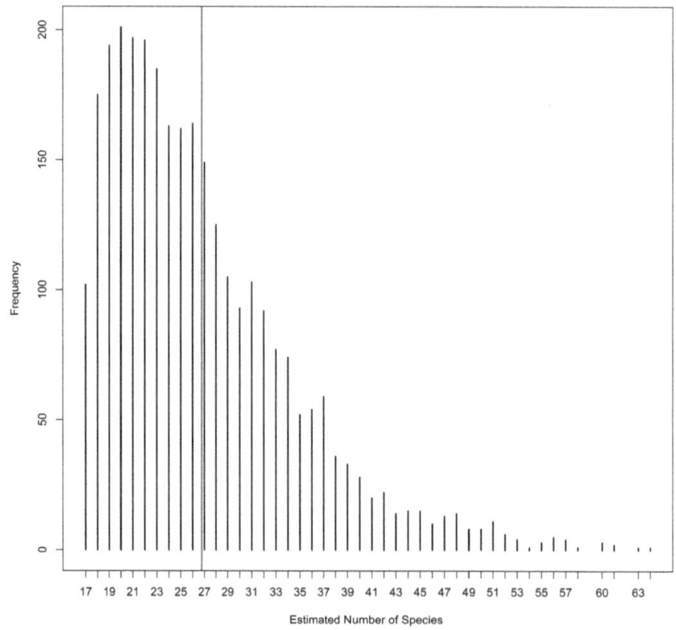

Fig. 17.2: Numbers of species estimated by hierarchical Bayesian procedure and frequencies of estimation. Mean estimated species are represented by the red line.

A total of 79 regression models had significant results ($p < 0.05$) and only the first four models with lower Delta AICc were selected. Results for the selected models are shown in Table 17.2. Models were not run for carnivorous and insectivorous numbers of occurrences, as both groups had small numbers of detections. Graphs showing fitted curves from the GLM models are in Figure17.3 for Total Species Numbers, Figure 17.4. for Total Occurrences, Figure 17.5. for Herbivorous Occurrences and Figure 17.6. for Omnivorous Occurrences.

Tab.17.1: Species detected and functional groups in the areas accessed in this study.

Taxonomy			**Area**		
Family	**Species**	**Functional group**	**A**	**B**	**C**
Dasypodidae	*Dasypus novemcinctus* (Linnaeus, 1758)	Omnivorous	X	X	X
Myrmecophagidae	*Myrmecophaga tridactyla* (Linnaeus, 1758)	Insectivorous	X		X
	Tamandua tetradactyla (Linnaeus, 1758)	Insectivorous			X
Dasyproctidae	*Dasyprocta azarae* (Lichtenstein, 1823)	Herbivorous			X
Cuniculidae	*Cuniculus paca* (Linnaeus, 1758)	Herbivorous		X	
Leporidae	*Sylvilagus brasiliensis* (Linnaeus, 1758)	Herbivorous			X
Tayassuidae	*Pecari tajacu* (Linnaeus, 1758)	Herbivorous	X		
Cervidae	*Mazama* sp. (Rafinesque, 1817)	Herbivorous	X	X	X
Tapiridae	*Tapirus terrestris* (Linnaeus, 1758)	Herbivorous	X		
Felidae	*Leopardus* sp. (Linnaeus, 1758)	Carnivorous	X	X	X
	Panthera onca (Linnaeus, 1758)	Carnivorous	X		
	Puma concolor (Linnaeus, 1771)	Carnivorous	X		X
	Puma yagouaroundi (É. Geoffroy Saint-Hilare, 1803)	Carnivorous		X	
Canidae	*Cerdocyon thous* (Linnaeus, 1766)	Omnivorous			X
Mustelidae	*Eira barbara* (Linnaeus, 1758)	Omnivorous	X		X
Procionidae	*Nasua nasua* (Linnaeus, 1766)	Omnivorous		X	X
Atelidae	*Alouatta guariba* (Humboldt, 1812)	Herbivorous	X		
Cebidae	*Cebus nigritus* (Goldfuss, 1809)	Herbivorous		X	

Results of single species probabilities of occurrence (psi) were selected on the basis of Delta AIC, although here, the first models of the set were not always selected as some sets had negative values of Delta AIC. Models were selected with a Delta AIC limit of ten, and by the observation of their fit and convergence. For every selected species 38 models were run with two covariates, correspondent landscape metrics for native vegetation and eucalypt, on each spatial scale. Graphs are shown as pairs for each functional group, except for *Leopardus* sp. and *Myrmecophaga tridactyla*, that were the single species with sufficient detections in their groups, carnivorous and insectivorous. Herbivorous species were represented by *Tapirus terrestris* (Delta AIC = 9.92, -2*log(Likelihood) = 9.8589) and *Mazama* sp. (Delta AIC=4.11, -2*log(Likelihood) = 50.3636), both fitted for Mean Fractal Dimension of Native Forest (FDNV) and

Tab. 17.2: Summary of GLM Poisson and Binomial-Negative models. Group sets are: TSpp = Total Species, Occ = Total Occurrences, Herb = Herbivorous Occurrences, Omn = Omnivorous Occurrences, Carn = Carnivorous Occurrences, Ins = Insectivorous Occurrences; Independent variables are written with prefixes: A = Area, PE = Perimeter, PA = Perimeter/Area Ratio, FD = Fractal Dimension, AWFD = Area Weighted fractal Dimension; sugixes: NV = Native Vegetation, EU = Eucalypt Plantation; and spatial scales: 05 = 500 m buffer ratio, 10 = 1000 m buffer ratio, 15 = 1500 m buffer ratio, 20 = 2000 m buffer ratio, 25 = 2500 m buffer ratio, 30 = 3000 m buffer ratio.

Group Set	Mean	Std. Dev.	Variance	Skewness	Kurtosis	Min	Max	Obs.	D.f.	Model	Family
TSpp	3.786	2.293	5.258	1.384	4.858	1	10	14	12	GLM	Poisson
Occ	6.429	5.360	28.725	1.514	4.867	1	21	14	12	GLM	Bin. Neg.
Herb	1.357	0.929	0.863	0.433	2.447	1	3	14	12	GLM	Poisson
Omn	1.429	1.089	1.187	0.559	3.352	1	4	14	12	GLM	Poisson
Carn	0.571	0.646	0.418	0.616	2.429	1	2	14	12	NA	NA
Ins	0.429	0.514	0.264	0.289	1.083	1	1	14	12	NA	NA
Dep. Var.	**Indep. Var.**	**P > z**	**AIC**	**AICc**	**DeltaAICc**						
TSpp	FDNV20	0.013	4.106	316.106	0.000						
	PENV05	< 0.001	4.261	316.261	0.154						
	FDEU25	< 0.001	4.290	316.290	0.184						
	PEEU25	< 0.001	4.323	316.323	0.217						
Occ	FDEU25	< 0.001	5.912	317.912	0.000						
	AWFDEU05	0.001	5.938	317.938	0.026						
	PAEU15	0.019	5.952	317.952	0.040						
	AWFDEU10	0.004	5.962	317.962	0.050						
Herb	PANV05	0.02	2.718	314.718	0.000						
	AWFDEU10	0.017	2.769	314.769	0.051						
	AWFDEU20	0.003	2.793	314.793	0.075						
	AWFDEU15	0.004	2.802	314.802	0.084						
Omn	ANV20	< 0.001	2.871	314.871	0.000						
	ANV15	< 0.001	2.920	314.920	0.049						
	AWFDNV30	0.048	2.949	314.949	0.078						
	ANV10	0.001	2.950	314.950	0.079						

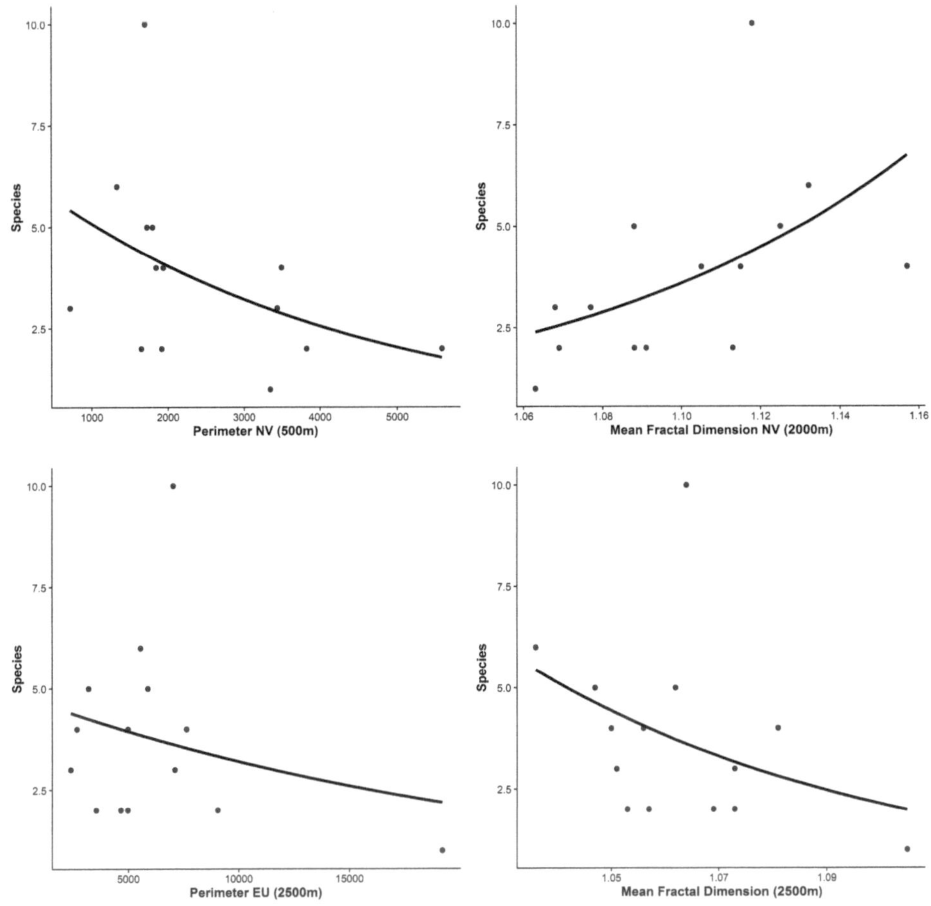

Fig. 17.3: Fitted curves of GLM models with Poisson identity function for Total Number of Species and Landscape metrics. Radius of buffer is referred in parentheses.

Eucalypt (FDEU) at 3000 m radius of buffers (Fig. 17.7). Omnivorous species were represented by *Dasypus novemcinctus* and *Cerdocyon thous*, fitted with Area Weighted Fractal Dimension at 1500 m of buffer radius (Delta AIC = 4.74, -2*log(Likelihood) = 52.71) and Percentage of Class Area at 2500 m of buffer radius (Delta AIC = 0.93, -2*log(Likelihood) = 21.32) respectively (Fig. 17.8). Carnivorous group was represented by *Leopardus* sp. and fitted in relation to Area Weighted Fractal Dimension at 2500 m of buffer radius (Delta AIC = 6.07, -2*log(Likelihood) = 28.49). *Myrmecophaga trydactila*, an insectivorous, was fitted with Perimeter of Classes Native Forest and Eucalypt at 500 m of buffer radius (Delta AIC = 1.51, -2*log(Likelihood) = 24.56). *Leopardus* sp. and *M. tridactyla* probabilities of occurrence are shown in Figure 17.9.

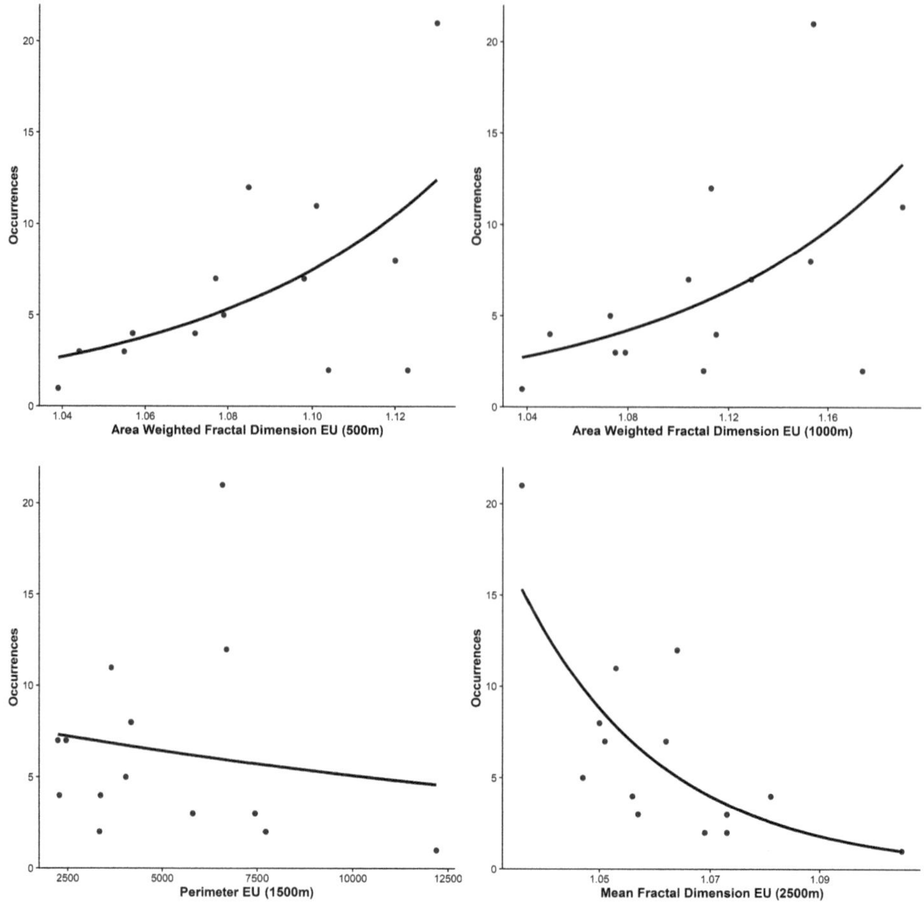

Fig. 17.4: Fitted curves of GLM models with Poisson identity function for Total Number of Occurrences and Landscape metrics. Radius of buffer is referred in parentheses.

17.3.5 Processes Inferred

17.3.5.1 Goals Achieved Hitherto: Riparian Forests Protection, Other Land Owned Protected Areas and Extant Wildlife Species

The total species numbers are negatively influenced by the increase of native vegetation perimeter at 500 m buffer radius and by increase in eucalypt perimeter at 2500 m buffer radius. In contrast, mean fractal dimension increase of native vegetation at 2000 m radius influences positively the number of species, as at 2500 m radius, eucalypt mean fractal dimension increase shows negative effect on species numbers. These results may imply that the increase of native vegetation perimeter in the immediate vicinities is favorable to the overall taxocenosis until some level when

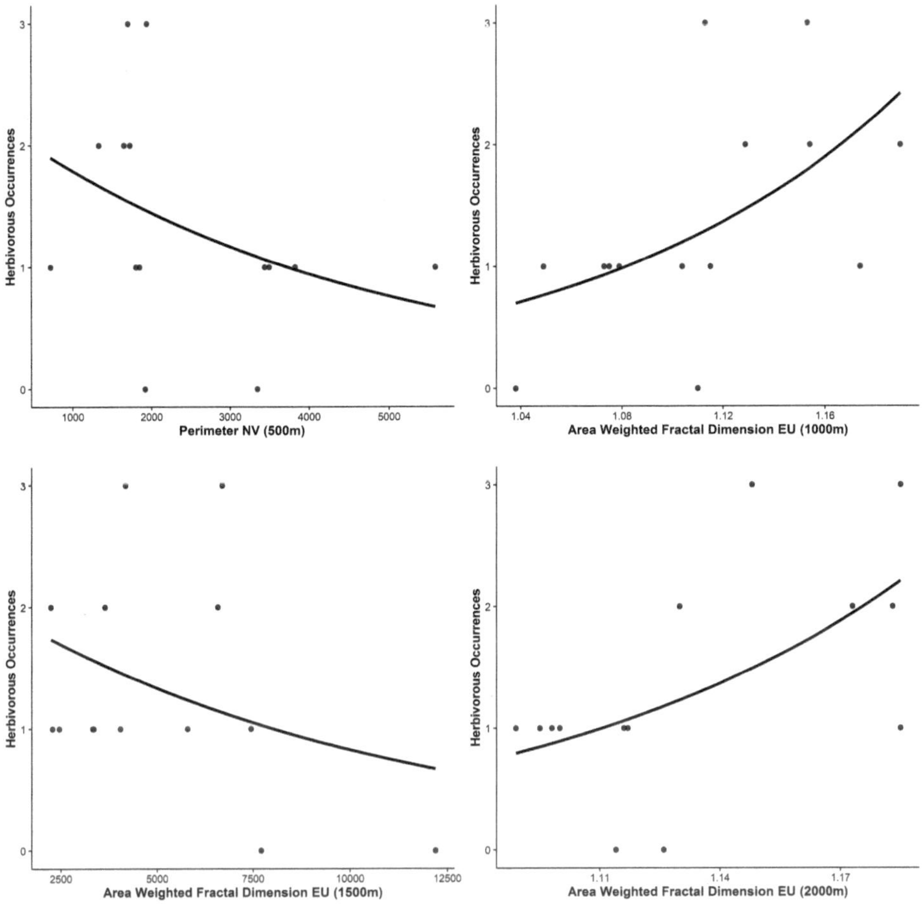

Fig. 17.5: Fitted curves of GLM models with Poisson identity function for Herbivorous Occurrences and Landscape metrics. Radius of buffer is referred in parentheses.

species numbers start to fall. This effect may be due to the ability of the buffer, as a window, in capturing complexity at this particular scale. The increase in perimeter of native vegetation in this scale means smaller dendritic shapes of this class that reflects the disruption of maintenance of more diverse floras able to sustain large herbivores (Milne et al., 1992). The same trend is shown for perimeter of eucalypt at 2500 m radius. We suggest that the increase in eucalypt plantation perimeters reflect the increase of their total area in the landscape, as the shape of stands is preferably regular for management reasons.

Concerning the mean fractal dimension, total species numbers are favored by the increase of this metric for native forest at 2000 m radius and disfavored by increase of FD at 2500 m in eucalypt class. These results are complementary as mean fractal dimension takes area into account and reflects not only the shape complexity, but

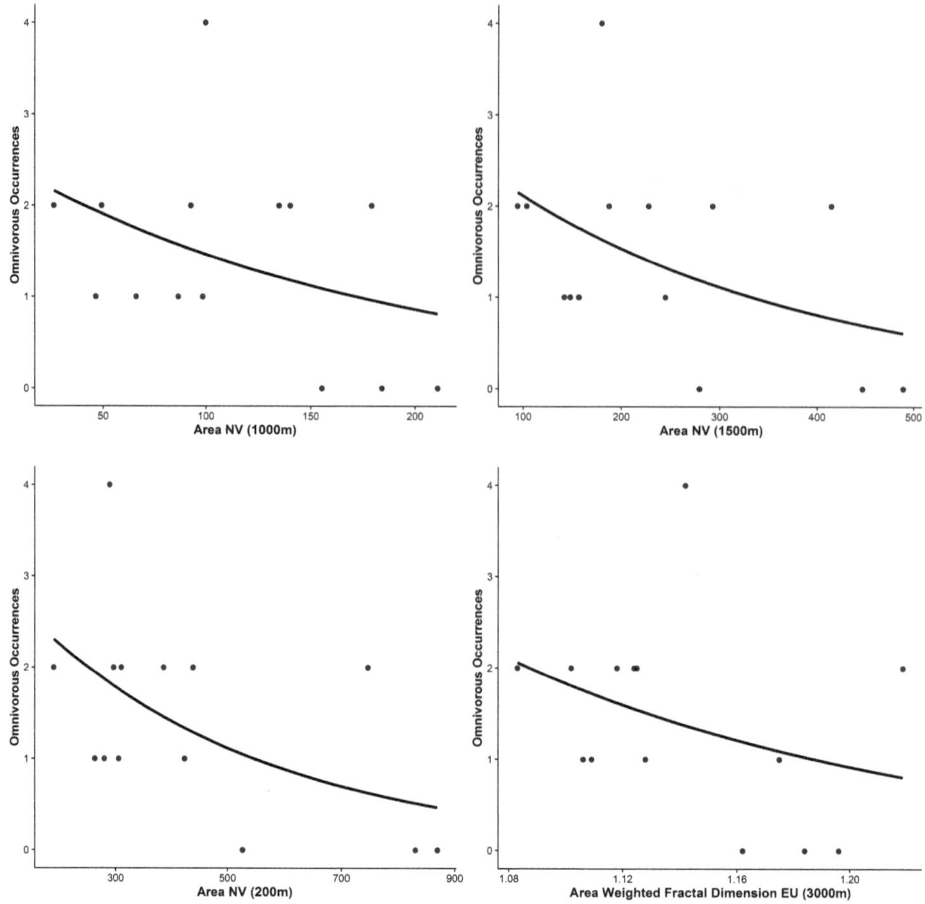

Fig. 17.6: Fitted curves of GLM models with Poisson identity function for Omnivorous Occurrences and Landscape metrics. Radius of buffer is referred in parentheses.

also the habitat amount (Theiler, 1990). In this context, species richness increases with the increase of mean fractal dimension of native vegetation and diminishes with the same increase for eucalypt, except that spatial scales are slightly different. However, the perception of such effect occurs at broad extensions what may suggest that area and its interaction with shape are factors that affect diversity at broad scale in the landscape. In addition, the pattern observed for species richness and fractal dimension points to a corroboration of the classic species-area relationship (Connor & Mccoy, 1979).

Area Weighted Mean Fractal Dimension (AWFD) is a metric of shape complexity, but as it is weighted for patch area, the effect of the specific habitat amount is annulated in the final output of the metric, making shape complexity the central measure of interest (Lausch & Herzog, 2002). Total Occurrences were increased with AWFD of

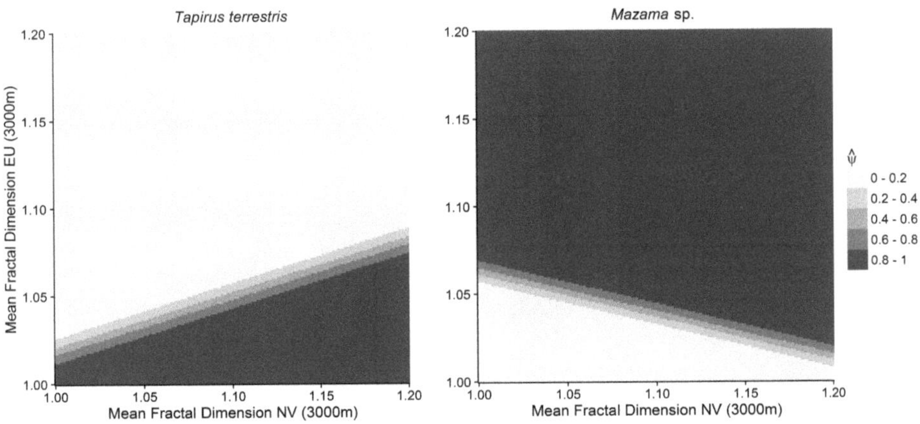

Fig. 17.7: Probability of Occurrence (Psi) of *Tapirus terrestris* and *Mazama* sp. in relation to Mean Fractal Dimension of Native Forest (FDNV) and Eucalypt (FDEU) at 3000 m of buffer radius.

Eucalypt at 500 and 1000 m buffer radiuses. The increase in eucalypt stands shape complexity may well reflect the increase of natural vegetation amount on landscape, as simpler shapes of stands are preferable for management reasons. Also, more complex shapes of eucalypt may represent higher length and complexity of contacts with native vegetation, which grants bigger ruderal species exchange and bigger turnover of plant matter on edges. This is likely to favor directly the occurrences of large herbivores and indirectly foster the subsistence of omnivorous and insectivorous species (Fletcher, 2005; Fleury & Galetti, 2006; Fischer & Lindenmayer, 2007; Taylor et al., 2008; Pawson et al., 2009).

On the other hand, Perimeter-Area Ratio (PA) at 1500 m and Mean Fractal Dimension (FD) at 2500 m radius for eucalypt show opposing trends. Numbers of occurrences are negatively influenced by FD (as mentioned above, this metric is area sensitive), yet show positive relation with PA. The pattern shown by these two metrics is, again, of the increase in eucalypt area as the limiting factor for number of mammal occurrences. As complexity of eucalypt increases (PA), so does the amount of natural vegetation, as the interaction of area and shape of eucalypt enlarges (FD), the area of natural vegetation shrinks.

The relationship of Herbivorous Occurrences with Perimeter-Area Ratio (at 500 m radius) illustrates the importance of native vegetation amount on the landscape and the role of shape complexity by showing a steeper increase of occurrences on initial levels of complexity followed by a mild decreasing trend when perimeter values surpass those of area for Native Vegetation. In contrast, complexity of eucalypt shape (AWFD) favors occurrence numbers of such group, corroborating the hypothesis that increased complexity is a positive factor in production of plant matter resources (Venevsky & Veneskaia, 2003; Anderson et al., 2004; De Jager et al., 2009).

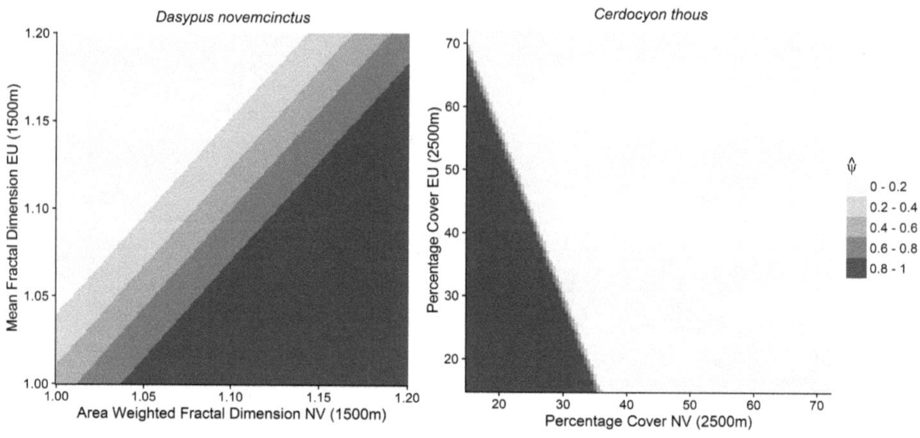

Fig. 17.8: Probability of Occurrence (Psi) of *Dasypus novemcinctus* in relation to Area Weighted Fractal Dimension of Native Forest (AWFDNV) and Eucalypt (AWFDEU) at 1500m of Buffer radius and, *Cerdocyon thous* in relation to Percentage of Area Class of Native Forest (AWFDNV) and Eucalypt (AWFDEU) at 2500 m of buffer radius.

Trends observed for occurrence numbers are confirmed by individual herbivorous species. *Tapirus terrestris* probability of occurrence is favored by the increase of native vegetation fractal dimension (FDNV) at 3000 m of buffer radius and shows a negative relationship with the corresponding metric of eucalypt. *Mazama* sp. is also favored by the increase of the interaction between area and shape of native vegetation, but shows a positive relation with eucalypt area as well (although a less steep influence than for FDNV). This result highlights that in the same foraging functional group there are differences particular to each species. We hypothesize that the reasons for such differences rely on foraging habits, as tapirs consume great quantities of plant matter from a large number of species every day (Bodmer, 1991; Padilla & Dowler, 1994; Salas & Fuller, 1996; Stoner et al., 2007), while deer are less specialized and have a more relaxed diet (Kufner et al., 2008). In addition, tapirs are larger animals and much more attractive to poachers and hunters, and have lower birth rates than deer (Padilla & Dowler, 1994; Hurtado-Gonzales & Bodmer, 2006; Rivero et al., 2007; Di Bitetti et al., 2008).

In contrast to Herbivorous Occurrences results, Omnivorous Species Occurrences are negatively influenced by the amount of native vegetation on landscape (here represented by Area of Native Vegetation – ANV). The increase in spatial scale of this relationship shows an interesting pattern- a slight increase in occurrence numbers happens at the lower levels of native vegetation area at 1000 m radius buffers, while this increase is diluted at broader scales (1500 and 2000 m). This trend suggests that landscape class area dependence is also scale dependent for numbers of occurrences of omnivorous. Moreover, these results show that these generalist and opportunist

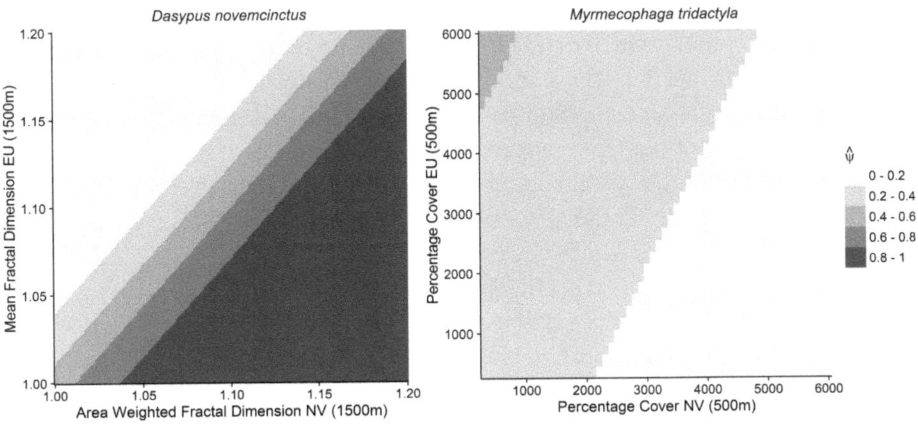

Fig. 17.9: Probability of Occurrence (Psi) of *Leopardus* sp. in relation to Area Weighted Fractal Dimension of Native Forest (AWFDNV) and Eucalypt (AWFDEU) at 2500 m of buffer radius and, *Myrmecophaga trydactila* in relation to Perimeter of Class of Native Forest (AWFDNV) and Eucalypt (AWFDEU) at 500 m of buffer radius.

species need, at least in some degree, that native vegetation is present in the landscape and they are not only favored by the native habitats scarcity, as they may be fostered by the absence of larger predators as well (Crooks & Soulé, 1999).

Dasypus novemcinctus had a positive relationship with higher complexities of native forest shapes, as well as it was not favored by the increase of eucalypt complexity at 1500m radius. Common Long-Nosed Armadillos are generalists, ancient (they were here when vegetation shifted with climate changes from Miocene, Pleistocene and Holocene, according to Flynn et al. (2012) and Goin et al. (2012)), resilient and highly successful reproductive organisms (Mcbee & Baker, 1982; Breece & Dusi, 1985). Their adaptability to new habitats can be illustrated by the recent range expansion to North American northern regions (Taulman & Robbins, 1996).

Cerdocyon thous presents a diverse pattern, as it is negatively influenced by native vegetation percentage of area in the landscape (ZANV) and positively influenced by the same metric for eucalypt (ZAEU) at 2500 m radius of buffers. As Eizirik (2012) points out, *C. thous* radiated from an ancestral species that reached South America during GABI, on the interface of Pleistocene and Holocene. Although this Canidae has its phylogeography strongly related to Mata Atlântica (Costa & Leite, 2012; Tchaicka et al., 2007), *C. thous* is a forest edge dweller with generalist diet (Berta, 1982; Arruda Bueno & Motta-Junior, 2004; Gatti et al., 2006; Di Bitetti et al., 2009).

Although insectivorous and carnivorous species were not evaluated as community groups, single species from both had enough detections that allowed us to model their relation with landscape. *Leopardus* sp. and *Myrmecophaga tridatyla* showed opposite patterns of probabilities of occurrence. While the first is favored by the increase of complexity of native vegetation patches' shapes at broad scale (AWFDNV at 2500 m buffers radiuses), the former benefits from eucalypt increasing in perimeter at 500 m radiuses. *Leopardus* sp. radiated in South America from an ancestor species after GABI. *Myrmecophaga* genus, in contrast, has a long history of presence in South America, as its own Order is of South American origin, the Xenarthra. Ocelots and margays are relatively small Neotropical cats in comparison to *Puma concolor* and *Panthera onca*, especially margay. These species probably benefit from the absence of larger predators and may also be favored by edge habitats (Gonz et al., 2003; Dillon & Kelly, 2008; Lyra-Jorge et al., 2008; Rocha-Mendes & Bianconi, 2009; de Oliveira Calleia et al., 2009) as more abundant small mammals are found in highly productive edges (Freemark, 1995; Bayne & Hobson, 1997; Christian et al., 1998; Gascon et al., 1999; Gheler-Costa, 2006). *M. tridactyla*, is influenced positively by narrow scale (500 m radius buffers) increase in eucalypt perimeter. This ancient edentate is fostered, probably, by the profusion of social insects that benefit from the high plant matter turnover of eucalypt edges. The *M. tridactyla* diet is based on insects, especially ants and termites, usually abundant in *Eucalyptus* plantations (Breece & Dusi, 1985; Medri et al., 2003). As a matter of fact, ants are considered the most important "plague" of *Eucalyptus* plantations in Southeastern Brazil demanding a rigorous control by the pulp and paper industry (Ramos et al., 2004).

Our results suggest a strong relationship between large to medium mammal species and the geometry of landscape, including the length of borders and the area of features. Although omnivorous species are less dependent on the amount of native vegetation than herbivores, they clearly need, at least, small areas of these vegetation types, even if in extensive spatial scales. Complexity of edges and the shifts between the intensity of influence of both tested vegetation types point to self-critical effects of landscape geometry on community structure (Holling, 1992; With & Crist, 1995; Hill & Caswell, 1999; Nikora et al., 1999) as exemplified by the relationship of species richness and perimeters and mean fractal dimensions of native vegetation and eucalypt.

Dependence on even small amounts of native vegetation stresses the need for proper protection of riparian vegetation stripes as long upland traits of native vegetation. More efficient designs of such protected areas may favor the occurrence and permanence of more than only generalist species. Furthermore, it may certainly aid in the fluxes of energy, matter and genetic information that favors intra and interspecific diversity (Taylor et al., 1993; With & Crist, 1995; Lindborg & Eriksson, 2004; Martins, 2005).

Legal environmental regulation and international trade certificates play an important role in determining the overall geometry of forest plantation farms, in addition to local socioeconomic characteristics and physical and biotic factors. Management decisions and environmental protection are, in final analyses, results of economic demands and governments concerns with natural resources. Although organized civil society affects the balance of this relation, most of the time, market driving forces surpass them in infinite fold, and the legal craftiness applied on such maneuvers is astonishing (e.g. Dalebout et al., 2002). While conservation of extant native Neotropical species demands a deeper analysis encompassing multiple dimensions, including time (as long-term inventories and historical accessory data), legal regulation relies on flexible theoretical concepts and legal requirements.

As emphasized previously in this chapter, Brazilian Environmental Laws require that landowners of São Paulo State protect riparian forests, sometimes very narrow stripes, and medium to large landowners protect at least 20% of native vegetation in relation to the total area of each property, riparian forests accounted. This does not guarantee the protection of native vegetation at the same locality, as compensation for incomplete reserves may be conducted by planting native forest tree species in other areas. The panorama presented above may not be enough to maintain many large and medium mammal species with the growing demand for food and commodities combined with the standardization of agricultural international trade cooperatives.

17.3.5.2 From Now On: Novel Analysis, Complex Geometries and Market Driven Review of Environmental Legal Regulation

The questions accessed herein are complex and, in some way, the fate of Neotropical extant faunas is unpredictable. Our ability to model their reactions to landscape changes and cultural shifts is limited to our comprehension of the factors that shape their biological characteristics and the influence of innumerous factors in their evolution. Our historical moment has been called the "Anthropocene" (Crutzen & Stoermer, 2000) for the impact of our production systems became much more intense and widespread than those observed for past lived cultures, and now we are comparable to a combination of geological scale forces that are used to name geological periods.

Long-term studies of biodiversity are still scarce and expensive but analyses are improving and, more models and modelling frameworks are becoming available that can address variables collected from diverse sources and account explicitly for error and biases, at least those prevenient from biological sampling schemes (e.g. Mackenzie et al., 2002; Dorazio & Royle, 2005; Stanley & Royle, 2005; Dorazio et al., 2006; Kéry & Royle, 2008; Russell et al., 2009; Zipkin et al., 2009).

Human constructed features such as roads, power-lines, ducts and many others impose an artificial sense of order to the landscape for our civilized point of view. Contrary to the first perception of beauty in symmetry, the anthropomorphized environments in western modern and post-modern market driven fashion disrupts several fluxes of a very complex web that maintains wildlife (Meine, 1997). The investigation of the hierarchical structure of wildlife communities poses a very difficult task given that, as the landscape shaped by ancient human inhabitants were substituted during the process of drastic reductions and extirpations of their populations, wildlife was also reshaped (Verdade et al., 2014a).

17.4 Wildlife Conservation and Silviculture

Mankind evolved in comprehension of the space where it dwells and developed as complex concepts such as self-affinity and self-criticality aided with the conquest of new technologies, the exploration of extra-terrestrial space and the shift of scientific paradigms that is leading to novel ways of perceiving processes and patterns in every knowledge field.

As culture spreads through novel kinds of media, less and less space is left to traditional knowledge and, despite efforts of rescue of such ways of life and production systems, globalized networks of food and commodity productions threaten the diversity of agricultural systems among others. Like other agricultural sectors, silvicultural practice is part of this panorama and is subjected to local environmental laws (Verdade et al., 2014b). Environmental law regulations, by their turn, are more and more subjected by thresholds imposed from market, as large corporations spread influence in every social stratum.

This time, the changes in production systems are much faster and wider than they used to be from the arrival of man in South America to the construction of the first railroads for coffee exportation in São Paulo State. As culture and perception drive decisions that, at final analyses, encompass even the widest and impersonal management schemes, ours is the decision of incorporating in education and formal/informal professions, the issues of complexity and their role in the "almost unpredictable" behavior of nature.

As the current magnitude of our impact on planet earth can be comparable to that of a geological force, our responsibility lies in choosing the path to follow concerning landscape and wildlife, and what landscape we may shape and who will share it with us.

Acknowledgements: We thank Eucatex Florestal and IPEF (Forestry Science and Research Institute), we also thank CNPq (National Counsel of Technological and Scientific Development) (Process: 42732/2007-9), for the funding of our research.

Bibliography

Abraf (2013). Anuário estatístico ABRAF 2013 ano base 2012. Brasília: ABRAF.

Anderson, D.R., Burnham, K.P., & White, G.C. (1998). Comparison of Akaike information criterion and consistent Akaike information criterion for model selection and statistical inference from capture-recapture studies. Journal of Applied Statistics, 25 (2), 263-282.

Anderson, T.M., Mcnaughton, S.J., & Ritchie, M.E. (2004). Scale-dependent relationships between the spatial distribution of a limiting resource and plant species diversity in an African grassland ecosystem. Oecologia, 139 (2), 277-287.

Antrop, M. (2005). From holistic landscape synthesis to transdisciplinary landscape management. In B. Tress (Ed.), From landscape research to landscape planning: aspects of integration, education and application (p.27-50). Wageningen: Springer.

Arruda Bueno, A.D., & Motta-Junior, J. C. (2004). Food habits of two syntopic canids, the maned wolf (*Chrysocyon brachyurus*) and the crab-eating fox (*Cerdocyon thous*), in southeastern Brazil. Revista Chilena de Historia Natural, 77, 5-14.

Atauri, J.A., & De Lucio, J.V. (2001). The role of landscape structure in species richness distribution of birds, amphibians, reptiles and lepidopterans in Mediterranean landscapes. Landscape Ecology, 16 (2), 147-159.

Bak, P. & Bak, P. (1996). How nature works: the science of self-organized criticality. Copernicus: New York.

Bayne, E.M. & Hobson, K.A. (1997). Comparing the effects of landscape fragmentation by forestry and agriculture on predation of artificial nests. Conservation Biology, 11 (6), 1418-1429.

Bazzaz, F.A. (1975). Plant species-diversity in old-field successional ecosystems in Southern Illinois. Ecology, 56 (2), 485-488.

Beisiegel, B.D.M. (2006). Shelter availability and use by mammals and birds in an Atlantic Forest area. Biota Neotropica, 6, 0-0.

Berges, S.A., Moore, L.A.S., Isenhart, T.M., & Schultz, R.C. (2010). Bird species diversity in riparian buffers, row crop fields, and grazed pastures within agriculturally dominated watersheds. Agroforestry Systems, 79 (1), 97-110.

Berta, A. (1982). *Cerdocyon thous*. Mammalian Species, (186), 1-4.

Bodmer, R.E. (1991). Strategies of seed dispersal and seed predation in Amazonian ungulates. Biotropica, 23 (3), 255-261.

Breece, G.A., & Dusi, J.L. (1985). Food habitats and home range of the common long-nosed armadillo *Dasypus novemcinctus* in Alabama. In G.G. Montgomery (Ed.), The evolution and ecology of armadillos, sloths and vermilinguas (p.419-427). Washington: Smithsonian Institution Press.

Brioschi, L.R. (1999). Caminhos do ouro. In C.A.P. Bacellar & L.R. Brioschi (Eds.), Na estrada do Anhengüera: uma visão regional da história Paulista (p.35-54). São Paulo: Humanitas FFLCH / USP.

Buarque De Holanda, S. (1990). Monções. São Paulo: Brasiliense.

Buarque De Holanda, S. (1994). Caminhos e Fronteiras. São Paulo: Companhia das Letras.

Burnham, K.P., & Anderson, D.R. (2004). Multimodel Inference: Understanding AIC and BIC in Model Selection. Sociological Methods Research, 33 (2), 261-304.

Butler, S., Vickery, J., & Norris, K. (2007). Farmland biodiversity and the footprint of agriculture. Science, 315 (5810), 381-384.

Chiarello, A.G. (1999). Effects of fragmentation of the Atlantic forest on mammal communities in Southeastern Brazil. Biological Conservation, 89 (1), 71-82.

Christian, D.P., Hoffman, W., Hanowski, J.M., Niemi, G.J., & Beyea, J. (1998). Bird and mammal diversity on woody biomass plantations in North America. Biomass and Bioenergy, 14 (4), 395-402.

Connor, E.F. & Mccoy, E.D. (1979). The statistics and biology of the species-area relationship. American Naturalist, 113 (6), 791.

Coppock, J.T. & Rhind, D.W. (1991). The history of GIS. In D.J. Maguire, M.F. Goodchild & D. Wrhind (Eds.), Geographical information systems: Principles and applications (p.21-43). Essex: Longman Scientific & Technical.

Costa, L.P. & Leite, Y.L. (2012). Historical Fragmentation Shaping Vertebrate Diversification in the Atlantic Forest Biodiversity Hotspot. In B.D. Patterson & L.P. Costa (Eds.), Bones, clones, and biomes: the history and geography of recent neotropical mammals (p.283-306). Chicago: The University of Chicago Press.

Crooks, K.R. & Soulé, M.E. (1999). Mesopredator release and avifaunal extinctions in a fragmented system. Nature, 400 (6744), 563-566.

Crutzen, P.J. & Stoermer, E.F. (2000). Global change newsletter. The Anthropocene, 41, 17-18.

Dalebout, M.L., Lento, G.M., Cipriano, F., Funahashi, N. & Baker, C.S. (2002). How many protected minke whales are sold in Japan and Korea? A census by microsatellite DNA profiling. Animal Conservation, 5 (2), 143-152.

Davidoff, C.H. (1986). Bandeirantismo: Verso e Reverso. São Paulo: Brasiliense.

De Abreu, J.C. (1999). O descobrimento do Brasil. Rio de Janeiro: Ed. Da Soc. Capistrano de Abreu.

De Heinzelin, J. (1962). Ishango. *Scientific American, 206,* 105-116.

De Jager, N.R., Pastor, J., & Hodgson, A.L. (2009). Scaling the effects of moose browsing on forage distribution, from the geometry of plant canopies to landscapes. Ecological Monographs, 79 (2), 281-297.

De Mello Araujo, A.G. (2007). A tradição cerâmica Itararé-Taquara: características, área de ocorrência e algumas hipóteses sobre a expansão dos grupos Jê no Sudeste do Brasil. Revista de Arqueologia, 20 (1), 9-38.

De Oliveira Calleia, F., Rohe, F., & Gordo, M. (2009). Hunting Strategy of the Margay (*Leopardus wiedii*) to Attract the Wild Pied Tamarin (*Saguinus bicolor*). Neotropical Primates, 16 (1), 32-34.

De Vivo, M., Carmignotto, A.P., Gregorin, R., et al.(2011). Checklist dos mamíferos do Estado de São Paulo, Brasil. Biota Neotropica, 11, 111-131.

Dean, W. (1997). With broadax and firebrand: the destruction of the Brazilian Atlantic Forest. Berkeley and Los Angeles: University of California Press.

Di Bitetti, M.S., Di Blanco, Y.E., Pereira, J.A., Paviolo, A.N., & PãRez, I.J.N. (2009). Time Partitioning Favors the Coexistence of Sympatric Crab-Eating Foxes (*Cerdocyon thous*) and Pampas Foxes (*Lycalopex gymnocercus*). Journal of Mammalogy, 90 (2), 479-490.

Di Bitetti, M.S., Paviolo, A., Ferrari, C.A., De Angelo, C., & Di Blanco, Y. (2008). Differential Responses to Hunting in Two Sympatric Species of Brocket Deer (*Mazama americana* and *M. nana*). Biotropica, 40 (5), 636-645.

Dillon, A., & Kelly, M.J. (2008). Ocelot home range, overlap and density: comparing radio telemetry with camera trapping. Journal of Zoology, 275 (4), 391-398.

Dorazio, R.M., & Royle, J.A. (2005). Estimating Size and Composition of Biological Communities by Modeling the Occurrence of Species. Journal of the American Statistical Association, 100 (470), 389-398.

Dorazio, R.M., Royle, J.A., Söderströ, B., & Glimskär, A. (2006). Estimating species richness and accumulation by modeling species occurrence and detectability. Ecology, 87 (4), 842-854.

Dotta, G. (2005). Diversidade de mamíferos de médio e grande porte em relação à paisagem da Bacia do Rio Passa-Cinco, São Paulo.. Piracicaba: Ciências Biológicas, Universidade de São Paulo.

Dotta, G., & Verdade, L.M. (2007). Trophic categories in a mammal assemblage: diversity in an agricultural landscape. Biota Neotropica, 7 (2), 287-292.

Dunning, J.B., Danielson, B.J. & Pulliam, H.R. (1992). Ecological processes that affect populations in complex landscapes. Oikos, 65 (1), 169-175.

Dunnington, G. W. (2004). Carl Friedrich Gauss: Titan of science. New York: Hafner.

Dyer, L.A. (2007). Tropical tritrophic interactions: Nasty hosts and ubiquitous cascades. In W.P. Carson & S.A. Schnitzer (Eds.), Tropical forest community ecology. Blackwell Science, Oxford (p.275-293). Chichester: John Wiley & Sons Ltd.

Edenius, L. & Sjoberg, K. (1997). Distribution of birds in natural landscape mosaics of old-growth forests in Northern Sweden: relations to habitat area and landscape context. Ecography, 20 (5), 425-431.

Eisenberg, J.F. (1981). The mammalian radiations : an analysis of trends in evolution, adaptation, and behaviour. London: Athlone Press.

Eizirik, E. (2012). A Molecular View on the Evolutionary History and Biogeography of Neotropical Carnivores (Mammalia, Carnivora). In B.D. Patterson & L.P. Costa (Eds.), Bones, Clones, and Biomes: The History and Geography of Recent Neotropical Mammals (p.123-142). Chicago: The University of Chicago Press.

Fahrig, L. (2003). Effects of habitat fragmentation on biodiversity. Annual Review of Ecology, Evolution, and Systematics, 34 (1), 487-515.

Falconer, K. (2013). Fractal geometry: mathematical foundations and applications. Chichester: John Wiley & Sons.

Fausto, C. (2008). Fragmentos de História e Cultura Tupinambá - Da etnologia como instrumento crítico de conhecimento etno-histórico. In M. Carneiro da Cunha (Ed.), História dos Índios no Brasil (p.388-393). São Paulo: Companhia das Letras, Secretaria Municipal de Cultura, FAPESP.

Fischer, J., & Lindenmayer, D.B. (2007). Landscape modification and habitat fragmentation: a synthesis. Global Ecology and Biogeography, 16 (3), 265-280.

Fletcher, R.J. (2005). Multiple edge effects and their implications in fragmented landscapes. Journal of Animal Ecology, 74 (2), 342-352.

Fleury, M., & Galetti, M. (2006). Forest fragment size and microhabitat effects on palm seed predation. Biological Conservation, 131 (1), 1-13.

Flynn, J., Charrier, R., Croft, D., Wyss, A., Patterson, B., & Costa, L. (2012). Cenozoic Andean faunas: shedding new light on South American mammal evolution, biogeography, environments, and tectonics. In B.D. Patterson & L.P. Costa (Eds.), Bones, Clones, and Biomes: The History and Geography of Recent Neotropical Mammals (p.51-75). Chicago: The University of Chicago Press.

Fonseca, G.A.B., Herrmann, G., Leite, Y.L.R., Meittermeier, R.A., Rylands, A.B., & Patton, J.L. (1996). Lista anotadados mamíferos do Brasil. Washington: Conservation International.

Freemark, K. (1995). Assessing effects of agriculture on terrestrial wildlife - developing a hierarchical approach for the United-States-EPA. Landscape and Urban Planning, 31 (1-3), 99-115.

Friendly, M. (2008). A Brief History of Data Visualization. In C.-H. Chen, W. Härdle, & A. Unwin (Eds.), Handbook of Data Visualization (p.15-56). Berlin: Springer Berlin Heidelberg.

Friendly, M. (2005). Milestones in the History of Data Visualization: A Case Study in Statistical Historiography. In C. Weihs & W. Gaul (Eds.), Classification — the Ubiquitous Challenge (p.34-52). Berlin: Springer Berlin Heidelberg.

Galliez, M., Leite, M.D., Queiroz, T.L., & Fernandez, F.A.D. (2009). Ecology of the water opossum *Chironectes minimus* in Atlantic Forest streams of Southeastern Brazil. Journal of Mammalogy, 90 (1), 93-103.

Gascon, C., Lovejoy, T.E., Bierregaard Jr, R.O., et al. (1999). Matrix habitat and species richness in tropical forest remnants. Biological Conservation, 91 (2-3), 223-229.

Gatti, A., Bianchi, R., Rosa, C.R.X., et al. (2006). Diet of two sympatric carnivores, *Cerdocyon thous* and *Procyon cancrivorus*, in a restinga area of Espirito Santo State, Brazil. Journal of Tropical Ecology, 22 (02), 227-230.

Gheler-Costa, C. (2006). Distribuição e abundância de pequenos mamíferos em relação à paisagem da bacia do Rio Passa-Cinco, São Paulo, Brasil. Piracicaba: Ciências Biológicas, Universidade de São Paulo.

Ghirardello, N. (2002). À beira da linha. Núcleos urbanos ao longo da Noroeste paulista. São Paulo: Editora Unesp.

Gleick, J. (2011). Chaos: Making a New Science. New York: Open Road Media.

Goin, F., Gelfo, J., Chornogubsky, L., Woodburne, M., & Martin, T. (2012). Origins, radiations, and distribution of South American mammals: from greenhouse to icehouse worlds. In B.D. Patterson & L.P. Costa (Eds.), Bones, Clones, and Biomes: The History and Geography of Recent Neotropical Mammals (p.20-50). Chicago: The University of Chicago Press.

Gonzales, C.A.L., Brown, D.E., & Gallo-Reynoso, J.P. (2003). The ocelot *Leopardus pardalis* in north-western Mexico: ecology, distribution and conservation status. Oryx, 37 (03), 358-364.

Gouveia, S., Pessenda, L., Aravena, R., Boulet, R., Scheel-Ybert, R., Bendassoli, J., Ribeiro, A. & Freitas, H. (2002). Carbon isotopes in charcoal and soils in studies of paleovegetation and climate changes during the late Pleistocene and the Holocene in the Southeast and Centerwest regions of Brazil. Global and Planetary Change, 33 (1), 95-106.

Granö, J.G. (2003). Origin of Landscape Science: JG Granö and a New Pure Geography for a New State: a Collection of Papers. Turku: Turku University Foundation.

Greenberg, M.J. (1993). Euclidean and non-Euclidean geometries: Development and history. New York: W.H. Freedman and Company.

Guthrie, W.K.C. (1962). A History of Greek Philosophy, Volume I: The Earlier PreSocratics and the Pythagoreans. Cambridge: Cambridge University Press.

Guthrie, W.K.C. (1967). A History of Greek Philosophy: Vol. 1 The Earlier Presocratics and the Pythagoreans. Cambridge: Cambridge University Press.

Guthrie, W.K.C. (1969). A history of Greek philosophy. Vol. 3: The fifth century enlightenment. Cambridge: Cambridge University Press.

Guthrie, W.K.C. (2012). The Greek Philosophers: From Thales to Aristotle. New York: Routledge.

Hanski, I., & Gilpin, M. (1991). Metapopulation dynamics - brief-history and conceptual domain. Biological Journal of the Linnean Society, 42 (1-2), 3-16.

Hawes, J., Barlow, J., Gardner, T.A., & Peres, C.A. (2008). The value of forest strips for understorey birds in an Amazonian plantation landscape. Biological Conservation, 141 (9), 2262-2278.

Heath, T. (1909). The thirteen books of Euclid's Elements translated from the text of Heiberg with introduction and commentary. Bulletin of American Mathematic Society, 15 (1909), 386-3.

Hill, M.F., & Caswell, H. (1999). Habitat fragmentation and extinction thresholds on fractal landscapes. Ecology Letters, 2 (2), 121-127.

Holling, C.S. (1986). Resilience of ecosystems; local surprise and global change. In W.C. Clark, & R.E. Munn (Eds.), Sustainable development of the biosphere (p.297-317). Cambridge: Cambridge University Press.

Holling, C.S. (1992). Cross-scale morphology, geometry, and dynamics of ecosystems. Ecological Monographs, 62 (4), 447-502.

Holling, C.S., & Allen, C.R. (2002). Adaptive Inference for Distinguishing Credible from Incredible Patterns in Nature. Ecosystems, 5 (4), 319-328.

Hurtado-Gonzales, J., & Bodmer, R. (2006). Reproductive biology of female Amazonian brocket deer in Northeastern Peru. European Journal of Wildlife Research, 52 (3), 171-177.

Jacquemyn, H., Butaye, J., & Hermy, M. (2003). Influence of environmental and spatial variables on regional distribution of forest plant species in a fragmented and changing landscape. Ecography, 26 (6), 768-776.

Jelinski, D., & Wu, J. (1996). The modifiable areal unit problem and implications for landscape ecology. Landscape Ecology, 11 (3), 129-140.

Jensen, J.R. (1996). Introductory digital image processing: a remote sensing perspective. Englewood: Prentice-Hall Inc.

Jiménez García, D., Martínez Pérez, J., & Peiró Clavell, V. (2006). Relationship between game species and landscape structure in the SE of Spain. Wildlife Biology in Practice, 2 (2), 48-62.

Johnson, C. (2002). Determinants of loss of mammal species during the Late Quaternary 'megafauna'extinctions: life history and ecology, but not body size. Proceedings of the Royal Society of London. Series B: Biological Sciences, 269 (1506), 2221-2227.

Kemp, T.S. (2005). The origin and evolution of mammals. Oxford: Oxford University Press.

Kéry, M., & Royle, J.A. (2008). Hierarchical Bayes estimation of species richness and occupancy in spatially replicated surveys. Journal of Applied Ecology, 45 (2), 589-598.

Keuroghlian, A., & Eaton, D.P. (2008). Importance of rare habitats and riparian zones in a tropical forest fragment: preferential use by *Tayassu pecari*, a wide-ranging frugivore. Journal of Zoology, 275 (3), 283-293.

Kruhl, J.H. (2013). Fractal-geometry techniques in the quantification of complex rock structures: A special view on scaling regimes, inhomogeneity and anisotropy. Journal of Structural Geology, 46, 2-21.

Kufner, M.B., Sepúlveda, L., Gavier, G., Madoery, L., & Giraudo, L. (2008). Is the native deer *Mazama gouazoubira* threatened by competition for food with the exotic hare *Lepus europaeus* in the degraded Chaco in Córdoba, Argentina? Journal of Arid Environments, 72 (12), 2159-2167.

Lausch, A., & Herzog, F. (2002). Applicability of landscape metrics for the monitoring of landscape change: issues of scale, resolution and interpretability. Ecological indicators, 2 (1), 3-15.

Lee, M., Fahrig, L., Freemark, K., & Currie, D.J. (2002). Importance of patch scale vs landscape scale on selected forest birds. Oikos, 96 (1), 110-118.

Lentz, D.L. (2000). Anthropocentric food webs in the Precolumbian Americas. In D. L. Lentz (Ed.), Imperfect balance: landscape transformations in the pre-Columbian Americas (p.89-119). New York: Columbia University Press.

Li, H., & Wu, J. (2004). Use and misuse of landscape indices. Landscape Ecology, 19 (4), 389-399.

Li, X., He, H. S., Bu, R., Wen, Q., Chang, Y., Hu, Y., & Li, Y. (2005). The adequacy of different landscape metrics for various landscape patterns. Pattern Recognition, 38 (12), 2626-2638.

Lindborg, R., & Eriksson, O. (2004). Historical landscape connectivity affects present plant species diversity. Ecology, 85 (7), 1840-1845.

Lindenmayer, D.B., & Hobbs, R. J. (2004). Fauna conservation in Australian plantation forests - a review. Biological Conservation, 119 (2), 151-168.

Lomolino, M.V., & Perault, D.R. (2001). Island biogeography and landscape ecology of mammals inhabiting fragmented, temperate rain forests. Global Ecology and Biogeography, 10 (2), 113-132.

Luck, G.W., & Korodaj, T.N. (2008). Stand and landscape-level factors related to bird assemblages in exotic pine plantations: Implications for forest management. Forest Ecology and Management, 255 (7), 2688-2697.

Lyra-Jorge, M., Ciocheti, G., & Pivello, V. (2008). Carnivore mammals in a fragmented landscape in Northeast of São Paulo State, Brazil. Biodiversity and Conservation, 17 (7), 1573-1580.

Mac Nally, R., Parkinson, A., Horrocks, G., Conole, L., & Tzaros, C. (2001). Relationships between terrestrial vertebrate diversity, abundance and availability of coarse woody debris on Southeastern Australian floodplains. Biological Conservation, 99 (2), 191-205.

Macdonald, S.E., Eaton, B., Machtans, C.S., Paszkowski, C., Hannon, S., & Boutin, S. (2006). Is forest close to lakes ecologically unique? Analysis of vegetation, small mammals, amphibians, and songbirds. Forest Ecology and Management, 223 (1-3), 1-17.

Mackenzie, D.I., Nichols, J.D., Lachman, G.B., Droege, S., Andrew Royle, J., & Langtimm, C.A. (2002). Estimating site occupancy rates when detection probabilities are less than one. Ecology, 83 (8), 2248-2255.

Mackenzie, D.I., Nichols, J.D., Royle, J.A., Pollock, K.H., Bailey, L.L., & Hines, J.E. (2006). Occupancy Estimation and Modeling. Burlington: Elsevier.

Mandelbrot, B. (1976). Fractal geometry of turbulence - hausdorff dimension, dispersion and nature of singularities of fluid motion. Comptes Rendus Hebdomadaires Des Seances De L Academie Des Sciences Serie A, 282 (2), 119-120.

Mandelbrot, B. (1982). The fractal geometry of nature. San Francisco: Freeman.

Mandelbrot, B. (1985). Self-Affine Fractals and Fractal Dimension. Physica Scripta, 32 (4), 257-260.

Mandelbrot, B. (1999). Multifractals and 1/f Noise: Wild Self-Affinity in Physics. New York: Springer.

Mandelbrot, B. (2002). Gaussian self-affinity and fractals: Globality, the earth, 1/f noise, and R/S. New York: Springer.

Marsh, C.J., & Ewers, R.M. (2013). A fractal-based sampling design for ecological surveys quantifying β-diversity. Methods in Ecology and Evolution, 4 (1), 63-72.

Marshall, D.W. (1972). A List of Manuscript Editions of Ptolemy's Geographia. Special Libraries Association, Geography and Map Division Bulletin, 87, 17-38.

Martins, M.M. (2005). Density of Primates in Four Semi-deciduous Forest Fragments of São Paulo, Brazil. Biodiversity and Conservation, 14 (10), 2321-2329.

Mcarthur, R.H., & Wilson, E. (1967). The equilibrium theory of island biogeography. New Jersey: Princeton University Press.

Mcbee, K., & Baker, R. J. (1982). *Dasypus novemcinctus*. Mammalian Species, (162), 1-9.

Mcgarigal, K., & Marks, B. (1995). Fragstats: spatial pattern analisys program for quantifying landscape structure. Portland: Department of Agriculture - Forest Service - Pacific Northwestern Research Station.

Medri, I.M., Mourão, G.M., & Harada, A.Y. (2003). Dieta de tamanduá-bandeira (*Myrmecophaga tridactyla*) no Pantanal da Nhecolândia, Brasil. Edentata, 5, 29-34.

Meine, C. (1997). Inherit the grid. Placing nature: Culture and landscape ecology, 45-62.

Mendonça, R.R. (2004). A história da ocupação do interior do Estado de São Paulo. In M.D. Bitencourt & R.R. Mendonça (Ed.), Viabilidade de Conservação dos Remanescentes de Cerrado no Estado de São Paulo (p.57-76). São Paulo: Annablume Editora.

Milne, B.T., Turner, M.G., Wiens, J.A., & Johnson, A.R. (1992). Interactions between the fractal geometry of landscapes and allometric herbivory. Theoretical Population Biology, 41 (3), 337-353.

Mlodinow, L. (2010). Euclid's window: the story of geometry from parallel lines to hyperspace. Simon and Schuster. New York: Touchstone.

Molland, G. (1995). Mathematics and the medieval ancestry of physics. Aldershot: Variorum.

Monbeig, P. (1984). Pioneiros e Fazendeiros de São Paulo. São Paulo: Hucitec.

Nielsen, S.E., Bayne, E.M., Schieck, J., Herbers, J., & Boutin, S. (2007). A new method to estimate species and biodiversity intactness using empirically derived reference conditions. Biological Conservation, 137 (3), 403-414.

Nikora, V.I., Pearson, C.P., & Shankar, U. (1999). Scaling properties in landscape patterns: New Zealand experience. Landscape Ecology, 14 (1), 17-33.

Nilsen, E.B., Finstad, A.G., Næsje, T.F., & Sverdrup-Thygeson, A. (2013). Using mass scaling of movement cost and resource encounter rate to predict animal body size–population density relationships. Theoretical Population Biology, 86, 23-28.

Noelli, F.S. (2008). The Tupi Expansion. In H. Silverman & W.H. Isbel (Eds.), Handbook of South American Archaeology (p.659-670) New York: Springer Science+Business Media.

Numa, C., Verdu, J.R., & Sanchez-Palomino, P. (2005). Phyllostomid bat diversity in a variegated coffee landscape. Biological Conservation, 122 (1), 151-158.

O'neill, R.V., Krummel, J.R., Gardner, et al. (1988). Indices of landscape pattern. Landscape Ecology, 1 (3), 153-162.

Padilla, M., & Dowler, R.C. (1994). *Tapirus terrestris*. Mammalian Species, (481), 1-8.

Palmer, G.C., & Bennett, A.F. (2006). Riparian zones provide for distinct bird assemblages in forest mosaics of south-east Australia. Biological Conservation, 130 (3), 447-457.

Pardini, R., Bueno, A.D.A., Gardner, T.A., Prado, P.I., & Metzger, J.P. (2010). Beyond the Fragmentation Threshold Hypothesis: Regime Shifts in Biodiversity Across Fragmented Landscapes. Plos One, 5 (10), e13666.

Parsons, J.J. (1972). Spread of African pasture grasses to America tropics. Journal of Range Management, 25 (1), 12-17.

Patterson, B.D., & Costa, L.P. (2012). Bones, Clones, and Biomes: The History and Geography of Recent Neotropical Mammals. Chicago: University of Chicago Press.

Pawson, S.M., Brockerhoff, E.G., & Didham, R.K. (2009). Native forest generalists dominate carabid assemblages along a stand age chronosequence in an exotic *Pinus radiata* plantation. Forest Ecology and Management, 258, S108-S116.

Persinger, M.A. & Koren, S.A. (2013). Dimensional Analyses of Geometric Products and the Boundary Conditions of the Universe: Implications for a Quantitative Value for the Latency to Display Entanglement. Open Astronomy Journal, 6, 10-13.
Peters, C.M. (2000). Precolumbian silviculture and indigenous management of neotropical forests. In D.L. Lentz (Ed.), Imperfect balance: Landscape transformations in the Precolumbian Americas (p.203-223) New York: Columbia University Press.
Primdahl, J., & Swaffield, S. (2010). Globalisation and agricultural landscapes: change patterns and policy trends in developed countries. New York: Cambridge University Press.
Ptolemy, C. (1525). Geographia. Strasbourg: Johannes Regiomontanus.
Ramos, L.D.S., Zanetti, R., Marinho, C.G.S., Delabie, J.H.C., Schlindwein, M.N., & Almado, R.D.P. (2004). Impacto das capinas mecânica e química do sub-bosque de Eucalyptus grandis sobre a comunidade de formigas (Hymenoptera: Formicidae). Revista Árvore, 28, 139-146.
Rempel, R., Kaukinen, D., & Carr, A. (2012). Patch analyst and patch grid. Thunder Bay, Ontario: Ontario Ministry of Natural Resources. Centre for Northern Forest Ecosystem Research.
Rivero, K., Rumiz, D.N.I., & Taber, A.B. (2007). Differential habitat use by two sympatric brocket deer species (*Mazama americana* and *M. gouazoubira*) in a seasonal Chiquitano Forest of Bolivia. Mammalia, 69 (2), 169-183.
Robinson, J.G., & Redford, K.H. (1986a). Body Size, Diet, and Population Density of Neotropical Forest Mammals. American Naturalist, 128 (5), 665.
Robinson, J.G., & Redford, K.H. (1986b). Intrinsic rate of natural increase in neotropical forest mammals - relationship to phylogeny and diet. Oecologia, 68 (4), 516-520.
Rocha-Mendes, F., & Bianconi, G.V. (2009). Opportunistic predatory behavior of margay, *Leopardus wiedii* (Schinz, 1821), in Brazil. Mammalia, 73 (2), 151-152.
Roosevelt, A.C., Lima Da Costa, M., Lopes Machado, C., et al. (1996). Paleoindian cave dwellers in the Amazon: the peopling of the Americas. Science, 272, 373-384.
Rosalino, L.M., Do Rosario, J., & Santos-Reis, M. (2009). The role of habitat patches on mammalian diversity in cork oak agroforestry systems. Acta Oecologica-International Journal of Ecology, 35 (4), 507-512.
Royle, J.A., Dorazio, R.M., & Link, W.A. (2007). Analysis of Multinomial Models With Unknown Index Using Data Augmentation. Journal of Computational and Graphical Statistics, 16 (1), 67-85.
Russell, R.E., Royle, J.A., Saab, V.A., Lehmkuhl, J.F., Block, W.M., & Sauer, J.R. (2009). Modeling the effects of environmental disturbance on wildlife communities: avian responses to prescribed fire. Ecological Applications, 19 (5), 1253-1263.
Sack, R.D. (1986). Human Territoriality: Its Theory and History. Cambridge: Cambridge University Press.
Saint-Hilaire, A.F.C.P. (1972). Viagem à Província de São Paulo e Resumo dasViagens ao Brasil, Província Cisplatina e Missões do Paraguai. São Paulo: Universidade de São Paulo.
Salas, L.A., & Fuller, T.K. (1996). Diet of the lowland tapir (*Tapirus terrestris* L.) in the Tabaro River Valley, Southern Venezuela. Canadian Journal of Zoology, 74 (8), 1444-1451.
Sampaio, A.N. (1949). Edmundo Navarro de Andrade: um pouco de sua vida e do seu trabalho. Anuário Brasileiro de Economia Florestal, 2 (2), 51-63.
Sampaio, A.N. (1957). Eucaliptos para o Brasil. Arquivos do Seviço Florestal, 12 (2), 3-17.
Scally, M., Madsen, O., Douady, C.J., De Jong, W.W., Stanhope, M.J., & Springer, M.S. (2001). Molecular evidence for the major clades of placental mammals. Journal of Mammalian Evolution, 8 (4), 239-277.
Scheel-Ybert, R., Gouveia, S.E.M., Pessenda, L.C.R., Aravena, R., Coutinho, L.M., & Boulet, R. (2003). Holocene palaeoenvironmental evolution in the São Paulo State (Brazil), based on anthracology and soil δ13C analysis. The Holocene, 13 (1), 73-81.
Schmidt, C.B. (1959). Tropas e Tropeiros. Journal of Inter-American Studies, 1 (2), 103-122.

Secretaria Do Meio Ambiente - Instituto Florestal (2005). Inventário Florestal da Vegetação Natural do Estado de São Paulo. São Paulo: Imprensa Oficial.
Secretaria Do Meio Ambiente (2007). Nos Caminhos da Biodiversidade Paulista. São Paulo: Imprensa Oficial.
Smith-Ramirez, C., Celis-Diez, J.L., Von Jenstchyk, E., Jimenez, J.E., & Armesto, J.J. (2010). Habitat use of remnant forest habitats by the threatened arboreal marsupial Dromiciops gliroides (Microbiotheria) in a rural landscape of southern Chile. Wildlife Research, 37 (3), 249-254.
Stanley, T.R., & Royle, J.A. (2005). Estimating site occupancy and abundance using indirect detection indices. Journal of Wildlife Management, 69 (3), 874-883.
Stoner, K.E., Riba-Hernández, P., Vulinec, K., & Lambert, J.E. (2007). The Role of Mammals in Creating and Modifying Seedshadows in Tropical Forests and Some Possible Consequences of Their Elimination. Biotropica, 39 (3), 316-327.
Tansley, A.G. (1935). The use and abuse of vegetational concepts and terms. Ecology, 16 (3), 284-307.
Tartaglia, J.C. (1988). Agricultura e interiorização do desenvolvimento em São Paulo. In J.C. Tartaglia & O.L. Oliveira (Eds.), Modernização e Desenvolvimento em São Paulo (p.63-75) São Paulo: Fundação para o Desenvolvimento da UNESP.
Taulman, J.F., & Robbins, L.W. (1996). Recent range expansion and distributional limits of the nine-banded armadillo (*Dasypus novemcinctus*) in the United States. Journal of Biogeography, 23 (5), 635-648.
Taylor, P.D., Fahrig, L., Henein, K., & Merriam, G. (1993). Connectivity is a vital element of landscape structure. Oikos, 68 (3), 571-573.
Taylor, R.S., Oldland, J.M., & Clarke, M.F. (2008). Edge geometry influences patch-level habitat use by an edge specialist in Southeastern Australia. Landscape Ecology, 23 (4), 377-389.
Tchaicka, L., Eizirik, E., Oliveira, T.G.D., Jr, J.F.C., & Freitas, T.R.O. (2007). Phylogeography and population history of the crab-eating fox (*Cerdocyon thous*). Molecular Ecology, 16 (4), 819-838.
Theiler, J. (1990). Estimating fractal dimension. Journal of the Optical Society of America, 7 (6), 1055-1073.
Tilman, D., Cassman, K.G., Matson, P.A., Naylor, R., & Polasky, S. (2002). Agricultural sustainability and intensive production practices. Nature, 418 (6898), 671-677.
Trzcinski, M.K., Fahrig, L., & Merriam, G. (1999). Independent effects of forest cover and fragmentation on the distribution of forest breeding birds. Ecological Applications, 9 (2), 586-593.
Turner, M.G. (1989). Landscape Ecology: The Effect of Pattern on Process. Annual Review of Ecology and Systematics, 20 (1), 171-197.
Turner, M.G., Gardner, R.H., & O'neill, R.V. (2001). Landscape Ecology in Theory and Practice: Pattern and Process. New York: Springer.
Venevsky, S., & Veneskaia, I. (2003). Large-scale energetic and landscape factors of vegetation diversity. Ecology Letters, 6 (11), 1004-1016.
Verdade, L.M., Gheler-Costa, C., Penteado, M., & Dotta, G. (2012). The Impacts of Sugarcane Expansion on Wildlife in the State of São Paulo, Brazil. Journal of Sustainable Bioenergy Systems, 2 (4), 138-144.
Verdade, L.M., Penteado, M., Gheler-Costa, C., Dotta, G., Rosalino, L.M., Pivello, V.R., Piña, C.I. & Lyra-Jorge, M.C. (2014). The conservation value of agricultural landscapes. In L.M. Verdade, M.C. Lyra-Jorge & C.I. Piña (Eds.), Applied Ecology and Human Dimensions in Biological Conservation (p.91-102). Berlin: Springer.
Verdade, L.M., Piña, C.I., & Lyra-Jorge, M.C. (2014). Redirections in conservation biology. In L.M. Verdade, M.C. Lyra-Jorge & C.I. Piña (Eds.), Applied Ecology and Human Dimensions in Biological Conservation (p.3-17). Berlin: Springer.
Vizcaino, S.F., Cassini, G.H., Toledo, N., & Bargo, M.S. (2012). On the Evolution of Large Size in Mammalian Herbivores of Cenozoic. In B.D. Patterson & L.P. Costa (Eds.), Bones, Clones, and

Biomes: The History and Geography of Recent Neotropical Mammals (p.76-101). Chicago: The University of Chicago Press.

Voss, R.F. (1989). Random fractals: Self-affinity in noise, music, mountains, and clouds. Physica D: Nonlinear Phenomena, 38 (1), 362-371.

Wilson, D.E., & Reeder, D.M. (2005). Mammal species of the world : a taxonomic and geographic reference. Baltimore: Johns Hopkins University Press.

With, K.A., & Crist, T.O. (1995). Critical thresholds in species responses to landscape structure. Ecology, 76 (8), 2446-2459.

Zipkin, E.F., Dewan, A., & Andrew Royle, J. (2009). Impacts of forest fragmentation on species richness: a hierarchical approach to community modelling. Journal of Applied Ecology, 46 (4), 815-822.

Zonneveld, I.S. (1995). Land ecology: an introduction to landscape ecology as a base for land evaluation, land management and conservation. Amsterdam: SPB Academic Publishing.

List of Figures

List of Tables

Index

FSC
www.fsc.org
MIX
Papier | Fördert gute Waldnutzung
FSC® C083411

Kolibri 360 GmbH
Ferdinand-Jühlke-Straße 7
99095 Erfurt, Deutschland
produktsicherheit@kolibri360.de